STOCHASTIC AND STATISTICAL METHODS
IN HYDROLOGY
AND ENVIRONMENTAL ENGINEERING

VOLUME 1

EXTREME VALUES: FLOODS AND DROUGHTS

Water Science and Technology Library

VOLUME 10/1

The titles published in this series are listed at the end of this volume.

STOCHASTIC AND STATISTICAL METHODS
IN HYDROLOGY
AND ENVIRONMENTAL ENGINEERING

Volume 1

EXTREME VALUES: FLOOD AND DROUGHTS

edited by

KEITH W. HIPEL

Departments of Systems Design Engineering and
Statistics and Actuarial Science,
University of Waterloo,
Waterloo, Ontario, Canada

KLUWER ACADEMIC PUBLISHERS
DORDRECHT / BOSTON / LONDON

Library of Congress Cataloging-in-Publication Data

Stochastic and statistical methods in hydrology and environmental engineering.
p. cm. -- (Water science and technology library ; v. 10)
Papers presented at an international conference held at the University of Waterloo, Canada, June 21-23, 1993.
Includes index.
Contents: v. 1. Extreme values: floods and droughts / edited by Keith W. Hipel -- v. 2. Stochastic and statistical modelling with groundwater and surface water applications / edited by Keith W. Hipel -- v. 3. Time series analysis in hydrology and environmental engineering / edited by Keith W. Hipel ... [et al.] -- v. 4. Effective environmental management for sustainable development / edited by Keith W. Hipel and Liping Fang.

1. Hydrology--Sratistical methods--Congresses. 2. Stochastic processes--Congresses. I. Series.
GB656.2.S7S815 1994
551.48'01'5195--dc20 94-27708

ISBN 978-90-481-4378-8

Published by Kluwer Academic Publishers,
P.O. Box 17, 3300 AA Dordrecht, The Netherlands.

Kluwer Academic Publishers incorporates
the publishing programmes of
D. Reidel, Martinus Nijhoff, Dr W. Junk and MTP Press.

Sold and distributed in the U.S.A. and Canada
by Kluwer Academic Publishers,
101 Philip Drive, Norwell, MA 02061, U.S.A.

In all other countries, sold and distributed
by Kluwer Academic Publishers Group,
P.O. Box 322, 3300 AH Dordrecht, The Netherlands.

Printed on acid-free paper

In Memory of Professor T.E. Unny
(1929 - 1991)

The Editor of this volume, Professor Keith W. Hipel (left), is showing Professor Unny posters for the International Conference on Stochastic and Statistical Methods in Hydrology and Environmental Engineering held in his honour June 21 to 23, 1993. The photograph was taken at the University of Waterloo on December 20, 1991, eight days before Professor Unny's untimely death.

TABLE OF CONTENTS

PREFACE

Objectives

The great American flood of 1993 will be remembered for generations to come. During the spring and summer of 1993, the Mississippi River and its tributaries exceeded the so-called 100-year flood levels and caused widespread devastation in the American mid-West. About 50 people lost their lives and damages were estimated to be more than 10 billion U.S. dollars. In addition to flooding in the United States and elsewhere, another of nature's lethal weapons was on the rampage – drought. In Africa, for example, the Sahara desert continued to spread southwards due to the lack of precipitation and over-grazing. As a matter of fact, the schizophrenic personality of water – floods and droughts – has been recorded and bemoaned since Biblical times. To better understand and learn how to properly deal with extreme hydrological events, extensive research is still required. Accordingly, the major objective of this volume is to present some of the latest and most advanced research results that can be employed in the ongoing struggle against water's erratic behaviour.

Contents

As shown in the Table of Contents, the book is subdivided into the following main parts:

PART I	**MAXIMUM ANNUAL SERIES**
PART II	**PARTIAL DURATION SERIES**
PART III	**DROUGHTS**
PART IV	**STORMS**
PART V	**FREQUENCY ANALYSIS AND FITTING DISTRIBUTIONS**
PART VI	**REGIONALIZATION TECHNIQUES AND DISTRIBUTED MODELS**
PART VII	**ENTROPY AND FRACTAL ANALYSIS**

Scientists and other decision makers often wish to estimate the probability of occurrence of an extreme event by fitting an appropriate probability distribution to the empirical data. Two techniques that are commonly employed for the statistical estimation of flood quantiles are the maximum annual series and the partial duration series approaches. Part I presents a set of interesting papers that are mainly concerned with modelling maximum annual series. For the case of flooding by a river, in the maximum annual flood approach a time series consisting of the maximum annual riverflows is modelled using a suitable distribution. The first paper in Part I furnishes a perspective on the annual maximum flood method for flood frequency analysis. Within the second paper, the partial duration series technique, which is commonly referred to as the peak over threshold method, is defined and then compared to the annual maximum flood method. In the peak over threshold technique, a partial series comprising data above a certain level is statistically modelled. The remaining papers in Part I deal largely with modelling maximum annual series while the papers in Part II are concerned with studying partial duration series.

The drought papers in Part III investigate the stochastic modelling of drought in Kenya as well as derived frequency distribution for low flow analysis. The stochastic modelling of storms is described in the papers published in Part IV.

Part V contains an intriguing array of papers that address a broad spectrum of problems on frequency analysis and fitting distributions. Subsequently, the papers in Part VI are concerned with regionalization techniques and distributed analysis, with applications in the Canadian province of Alberta and in Japan. Finally, techniques used in the papers in Part VII include entropy and fractal analysis.

Audience

This book should be of direct interest to anyone who is concerned with the latest developments in the systematic study of floods and droughts. Accordingly, the types of **Professionals** who may wish to use this book include:

Water Resources Engineers
Environmental Scientists
Civil Engineers
Hydrologists
Geographers
Planners
Systems Engineers
Management Scientists
Statisticians
Meteorologists

Within each professional group, the book should provide useful information for:

Researchers
Teachers
Students
Practitioners and Consultants

When utilized for teaching purposes, the book could serve as a complementary text at the upper undergraduate and graduate levels. As a matter of fact, authors who have written papers for this volume, have also written earlier books on flood frequency analysis that could also be utilized in courses (see *The Gamma Family and Derived Distributions Applied in Hydrology* by B. Bobée and F. Ashkar, Water Resources Publication, Littleton, Colorado, 1991, as well as *Partial Duration Series in Water Resources* by Dan Rosbjerg, Institute of Hydrodynamics and Hydraulic Engineering, Technical University of Denmark, DK-2800 Lyngby, Denmark, 1993). Researchers should obtain guidance and background material for carrying out worthwhile research projects in flood and drought frequency analyses. Consultants who wish to keep their companies at the leading edge of activities in floods and droughts and thereby serve their clients in the best possible ways will find this book to be an indispensable resource.

AN INTERNATIONAL CELEBRATION

Dedication

The papers contained in this book were originally presented at the international conference on **Stochastic and Statistical Methods in Hydrology and Environmental Engineering** that took place at the University of Waterloo, Waterloo, Ontario, Canada, from June 21 to 23, 1993. This international gathering was held in honour and memory of the late **Professor T.E. Unny** in order to celebrate his lifelong accomplishments in many of the important environmental topics falling within the overall conference theme. When he passed away in late December, 1991, Professor T.E. Unny was Professor of Systems Design Engineering at the University of Waterloo and Editor-in-Chief of the international journal entitled Stochastic Hydrology and Hydraulics.

About 250 scientists from around the world attended the Waterloo conference in June, 1993. At the conference, each participant was given a **Pre-Conference Proceedings**, published by the University of Waterloo and edited by K.W. Hipel. This 584 page volume contains the detailed conference program as well as the refereed extended abstracts for the 234 papers presented at the conference. Subsequent to the conference, full length papers submitted for publication by presenters were mailed to international experts who kindly carried out thorough reviews. Accepted papers were returned to authors for revisions and the final manuscripts were then published by Kluwer according to topics in the following four volumes:

TIME SERIES ANALYSIS IN HYDROLOGY AND ENVIRONMENTAL ENGINEERING
edited by
Keith W. Hipel, A. Ian McLeod, U. S. Panu and Vijay P. Singh

STOCHASTIC AND STATISTICAL MODELLING WITH GROUNDWATER AND SURFACE WATER APPLICATIONS
edited by
Keith W. Hipel

EFFECTIVE ENVIRONMENTAL MANAGEMENT FOR SUSTAINABLE DEVELOPMENT
edited by
Keith W. Hipel and Liping Fang

as well as the current book on:

EXTREME VALUES: FLOODS AND DROUGHTS
edited by
Keith W. Hipel

The Editors of the volumes as well as Professor Unny's many friends and colleagues from around the globe who wrote excellent research papers for publication in these four volumes, would like to dedicate their work as a lasting memorial to Professor T. E. Unny. In addition to his intellectual accomplishments, Professor Unny will be fondly remembered for his warmth, humour and thoughtful consideration of others.

Conference Organization and Sponsorships

The many colleagues and sponsors who took part in the planning and execution of the international conference on **Stochastic and Statistical Methods in Hydrology and Environmental Engineering** are given below.

Organizing Committee

K. W. Hipel (Chairman)	A. I. McLeod
U. S. Panu	V. P. Singh

International Programme Committee

S. Al-Nassri (Malaysia)	Z. Kundzewicz (Poland)
H. Bergmann (Austria)	Gwo-Fong Lin (Taiwan)
J. Bernier (France)	C. Lemaréchal (France)
B. Bobée (Canada)	L. Logan (Canada)
B. Bodo (Canada)	D. P. Loucks (U.S.A.)
D. S. Bowles (U.S.A.)	I. B. MacNeill (Canada)
W. P. Budgell (Norway)	A. Musy (Switzerland)
S. J. Burges (U.S.A.)	P. Nachtnebel (Austria)
F. Camacho (Canada)	D. J. Noakes (Canada)
S. Chandra (India)	N. Okada (Japan)
C-L. Chiu (U.S.A.)	R. M. Phatarford (Australia)
J. Ding (China)	V. Privalsky (U.S.S.R.)
L. Duckstein (U.S.A.)	D. Rosbjerg (Denmark)
A. H. El-Shaarawi (Canada)	J. D. Salas (U.S.A)
M. Fiorentino (Italy)	G. A. Schultz (Germany)
E. Foufoula (U.S.A.)	S. Serrano (U.S.A.)
I. C. Goulter (Australia)	U. Shamir (Israel)
Y. Y. Haimes (U.S.A.)	S. P. Simonovic (Canada)
N. Harmancioglu (Turkey)	S. Sorooshian (U.S.A.)
S. Ikebuchi (Japan)	A. Szollosi-Nagy (France)
Karmeshu (India)	C. Thirriot (France)
M. L. Kavvas (U.S.A.)	W. E. Watt (Canada)
J. Kelman (Brazil)	S. J. Yakowitz (U.S.A.)
J. Kindler (Poland)	V. Yevjevich (U.S.A.)
G. Kite (Canada)	Y. C. Zhang (China)
T. Kojiri (Japan)	P. Zielinski (Canada)
R. Krzysztofowicz (U.S.A.)	

University of Waterloo Committee

A. Bogobowicz	T. Hollands
S. Brown	J. D. Kalbfleisch
D. Burns	E. LeDrew
C. Dufournaud	E. A. McBean
L. Fang	K. Ponnambalam
G. Farquhar	E. Sudicky

Financial Support

Conestoga/Rovers and Associates
Cumming Cockburn Limited
Department of Systems Design Engineering, University of Waterloo
Faculty of Engineering, University of Waterloo
Natural Sciences and Engineering Research Council (NSERC) of Canada

Sponsors

American Geophysical Union
American Water Resources Association
Association of State Floodplain Managers
Canadian Society for Civil Engineering
Canadian Society for Hydrological Sciences
IEEE Systems, Man and Cybernetics Society
Instituto Panamericano de Geografia e Historia
International Association for Hydraulic Research
International Association of Hydrological Sciences
International Commission of Theoretical and Applied Limnology
International Commission on Irrigation and Drainage
International Institute for Applied Systems Analysis
International Statistical Institute
International Water Resources Association
Lakehead University
Louisiana State University
North American Lake Management Society
The International Environmetrics Society
The Pattern Recognition Society
The University of Western Ontario
University of Waterloo

University of Waterloo

President James Downey, Opening and Banquet Addresses
D. Bartholomew, Graphic Services
Danny Lee, Catering and Bar Services Manager
D. E. Reynolds, Manager, Village 2 Conference Centre
T. Schmidt, Engineering Photographer
Audio Visual Centre
Food Services
Graduate Students in Systems Design Engineering

Technical Assistance

Mrs. Sharon Bolender
Mr. Steve Fletcher
Mr. Kei Fukuyama
Ms. Hong Gao
Ms. Wendy Stoneman
Mr. Roy Unny

ACKNOWLEDGEMENTS

The Editor would like to sincerely thank the authors for writing such excellent papers for publication in this as well as the other three volumes. The thoughtful reviews of the many anonymous referees are also gratefully acknowledged. Moreover, the Editor appreciates the fine contributions by everyone who attended the Waterloo conference in June, 1993, and actively took part in the many interesting discussions at the paper presentations.

Additionally, the Editor would like to say *merci beaucoup* to the committee members and sponsors of the Waterloo conference listed in the previous section. Dr. Bernard Bobée, NSERC Chair, University of Quebec, Dr. Fahim Ashkar, University of Moncton, Dr. Peter Rasmussen, University of Quebec, Dr. Dan Rosbjerg, Technical University of Denmark and Dr. Kaz Adamowski, University of Ottawa, kindly assisted in organizing sessions for papers contained in this volume. Dr. Van Nguygen, McGill University, and Dr. Jean Rouselle, École Polytechnique, were involved in joint research projects with the aforementioned authors and thereby also contributed to the success of the conference. Finally, the Editor sincerely appreciates all the thoughtful personnel at Kluwer who assisted in the publication of the volumes. In particular, he is grateful to Dr. Petra D. Van Steenbergen, Acquisition Editor, and Professor V.P. Singh who is the Editor of the Water Science and Technology Library book series in which this volume is published.

Keith W. Hipel
Professor and Chair
Department of Systems Design Engineering

Cross Appointed Professor to
Department of Statistics and Actuarial Science
University of Waterloo

April, 1994

PART I

MAXIMUM ANNUAL SERIES

A PERSPECTIVE ON THE ANNUAL MAXIMUM FLOOD APPROACH TO FLOOD FREQUENCY ANALYSIS

F. ASHKAR[1], B. BOBÉE[2], P. RASMUSSEN[2], and D. ROSBJERG[3]
[1]Department of Mathematics, University of Moncton, Moncton, NB, Canada E1A 3E9
[2]INRS-Eau, University of Quebec, 2800 Einstein St., C.P. 7500, Sainte-Foy, Quebec, Canada G1V 4C7
[3]Institute of Hydrodynamics and Hydraulic Engineering, Technical University of Denmark, 2800 Lyngby, Denmark

A reflection is made on certain aspects of the annual maximum flood (AMF) approach to flood frequency analysis, which we consider important from both a practical and a scientific viewpoint. A large number of D/E procedures (Distribution, D, and Method of parameter estimation, E) for analyzing AMF series have been proposed, but their lack of physical justification has led to a state of confusion among practitioners. Various regional flood frequency procedures have been proposed to reduce model and sampling uncertainties inherent in flood frequency estimation at sites with only short flood records, or to estimate flood quantiles at ungaged sites. However, the state of confusion that has been common in at-site AMF procedures, has also been so in regional methods. Past attempts at surveying and comparing the various AMF frequency estimation methodologies have failed to satisfy the needs of practitioners. We discuss some misleading or imprecise concepts such as "robustness to the distribution choice", "sensitivity to outliers", and the notion of "separation of skewness", whose use in the area of AMF-series analysis has added to the state of confusion. A large number of partial comparisons among AMF procedures has been conducted, but these comparisons are limited in use. There are compelling reasons to reconsider carefully the problem of flood frequency analysis and define comparison schemes and criteria that are more broadly agreed upon in the hydrological community.

INTRODUCTION

Statistical estimation of flood quantiles has been a topic of extensive research during the last decades. Two methods have been extensively used for this purpose: the annual maximum flood (AMF) approach, and the peaks over threshold (POT) approach. The present study is concerned with the former, which has been the more widely used of the two methods. It is not our intention to make an extensive review of AMF procedures, but to reflect on certain aspects of this topic which we consider important from both a

K. W. Hipel (ed.),
Stochastic and Statistical Methods in Hydrology and Environmental Engineering, Vol. 1, 3–14.

practical and a scientific viewpoint.

A sample of size n of AMF series constitutes the basis for the AMF method. After checking the sample for the conditions of randomness, independence, homogeneity, stationarity, and the absence of "outliers", the sample is then fitted by an "appropriate" distribution (D), using a method of parameter estimation (E). This D/E-procedure is then used to make inferences about flood quantiles, and their associated confidence intervals.

A large number of D/E procedures for analyzing AMF series have been recommended, of which some have been claimed physically justified, but in reality, none of these can be *completely* defended on a physical basis. The large number of proposed D/E procedures has led to a state of confusion among practitioners as to which one is the best (Bobée et al., 1993a). Many questions on this subject also remain unanswered in the minds of researchers. The present paper attempts to identify and discuss some of the research controversies, and some of the reasons for the state of confusion.

According to Potter (1987), much of the research on flood frequency analysis has "lacked balance", and has not put sufficient emphasis on the physical characteristics of the problem. Researchers have spent more time on developing new methods than on testing existing ones (Lettenmaier, 1988), which has added to the current state of confusion.

Attempts have been made to survey and compare the various flood frequency estimation methodologies (WMO, 1989, for example), but generally they fail to satisfy the needs of practitioners (Bobée et al., 1993a). However, by focusing on the flood frequency analysis problem as a whole, these attempts have revealed, in a clearer manner, the complexity of the issues under consideration, and have confirmed the lack of statistical and physical justification of many elements in today's flood-frequency-analysis techniques.

The flood frequency analysis problem is complex: it requires a thorough knowledge of statistics, as well as an understanding of the physical phenomena involved. The statistical analysis is made especially complicated by the fact that available data sets are too short to provide a good basis for selecting the "appropriate" D/E procedure.

Statistical inference involves two types of uncertainty which must be investigated and understood: (1) sampling uncertainty and (2) model uncertainty. The practitioner is interested in the global uncertainty of the final results, which is a combination of these two. The global uncertainty is especially pronounced when flood quantiles beyond the observed sample need to be estimated ("extrapolation"; Ashkar and Bobée, 1987). The model error is relatively more important in extrapolations, whereas for small return periods ("interpolation") the global uncertainty is mainly due to sampling. Both sources of uncertainty are highly dependent on the amount of data available.

CHOICE OF DISTRIBUTION FOR FITTING AMF SERIES

The choice of distributions (D) for fitting AMF series involves some complex considerations: (1) choice of the number of distribution parameters; (2) specification of general "desirable" properties that an AMF distribution should possess; (3) choice of criteria to be used for selecting between two or more distributions (with the same, or different number of parameters); and (4) identification of the factors that determine the "descriptive" and "predictive" abilities of distributions (Cunnane, 1987).

For descriptive purposes, it is convenient to have at least three distribution parameters in order to provide sufficient flexibility to the large variety of flood data found in practice. However, with typical sample sizes available, two-parameter distributions should in many cases be preferred for prediction in order to reduce the sampling variance. The selection between two distributions with the same, or different, number of parameters can pose major problems. Statistical goodness-of-fit tests are not a powerful tool for choosing between distribution candidates when small samples are involved, as is commonly the case in flood frequency analysis. In addition, when choosing between distributions with different numbers of parameters, these tests generally tend to put more emphasis on goodness of fit than on the uncertainty due to additional parameters. In this way, they tend to be biased towards the distribution with the largest number of parameters, as noted by Chow and Watt (1992). These authors propose the Akaike information criterion (AIC) (Akaike, 1974) as a measure to evaluate the "benefit" of goodness of fit and the "cost" of parameter uncertainty, but much research remains to be done to test the practical usefulness of this approach.

It has been suggested by some investigators that distributions should have both a cdf (cumulative distribution function) and an inverse cdf expressable in closed form, and that distributions should not be computationally complex. However, although desirable, these characteristics should not form a strict basis for choosing one type of AMF distribution over another. With the advancement of computer technology, mathematically complex pdf's can no longer be dismissed simply on the grounds of analytical intractability, especially when they may have other desirable characteristics that compensate for their mathematical complexity.

METHODS OF PARAMETER ESTIMATION (E)

For many AMF distributions, the maximum likelihood (ML) method often yields asymptotically optimal estimators of the parameters (unbiased, minimum variance). However, the ML method may produce bad estimators in small samples, especially when the random variable is restricted to an interval that depends on the parameters (e.g., distributions with a parameter of origin; Bobée, 1979). For this reason, hydrologists have often opted for other fitting methods, among which the method of moments has been greatly favoured for its simplicity. In the case of a three-parameter distribution, the classical use of this method is based on the sample mean, variance, and skewness. However, recent tendencies have been to avoid the use of the coefficient of skewness in favour of moments that are less variable such as the first and/or second order moments in log space ("method of mixed moments"; Rao, 1980; Phien and Hira, 1983) or the geometric and/or harmonic mean in real space ("generalized method of moments"; Ashkar et al., 1988). Greenwood et al. (1979) introduced the concept of probability weighted moments (PWM), which are linear combinations of order statistics. PWM's have gained wide popularity, and they have been used in many studies (Landwehr et al., 1979; Hosking et al., 1985b, Hosking and Wallis, 1987). Hosking (1990) suggested the application of L-moments, which are similar to PWM's but with a clearer statistical interpretation.

A key point concerning all these different parameter estimation methods is that they can differ greatly in the weights that they give to the different elements of the data set (sample values). For example, the method proposed by the U.S. Water Resources Council (WRC, 1967) for fitting the LP3 distribution uses moments in log space to fit the distribution, whereas another method proposed by Bobée (1975) uses moments in real space. Obviously, the two methods differ significantly in the weights they assign to the different elements of the data series.

Hydrologists have consecrated much research to higher order moments of distributions (particularly moments of order 3 and higher) because of their usefulness as shape indices for AMF distributions, and due to their large influence on the distribution tails. The sample coefficient of skewness has been given special attention in the hydrological literature. Some investigators have argued against the use of such higher order moments in fitting statistical distributions to flood data, because of the large variability of these sample moments. However, it should be remembered that the ultimate interest of flood frequency procedures is in the estimation of flood quantiles, and not in the estimation of the distribution moments themselves. Therefore, this argument against the use of higher sample moments needs to be examined more critically, with emphasis being put on the flood characteristics being estimated, which are flood quantiles. In reality, the argument against the use of higher sample moments for estimating flood frequency overlooks the very important fact that the variance of the estimated flood quantiles is a function not only of the *variance* of the sample moments that are used in the estimation, but also of the degree of *correlation* (*covariance*) that exists between these moments. This *covariance* problem resembles that of multicollinearity of the explanatory variables which is often encountered in regression analysis, where the collinear variables do not provide new information, and their effect on the dependent or response variable is difficult to separate. The effect of covariance between sample moments on flood-quantile estimators is a problem that needs further investigation in the hydrological literature, as proposed by Ashkar and Bobée (1987).

Another aspect of the (generalized) method of moments that needs to be emphasized is that the optimal choice of moments to be used in flood quantile estimation *depends on the distribution considered*, such as the normal, gamma, Weibull, etc. Statistical theory based on sufficient statistics shows that to estimate normal (Gaussian) distribution parameters, the sample arithmetic mean, and variance (moments of order 1 and 2) are the "optimal" choice (in the sense that they lead to unbiased minimum variance estimators of the distribution parameters). On the other hand, in the case of the gamma distribution, this same theory shows that the geometric and arithmetic means (respectively moments of order "quasi-zero" (Ashkar and Bobée, 1986; Bobée and Ashkar, 1991), and 1) are the best choice. Therefore, it is important not to lose sight of the fact that the optimal choice of moments to be used in fitting AMF series is closely tied to the type of distribution under consideration.

For estimation of quantiles from AMF series, the following two steps are usually carried out:

1. look for the best possible distribution (D) to fit the data set from the group of (best known) candidates; and

2. search for the best estimation method (E) for the parameters of the selected distribution, either on the basis of statistical theory, or through the help of simulation.

Very seldom have hydrologists taken the opposite route to this problem, by fixing a parameter-estimation method with desirable statistical properties, and then analytically attempting to develop a pdf that takes advantage of these properties. However, some interesting work along this path has been done by the French hydrologist Etienne Halphen, in the 1940's and 1950's. Starting from the desirable statistical properties of the ML method, and the related statistical theory of sufficient statistics, Halphen developed an interesting system of probability distributions comparable in richness (pdf shape diversity and flexibility) to the Pearson System, but with some favourable theoretical characteristics of its own, which are particularly useful in small-sample applications. Halphen's pdf's belong to the exponential class of pdf's with sufficient statistics, for which unbiased minimum variance estimators of the parameters can be found, based on ML-estimators (Lehmann, 1983). Other examples can be cited from hydrology of innovative work similar to Halphen's, which can be very beneficial to any scientific field. For a discussion of the Halphen System of distributions, the reader is referred to Morlat (1956) or to Bobée et al. (1993b).

SENSITIVITY TO "OUTLIERS"

Statisticians have often recommended the elimination of "outliers" before doing any parametric analysis to a data set. In hydrology, some authors have suggested that an AMF distribution must be insensitive to outliers, especially in the upper tail. Two fundamental questions arise:

1. how do we define an outlier? (In fact, there is no generally accepted definition of this term that bears special interest to the flood frequency analysis problem); and

2. what do we do when an outlier exists?

The answers to these two questions are by no means simple. The treatment of outliers is, in fact, one of the most complex problems of flood frequency analysis which demands a special understanding of both statistics and the physical problem at hand. In fact, statisticians who have recommended the elimination of outliers from data sets have not set generally accepted procedures for rejecting them.

It is important to stress that the sensitivity to outliers is dependent on both the statistical distribution (D) being used, as well as on the method of parameter estimation (E) (order of the moments being used with the method of moments, for example). Therefore, it is more appropriate to talk about the sensitivity of a D/E procedure to outliers, rather than the sensitivity of a pdf.

It was mentioned earlier that different parameter estimation methods can differ greatly in the weights they give to the different elements of the sample. This difference can become particularly pronounced when outliers are present in the data set. Some hydrological studies have suggested that a D/E procedure that is sensitive to outliers is inadequate for flood frequency analysis. However, for the purpose of estimating rare flood events, a procedure's sensitivity to outliers needs to be balanced against its "insensitivity" to them, in the same way that a statistical distribution's descriptive ability needs to be balanced against its predictive ability (Cunnane, 1987; Ashkar and Bobée, 1987). In other words, just as a pdf's descriptive ability should be allowed to overly dominate its predictive ability in estimating flood quantiles by extrapolation, a D/E-procedure's insensitivity to outliers should not be allowed to overly dominate its sensitivity to them, for the same purpose.

"FLOOD-LIKE" DISTRIBUTIONS AND THE ISSUE OF "ROBUSTNESS"

In statistics, the robustness concept is used to characterize a method of parameter estimation or a test of hypothesis. A parameter estimator is considered robust if small departures from the assumptions under which the estimator is derived have only minor influence on properties of the estimator, such as bias, variance, etc. For instance, an estimator of the population mean can be considered robust if its properties are not sensitive to small differences between the assumed model and the parent distribution.

The same concept has been used in hydrology for quantile (Q_T) estimators. So-called "Robustness studies" have been conducted by some investigators using Monte Carlo simulation to test and compare different regional flood-estimation procedures (e.g., Wallis and Wood, 1985; Hosking et al., 1985a). In these studies, a D/E procedure was considered "robust" if Q_T(D/E) is a good estimate of Q_T even if the procedure is based on an assumption which is not true (WMO, 1989). However, a major difficulty arises here, as to the interpretation of the word "robust".

In fact, to conclude what methods are "robust", the studies mentioned above made the explicit assumption of particular parent distribution families, such as the Two-Component Extreme Value (TCEV) and Wakeby, that they considered more "flood-like" than other families. However, a D/E-procedure that is robust with respect to one parent distribution may not be robust with respect to others, and the above families have not been demonstrated more "flood-like" than other families. Numerous multi-parameter distributions could, in fact, be found (which can be quite distinct from the TCEV and Wakeby), whose large flexibility could earn them the title "flood-like". Before one is able to clearly define what this label really means, clear definition of the term "robust D/E procedure" cannot be given.

When the problem is to estimate a quantile by extrapolation, the performances of the estimator will always be very sensitive to the choice of the model. If a sequence of quantiles is to be estimated, then the basic problem is precisely that of the choice of distribution model; therefore attempting to find a robust solution to this problem with respect to some external "flood-like" distribution is logically incoherent.

Basically, one seeks the "best" distribution for estimating flood quantiles at a given site (i.e. the one most consistent with regional and at-site information), and this constitutes the most "flood-like" distribution for the given site.

USE OF REGIONAL INFORMATION

It was mentioned earlier that model and sampling uncertainties inherent in the flood frequency estimation process are highly dependent on the amount of data available. To reduce these uncertainties at sites with only short flood records, or to estimate flood quantiles at ungaged sites, regional flood estimation has been recommended. The purpose of incorporating such information is to allow a reduction of sampling uncertainty by introducing more data, as well as a reduction of model uncertainty by facilitating the choice of distribution.

To facilitate the use of regional flood information for the above purposes, hydrologists have introduced the vague notion of "homogeneous regions" which, according to WMO (1989), is a "collection of catchments whose flood statistics are homogeneous". However, this definition is meaningless unless it is clear what flood statistics are referred to, and what is meant by the term "homogeneous".

As noted above, hydrologists have regarded higher order moments of distributions, such as the coefficient of skewness, as important shape indices of flood frequency distributions. Consequently, one of the most frequently used definitions of "homogeneous region" in hydrology, has been based on the coefficient of skewness (Cs). A homogeneous region, according to this definition, is a collection of catchments within which: (1) all floods follow the same type of distribution, and (2) all floods have the same Cs value.

Strictly speaking, however, it is contrary to physical logic to combine several basins having different hydrological characteristics (especially different sizes) into one region to be called "homogeneous" (as argued by Klemes, 1976). On the other hand, one can also argue that, although this homogeneity does not exist in the strict sense, it can still serve its useful role in improving the accuracy of flood estimates at ungaged sites, or at sites with short record by including information from "neighboring stations", or stations within the so-called "homogeneous region". This is a sound argument, but it is still not clear whether the best way to incorporate this kind of regional information is by using one single type of probability distribution (or one single Cs value or any other distributional shape characteristic) to represent a homogeneous region. Moreover, the problem of how to divide a large group of catchments into homogeneous subgroups is one that still requires much research.

"SEPARATION OF SKEWNESS" OR "SEPARATION EFFECT"

Hydrologists wanting to use the definition of homogeneous region as presented in the previous section, are faced with the question of what distribution to choose to represent the homogeneous region. A study that attempted to address this question was published by Matalas et al. (1975). This study examined the properties of sample coefficient of skewness for 1351 AMF flood samples in fourteen hydrologically "homogeneous" regions in the USA and concluded that the dispersion in skewness values among the various series in each region was larger than what is expected from samples randomly generated from a specific group of distributions (normal, log-normal, Gumbel, Pearson Type 3, Weibull, Pareto and Uniform). The authors called this phenomenon the "condition of separation (of skewness)" or "separation effect" and considered it an important criterion for choosing flood frequency distributions. The tendency for historical flood sequences in a region to exhibit greater variability in sample skewness than samples drawn from a particular group of distributions was considered by Matalas et al. (1975) to imply that no member of this specific group was capable of adequately estimating flood probabilities in the regions under consideration. The U.S. study was repeated with flood data from other countries (Rossi et al., 1984; Cunnane, 1984) with similar results (WMO, 1989). The above findings have been used by some hydrologists to establish the requirement that for a distribution to be adequate for flood frequency analysis, it must satisfy the "reproductive criterion", i.e. it must reproduce at least as much variability in flood characteristics (particularly skewness) as is observed in flood data sets (WMO, 1989; Ahmad et al. 1988).

There are at least two serious concerns about the way "separation studies" have been conducted and used in hydrology.

1. The regions considered by Matalas et al. (1975; Fig. 1) are identical to the "parts" used by the U.S.G.S. for publishing Water Supply Papers, which were determined at least partly for administrative reasons (Bobée et al., 1993a). It is very difficult to justify from the physical standpoint how such *large geographical regions* could be considered as hydrologically homogeneous. Matalas et al. (1975), and subsequent "separation studies", failed to show whether the actual cause of the separation phenomenon was the type of probability distribution used to fit the flood data, or whether a more natural cause was the combining in the same region of flood samples with different skewness.

2. It is of concern to note that, over time, a shift has taken place in hydrology from considering the condition of separation as a *regional attribute*, to one that is *site-specific*.

Serious concerns have been expressed by some hydrologists about the separation phenomenon (Klemes, 1976; Ashkar and Bobée, 1989; Bobée and Ashkar, 1989), but until recently no comprehensive study using real data had been conducted to determine, in a clear manner, whether the actual cause of the separation phenomenon was the type of probability distribution used to fit the flood data or whether a more important factor was the combining in the same region of flood samples with different skewness. Ashkar et al. (1992) recently examined the problem and were able to present evidence that the real cause of separation was spatial mixing of different skewness values. The conclusion of that study was that it is absolutely not justified to use the separation of skewness as a criterion for comparing probability distributions for flood frequency analysis.

NEED FOR SYSTEMATIC APPROACH TO COMPARE D/E PROCEDURES

The user who works with practical problems of flood estimation needs objective guidelines to avoid relying on his or her own subjective judgment, which ultimately could make him or her responsible for the failure or over-design of a hydraulic structure. In the 1960's and 1970's several countries embarked on large-scale studies of their own flood data which aimed at establishing recommendations of standard methods to be used by their government or private agencies in order to achieve uniformity in flood frequency estimation. An extensive study in the USA (WRC, 1967; Benson, 1968) led to the recommendation of the log-Pearson Type 3 (LP3) distribution, and a similar study in the UK (NERC, 1975) led to the recommendation of the Generalized Extreme Value (GEV) distribution as standard. The Pearson Type 3 (P3) and LP3 distributions were generally recommended in West Germany (Mosonyi et al., 1975), and the LP3 distribution was proposed by the Institution of Engineers in Australia (IEA, 1977).

The diversity of "standard methods" recommended in different countries implies the question: "Are we using the 'best' method?" Another question to be asked is: given the need for uniform practices, do we look for a "global" distribution to be used everywhere as standard, or should the data observed at a given site influence the choice of distribution?

Until now, no comprehensive comparison of distributions and estimation methods has been made, although a large number of partial comparisons based on either simulation or real data have been conducted. Unfortunately, the usefulness of these comparisons is limited because they are based on different simulation schemes and different comparison criteria. As a consequence, it is not possible to combine them to obtain a clear picture of which procedures are the most adequate.

Hence, there are compelling reasons to reconsider more carefully the problem of flood frequency analysis and to define comparison schemes and criteria which can be broadly agreed upon. What is needed is a systematic approach to this problem that will allow a more comprehensive comparison of the various D/E procedures (Bobée et al., 1993a). In this comparison, it would be beneficial not to rely entirely on statistical criteria (mean square error, bias, etc., of quantile estimates), but to consider a more general multiple-criteria-decision approach that allows other criteria to be incorporated into the comparison, such as computational ones (e.g., computer program availability, ease of use) or a user-related criterion, such as acceptability (Duckstein et al., 1991).

CONCLUSION

The problem of flood frequency estimation is one of the most challenging in hydrology, and one of great practical importance. However, it is filled with controversies that need to be sorted out and dealt with. Misleading or imprecise concepts such as "robustness to the distribution choice", "sensitivity to outliers", and the notion of "separation of skewness" have been used in the area of AMF-series analysis, and have led to a state of confusion among practitioners. This state is common to both at-site and regional AMF procedures. Efforts to clarify some of the controversies have not been successful in the past. One wonders whether time is more ripe now for a serious discussion of what should be done to resolve some of the outstanding dilemmas. It is hoped that as long as interest in sorting out these controversies is alive, there is hope that a systematic approach to handling the problem could be agreed upon as called for by Bobée et al. (1993a).

ACKNOWLEDGMENTS

The financial support of the Natural Sciences and Engineering Research Council of Canada, and of the Ministry of Intergovernmental affairs of New Brunswick (Cooperation N.B.-Québec) is gratefully acknowledged.

REFERENCES

Ahmad, M.I., Sinclair, C.D. and Werritty, A. (1988) "Log-logistic flood frequency analysis", J. Hydrol., 98, 205-224.

Akaike, H. (1974) "A new look at the statistical model identification", IEEE Transactions on Automatic Control, AC-19(60), 716-721.

Ashkar, F. and Bobée, B. (1986) "Comment on: variance of the T-Year event in the log-Pearson Type 3 distribution", J. Hydrol., 84(1), 181-187.

Ashkar, F. and Bobée, B. (1987) "The generalized method of moments as applied to problems of flood frequency analysis: some practical results for the log-Pearson Type 3 distribution", J. Hydrol., 90(3-4), 199-217.

Ashkar, F. and Bobée, B. (1989) "Reflection on some problems of flood frequency analysis", EOS: Trans., American Geophysical Union 70(15), 325 (abstract).

Ashkar, F. and Bobée, B. (1990) "The Halphen system of distributions for flood frequency analysis", EOS: Trans., American Geophysical Union 71(28), 871 (abstract).

Ashkar, F., Bobée, B., Leroux, D. and Morisette, D. (1988) "The generalized method of moments applied to the generalized gamma distribution", Stochastic Hydrol. Hydraul., 2, 161-174.

Ashkar, F., Bobée, B. and Bernier, J. (1992) "Separation of skewness: reality or regional artifact?" J. Hydraul. Eng., ASCE, 118(3), 460-475 (with discussion).

Benson, M.A. (1968) "Uniform flood frequency estimating methods for federal agencies", Water Resour. Res., 4(5), 891-908.
Bobée, B. (1975) "The log-Pearson Type 3 distribution and its application in hydrology", Water Resour. Res., 11(5), 681-689.
Bobée, B. (1979) "Comment on: The log Pearson type 3 distribution, the T-year event and its asymptotic standard error by maximum likelihood theory" by R. Condie, Water Resour. Res., 15(1), 189-90.
Bobée, B. and Ashkar, F. (1989) "Comment on: Log-logistic flood frequency analysis" by M.I. Ahmad, C.D. Sinclair and A. Werritty, J. Hydrol., 107, 367-372.
Bobée, B. and Ashkar, F. (1991) The Gamma Family and Derived Distributions Applied in Hydrology, Water Resources Publications, Littleton, Colo., 217 p.
Bobée, B., Cavadias, G., Ashkar, F., Bernier, J. and Rasmussen, P. (1993a) "Towards a systematic approach to the comparison of distributions used in flood frequency analysis", J. Hydrol., 142, 121-136.
Bobée, B., Rasmussen, P., Perrault, L. and Ashkar, F. (1993b) "Risk analysis of hydrological data: Review and new developments", To appear in NATO ASI Series (NATO ASI "Engineering Risk and Reliability in a Changing Physical Environment"; 24 May - 4 June; Deauville, France).
Chow, K.C.A. and Watt, W.E. (1992) "Use of Akaike information criterion for selection of flood frequency distribution", Canadian Journal of Civil Engineering, 19, 616-626.
Cunnane, C. (1984) "Condition of separation of skewness of random samples from the General Extreme Value distribution", Unpublished report, Department of Engineering Hydrology, University College, Galway, 7 p.
Cunnane, C. (1987) "Review of statistical models for flood frequency estimation", In V.P. Singh (ed.), Hydrologic Frequency Modeling, D. Reidel Publishing Company, pp. 49-95.
Duckstein, L., Bobée, B. and Ashkar, F. (1991) "A multiple criteria decision modelling approach to selection of estimation techniques for fitting extreme floods", Stochastic Hydrology and Hydraulics, 5(3), 227-238.
Greenwood, J.A., Landwehr, J.M., Matalas, N.C. and Wallis, J.R. (1979) "Probability weighted moments: Definition and relation to parameters of several distributions expressable in inverse form", Water Resour. Res., 15(5), 1049-1054.
Hosking, J.R.M. (1990) "L-moments: Analysis and estimation of distributions using linear combination of order statistics", J. Roy. Stat. Soc., Series B, 52(1), 105-124.
Hosking, J.R.M. and Wallis, J.R. (1987) "Parameter and quantile estimation for the generalized Pareto distribution", Technometrics, 29(3), 339-349.
Hosking, J.R.M., Wallis, J.R. and Wood, E.F. (1985a) "An appraisal of the regional flood frequency procedure in the UK flood studies report", Hydrol. Sci. J., 30, 85-109.
Hosking, J.R.M., Wallis, J.R. and Wood, E.F. (1985b) "Estimation of the generalized extreme value distribution by the method of probability weighted moments", Technometrics, 27(3), 251-261.
IEA (Institution of Engineers, Australia) (1977) "Australian rainfall and runoff: flood analysis and design",

Klemes, V. (1976) "Comment on: Regional skew in search for a parent" by N.C. Matalas, J.R. Slack and J.R. Wallis, Water Resour. Res., 12(6), 1325-1326.
Landwehr, J.M., Matalas, N.C. and Wallis, J.R. (1979) "Probability weighted moments compared with some traditional techniques in estimating Gumbel parameters and quantiles", Water Resour. Res., 15(5), 1055-1064.
Lehmann, E. L. (1983) Theory of point estimation, John Wiley & sons, New York.
Lettenmaier, D.P. (1988) "Evaluation and testing of flood frequency estimation methods", Proceedings, International Conference on Natural Hazards in European-Mediterranean Countries, Perugia, Italy, June.
Matalas, N.C., Slack, J.R. and Wallis, J.R. (1975) "Regional skew in search of a parent", Water Resour. Res., 11(6), 815-826.
Morlat, G. (1956) "Les lois de probabilités de Halphen", Revue de statistique appliquée, 4(3), 21-46.
Mosonyi, E.F., Eggers, H. and Koberg, D. (1975) "Influence of probability analysis on selection of design flood", II. World Congress of IWRA, New Delhi.
NERC (Natural Environment Research Council, London) (1975) Flood Studies Report, Vols. 1-5, 1100 p.
Phien, H.N. and Hira, M.A. (1983) "Log-Pearson type 3 distribution: parameter estimation", J. Hydrol., 64, 25-37.
Potter, K.W. (1987) "Research in flood frequency analysis: 1983-86", in J.R. Wallis (ed.), U.S. National Report to IUGG 1983-86, Reviews of Geophysics, 25(2), pp. 113-118 .
Rao, D.V. (1980) "Log Pearson type 3 distribution: method of mixed moments", J. Hydraul. Div., ASCE, 106(6), 999-1019.
Rossi, F., Fiorentino, M. and Versace, P. (1984) "Two-component extreme value distribution for flood frequency analysis", Water Resour. Res., 20(7), 847-856.
Wallis, J.R. and Wood, E.F. (1985) "Relative accuracy of Log Pearson 3 procedures", J. Hydr. Eng., ASCE, 111(7), 1043-1057.
WMO (World Meteorological Organization) (1989) "Statistical distributions for flood frequency analysis", Operational Hydrology, Report N° 33. WMO Publ. N° 718, WMO, Geneva, Switzerland.
WRC (Water Resources Council; USA) (1967) "A uniform technique for determining flood flow frequencies", Bull. No. 15, Washington, D.C.

THE POT METHOD FOR FLOOD ESTIMATION: A REVIEW

P. F. RASMUSSEN[1], F. ASHKAR[2], D. ROSBJERG[3], and B. BOBÉE[1]
[1]INRS-Eau, University of Quebec, 2800 rue Einstein, Sainte-Foy (PQ), G1V 4C7, Canada
[2]Dept. of Mathematics, University of Moncton, Moncton (N-B), E1A 3E9, Canada
[3]Institute of Hydrodynamics and Hydraulic Engineering, Technical University of Denmark, 2800 Lyngby, Denmark

Elements of the peaks over threshold (POT) method are reviewed. We discuss the need for statistically objective criteria to determine which flood peaks should be used for extreme value analyses, and to select appropriate thresholds. Attempts have been made to define such criteria, but none of those have yet gained broad acceptance. Distributions of exceedances are discussed, and the advantages of the generalized Pareto (GP) distribution (with the exponential as a special case) are pointed out. However, although the GP distribution is more flexible than the exponential and has been demonstrated to fit observed flood data well, the exponential distribution may be quite appropriate for flood frequency analysis. Methods for including seasonality exist, but for prediction purposes a non-seasonal model is usually the best choice. Comparisons of POT analysis and annual maximum flood (AMF) analysis have demonstrated the advantages of the former, especially when data series are short.

DATA SERIES FOR FLOOD FREQUENCY ANALYSIS

Flood frequency analysis is the hydrologic field dealing with estimation of extreme flood events. Based on experience, people have some ideas as to the probability of occurrence of floods of given size at given places. Hydrologists attempt to formalize these ideas by establishing networks of gauging stations and to analyze the recorded data. According to Kirby and Moss (1987), the US Geological Survey started its gauging program in 1888, but the early data are not reliable, and the typical record of discharge measurements rarely exceeds fifty years. The objective of flood frequency analysis is to extract as much information as possible from such short data series of daily discharges in order to furnish an estimate of a flood corresponding to a specified exceedance probability (risk). Basically, this kind of analysis can be accomplished in three different ways:

K. W. Hipel (ed.),
Stochastic and Statistical Methods in Hydrology and Environmental Engineering, Vol. 1, 15–26.

Type 1: Analysis of the complete data series of mean daily discharges
Type 2: Analysis of partial series comprising data above a certain level
Type 3: Analysis of annual series comprising the largest event in each year

Because of the nonstationarity and dependence structure in mean daily discharge series, a fairly complex model with many parameters is needed if a type 1 analysis is to be used. In reservoir management and other problems related to water yield analysis, it is frequently necessary to consider daily discharges and describe them for instance by ARIMA-models or periodic ARMA models. Such methods are, however, less useful in flood frequency analysis, because of the lack of emphasis put on extreme events. On the other hand, type 3 analysis is based on only the largest mean daily discharge within each year (annual maximum flood, AMF). It is common (and generally reasonable) to assume that annual floods in two consecutive years are independent events, and that series of AMF's are stationary. The type 2 analysis is a compromise between types 1 and 3. In this approach, which is denoted the peaks over threshold (POT) method (or partial duration series method), all flood peaks exceeding the threshold are considered in the analysis. In contrast to the type 1 analysis, it puts emphasis on flood events by discarding the daily discharge values which do not contain information about the extreme value process, and in contrast to the type 3 analysis, it permits to consider a wider spectrum of floods, not just annual floods. In contemporary frequency analysis, the competing alternatives seem to be the types 2 and 3. The POT method provides a more complete description of floods than does the AMF method, but it also gives rise to a number of difficulties which have restricted its popularity among practitioners. Perhaps the greatest difficulty concerns the very definition of POT series, or more specifically, the choice of threshold discharge and the criteria used to retain exceeding flood peaks for the subsequent statistical analysis. This is in clear contrast to the AMF method which is easily defined by simply selecting the largest daily discharge in each year (although in some cases, it may be necessary to choose the largest event in a particular season in order to justify the assumption that flood event are identically distributed). Much of the classical theoretical work on the POT approach (eg. Borgman, 1963; Shane and Lynn, 1964; Bernier, 1967; Todorovic, 1970) merely assumed that a sequence of peaks was available. While such an assumption may be quite appropriate in theoretical studies, the practitioner needs clear guidelines to define data series. In spite of its importance, the literature on this fundamental problem is surprisingly sparse. For instance, the NERC (1975) report, which contains a whole chapter on the POT method, does not directly address the issue. An approach which could be considered a compromise between the POT and AMF methods is to take a fixed number of flood peaks (Cunnane, 1973; Buishand, 1989). Thereby, one circumvents, at least to a certain extent, the problems of defining the POT series and gets a sequence of the most important flood events rather than of annual floods, of which some may not be real extreme events. A series with the N largest flood peaks in a record of N years is called an annual exceedance series. However, fixing a priori the number of peaks has the disadvantage of reducing the flexibility of the POT model. The main strength of the POT model is namely that, by appropriate selection of the threshold, one can include all extreme events in the analysis and only those events. This property is sacrificed if one fixes a priori the number of exceedances to be included in the analysis.

Two approaches to defining peaks can be distinguished:

- methods related to a predetermined threshold
- methods which do not involve a threshold

In the former, a threshold is selected, for example by visual inspection of the hydrograph, and peaks above it are retained. The discharge between two adjacent peaks must drop below the threshold and the peaks must be properly separated in time to justify the assumption of independence. However, there is a certain advantage in extracting peaks without reference to a threshold discharge. Such an approach permits to reduce substantially the data bank of daily discharges, thereby facilitating the subsequent choice of threshold and analysis of peaks. It is of course necessary to impose restrictions on the time interval between successive events and on the minimum flow between two peaks to ensure independence. The USWRC (1976) recommended that exceedances be separated in time by at least 5+ln(A) days, where A is the catchment area in square miles, with the additional requirement that the intermediate flow between two peaks drop below 75% of the smaller. However, such guidelines seem somewhat ad hoc, and there is an obvious need for more research on that issue, with emphasis on both physical and statistical characteristics of floods.

DISTRIBUTION ASSUMPTIONS

A POT analysis requires that flood peaks be modeled in both time and magnitude domains. This type of process is called a marked point process. Several papers in the hydrologic literature have specifically addressed the joint distribution of floods in time and magnitude (Todorovic and Woolhiser, 1972; Gupta et al., 1976; Karr, 1976; Todorovic, 1978). The Poisson model has been extensively used for describing the occurrence of exceedances, and its theoretical justification is demonstrated by Kirby (1969). However, the Poisson assumption is not crucial for flood frequency analysis, since explicit expressions for flood quantile estimators involve the expected annual number of exceedances, but not their distribution. In a study of data from 26 gauging stations in Great Britain, Cunnane (1979) found that the hypothesis that the annual number of exceedances was Poisson distributed had to be rejected when all the stations were considered jointly. The mean and the variance of a Poisson distributed variable are identical, but Cunnane found that there is a tendency for the variance of the annual number of exceedances to be greater than the mean. A potential alternative to the Poisson distribution is the negative Binomial distribution, for which the variance is greater than the mean (Ben-Zvi, 1991). However, Cunnane (1979) found that the negative Binomial distribution provides only a slight improvement over the Poisson model. The reason why the variance is often found to be greater than the mean is that flood peaks tend to occur in bunches. Kavvas (1982a,b) suggested a two-level trigger process where, at the first level, an event, say the spring snow melt, triggers a possible sequence of floods at the second level. The trigger process as well as the occurrence process at the second level can be modeled by Poisson processes. Kavvas derived analytical expressions for the mean number of exceedances in a given time interval, and for the covariance of numbers in two different intervals. The model, however, is complicated and has mainly theoretical interest.

The most important element in the POT method is the modeling of exceedances, which is closely related to the choice of threshold. Some distributions are sensitive to the choice of truncation, while others leave a certain degree of freedom. Ashkar and Rousselle (1983a) showed that the exponential distribution belongs to the latter group, because, if it applies theoretically to a certain threshold, then it also applies to any higher threshold. Among the well-known statistical distributions, the generalized Pareto (GP) distribution seems to be the most general and flexible distribution possessing this property. It contains the exponential, the uniform, and the triangular distributions as special cases (Hosking and Wallis, 1987). The GP distribution was introduced by Pickands (1975), and it has in recent years become very popular in POT analysis (Smith, 1984; Van Monfort and Witter, 1985, 1986; Fitzgerald, 1989; Jin and Stedinger, 1989; Rosbjerg et al., 1992). Another generally desirable feature of an exceedance distribution is the monotonic decrease of the density function (no mode). The GP distribution has that characteristic, but other common distributions have it as well, for example the Weibull and the gamma distributions when their shape parameter is between zero and one. The latter distributions have been used in POT analysis by Miquel (1984), Zelenhasic (1970), and Taesombut and Yevjevich (1978), but their application in practice seems so far to have been limited. However, when dealing with design floods of small return period, a low threshold must be used, and the empirical distribution of exceedances may exhibit modality. Therefore, it may be necessary to consider modal distributions. In addition to the Weibull and gamma distributions with no restriction on the shape parameter, Taesombut and Yevjevich (1978) also considered the log-normal and the Pearson type 3 distributions. Rosbjerg (1987b) specifically examined the use of the log-normal distribution in a POT context. For descriptive purposes, complex distribution assumptions may be appropriate, but for prediction of extreme events, it often turns out that the simple one-parameter exponential distribution outranks more sophisticated two-parameter distributions. Particular studies of the robustness of the exponential assumption have been presented by Rosbjerg et al. (1991) in case of log-normal distributed exceedances, and by Rosbjerg et al. (1992) in case of generalized Pareto distributed exceedances.

CHOICE OF THRESHOLD

No matter what method is used to extract flood peaks, the choice of threshold remains a crucial element in the frequency analysis. Through the selection, one should attempt to satisfy certain distribution criteria such as the Poissonian occurrence of exceedances and exponentiality of flood peaks. In fact, the subjective choice of threshold is a potential advantage of the POT method as compared to the AMF method, where data series are well defined, and the use of a particular distribution often prescribed (e.g. Log Pearson type 3 in the USA, USWRC, 1976). It seems that this important practical problem has been treated only superficially in the scientific literature. Practitioners often choose a threshold corresponding to an average of 2-3 exceedances per year or to a given flood quantile calculated from the AMF series, without paying particular attention to, for example, the fit of exceedances to a given distribution type. It is important, however, to ensure that the threshold be selected such as to separate informative events from those of no relevance for the extreme value process.

Ashkar and Rousselle (1987) exploited the fact that, if floods occur according to a Poisson process, the mean and the variance of the annual number of exceedances are identical. They constructed graphics showing the mean-to-variance ratio as a function of the threshold, and chose the threshold for which the ratio was equal to one. Good results were obtain in that particular study, but the method has the disadvantage of paying no attention to the distribution of exceedances which is the most important part of the modeling, and it may lead to unreasonable results.

Several other studies have focused on the Poissonian character of exceedances, e.g. Cunnane (1979) and Ashkar and Rousselle (1983b). Based on those, it has been generally recognized that, for estimating large floods, the Poisson assumption is less critical than the assumption of the distribution of exceedances, and thresholds should therefore be selected with special focus on the distribution of exceedances. As mentioned above, candidate distributions can be grouped in two categories:

- those which are sensitive to the choice of threshold (e.g. the Weibull distribution)
- those which are not sensitive to the choice of threshold (e.g. the exponential distribution)

Considerations regarding the distribution of exceedances should play an important role in selecting an appropriate threshold.

A common reasoning is that the threshold should be chosen as low as possible (with proper attention paid to the various distribution assumptions) in order to include as many exceedances as possible in the analysis. This is presumed to increase the reliability of flood estimates. However, in a recent study, Wang (1991) found that for certain models the variance of flood estimates actually decreases when the threshold is increased. Smith (1987) noted that due to difficulties in specifying the correct parametric form of probability distributions, it could be advantageous to select a high threshold and focus only on few of the largest flood peaks.

Another important consideration is the dependence among exceedances that may arise if the threshold is chosen too low. Yevjevich and Taesombut (1979) examined data for a particular river in the USA and found that the first-order serial correlation coefficient between peaks increased with the average number of exceedances per year, reaching a value of about 0.15 when the annual mean number of peaks was above 5-6. Theoretical studies of the implications of correlation have been published by Rosbjerg (1985), who assumed that exceedances could be modeled as a first-order autoregressive process. The meaning of serial correlation in POT series is, however, not obvious, because the data are not equally separated in time. NERC (1975) and Cunnane (1979) chose to plot peak values (Q1, Q2) separated by less than ten days in a scatter diagram in order to detect correlation between peaks separated by small time intervals. Cunnane (1979) concluded that correlation is not a major problem in POT analysis.

If a distribution is selected a priori from the class of distributions which are not sensitive to the threshold, then a possible approach is to compute a flood estimate (or moment or parameter estimate) and to depict it as function of different thresholds. From a high threshold and downwards, the quantile estimate should theoretically remain constant except for sampling variability until the point is reached where the distribution assumptions no longer apply. This could then be taken as an appropriate threshold for the analysis.

A review of various methods for threshold selection is provided by Rosbjerg and Madsen (1992). They recommended either the use of a statistical test for separation of the extreme region, or use of a threshold corresponding to the mean plus three times the standard deviation of daily discharges.

SEASONALITY OF EXCEEDANCES

Supporters of the AMF method often claim that stationarity conditions cannot be maintained in the POT analysis, because floods in different seasons have different statistical properties. To this, partisans of the POT method will argue that AMF analysis is hampered by the same kind of problems, and that the POT analysis provides the tools to cope with seasonality. For example, in a case study, Waylen and Woo (1982) clearly identified two distinct flood generating mechanisms and showed that, to obtain a good fit to the AMF series, it was necessary to consider a combination of two different distributions. On the other hand, Cruise and Arora (1990) also partitioned data according to climatic causes and found that, although significant differences existed between the arrival structure of floods caused by different climatic mechanisms, differences did not exist in the distribution of flood magnitudes.

No hydrologist will deny the seasonality of flood generating processes. In the classical work of Todorovic and Zelenhasic (1970), it was assumed that the occurrences of exceedances could be described by a nonhomogeneous Poisson process, whereas the magnitudes of exceedances were assumed identically distributed over the year. Todorovic and Rousselle (1971) generalized these assumptions by allowing a seasonal variation in the distribution of peak magnitudes. More specifically, they assumed that within a given season of the year exceedances were identically distributed. A further sophistication was presented by North (1980), who considered a continuous variation of the distributions of exceedances and of the occurrence rate. Recently, Rasmussen and Rosbjerg (1991b) examined the seasonality issue from a purely theoretical point of view, and concluded that for prediction purposes, it is rarely necessary to use the model of Todorovic and Rousselle (1971), since one of the following two situations generally prevails: (1) the majority of large floods occur within a given season of the year while the other seasons contain only minor floods; the frequency analysis could then be performed using floods only from the dominant season (hence no need for a seasonal model); or (2) the statistical difference between floods in two seasons is so small that, although a certain seasonality must be assumed, more accurate estimates of floods can be obtained by pooling together data from the two seasons. This reduces the number of model parameters and leads to a decrease in mean square error of flood estimates. The seasonal model may, however, be quite appropriate for descriptive purposes, and also in cases where floods of low return period are to be estimated.

QUANTILE ESTIMATION

The ultimate objective of flood frequency analysis is to estimate floods with given return period (i.e. average time between occurrences). It can be shown that, if λ is the average number of exceedances of the threshold, and T is the return period, then the theoretical T-year event is the $(1-1/\lambda T)$-quantile in the distribution of exceedances. From observed flood peaks one can estimate the model parameters, and thereby the T-year event. Such estimates are affected by errors due to incorrect model specification and to sampling variability (plus a number of other sources like measurement errors and inaccurate rating curve at high discharges). The scientific literature tends to focus on the errors caused by the inherent randomness in flood flows, whereas not much attention has been paid to model error. This is not because the latter is less important, but rather because the former is easier to treat statistically in a rigorous way. Quantile estimators and their precision in the classical model with exponentially distributed exceedances have been thoroughly examined by many researchers, for instance Cunnane (1973), Ashkar and Rousselle (1981), Konecny and Nachtnebel (1985), Nachtnebel and Konecny (1987), Rasmussen and Rosbjerg (1989), and Buishand (1990).

As an alternative to classical statistical estimation of confidence intervals, Bayesian analysis can be used to establish so-called credible regions. For the POT model with exponentially distributed peaks, simple analytical results can be obtained (Rousselle and Hindie, 1976; Rasmussen and Rosbjerg 1991a,c). For more complicated POT models, such as those based on the GP distribution, the method used to estimate the model parameters may significantly affect the performance of the T-year event estimator (for the exponential distribution there is essentially only one estimation method.) Hosking and Wallis (1987) compared the methods of moments, probability weighted moments, and maximum likelihood for estimating the parameters of the GP distribution. A similar study with specific emphasis on the application of the GP distribution in POT analysis was published by Rosbjerg et al. (1992). Wang (1991) compared variations of the method of probability weighted moments in the case of GP distributed exceedances, and found that, by estimating a GP distribution with unknown, rather than fixed, location parameter, one can in some cases reduce the variance of quantile estimators.

COMPARISON OF POT AND AMF MODELS

A question that naturally arises in flood frequency analysis is which methodology, POT or AMF, yields the most precise T-year flood estimates. In order to compare the two approaches, it is necessary to establish a mathematical link between the POT series and the

AMF series. One of the very first papers on the POT method, published by Langbein (1949), actually addressed this problem. Todorovic and Zelenhasic (1970) showed that the POT model with exponentially distributed peaks leads to a Gumbel distribution of annual floods (at least those above the threshold level). Cunnane (1973) examined the performance of the two models and found that for reasonably long records, the POT models yields a more precise estimate of the T-year flood when the average number of peaks per year is greater than 1.65. The annual exceedance model generally performs worse than the AMF model. In a simulation study, Tavares and Da Silva (1983) found a pronounced discrepancy between Cunnane's asymptotic variance expression and the variance obtained by Monte Carlo simulation. Rosbjerg (1985) demonstrated that this was because the randomness in the number of exceedances was only partly accounted for in Cunnane's expression, and he deduced a correction factor, which improves the accuracy of the asymptotic variance expression. Rasmussen and Rosbjerg (1989) calculated the exact bias and variance of the T-year flood estimator and used this result to show that the approximate formulae based on asymptotic theory are quite precise. Yevjevich and Taesombut (1979) found that the simple POT model may give much better results than the AMF model (Gumbel) when the average number of exceedances is large (3-5). However, a high number of peaks can cause dependence between successive peaks and thus reduce the gain of the POT method (Yevjevich and Taesombut, 1979; Rosbjerg, 1985, 1987a). Ashkar et al. (1987) also examined the validity of Cunnane's result, and found that for typical sample sizes in hydrology (20-40 years), the average number of exceedances needed for the POT model to perform better than the Gumbel distribution is markedly less than the asymptotic 1.65 deduced theoretically by Cunnane.

An interesting property of the GP distribution in POT analysis is that the derived AMF distribution is generalized extreme value (GEV) (Jenkinson, 1955). The GEV distribution is extensively used for modeling annual floods, and has been recommended as standard in the UK (NERC, 1975) - an additional reason for the popularity of the GP distribution in POT analysis. Just as the exponential distribution is a limiting form of the GP distribution obtained when the shape parameter tends to zero, so the Gumbel distribution is a particular case of the GEV, also obtained as the shape parameter tends to zero. There is thus consistency with previous results. The GEV/AMF model can be compared to the GP/POT model as done by Jin and Stedinger (1989) and Wang (1991). Both of these studies conclude that the two models perform equally well when the average number of exceedances is about one. The comparison is more complicated, however, because a number of estimation methods have to be considered. Jin and Stedinger used the maximum likelihood method for parameter estimation, while Wang used probability weighted moments, but other studies have concluded that the method of moments generally is the best method for estimating the parameters of the GP distribution (Hosking and Wallis, 1987; Rosbjerg et al., 1992).

CONCLUSIONS

Compared to the AMF method, the POT approach provides a more complete description of flood generating processes. The former, however, still maintains a leading role in contemporary flood frequency analysis, primarily because of its simplicity. During the last two decades, much research has been done on the POT method, but recent tendencies have been to focus on theoretical aspects rather than to solve more fundamental problems pertaining to the practical use of the method. In fact, this is the general situation in flood frequency analysis as pointed out by Bobée et al. (1993). The development of guidelines for an optimal choice of threshold is probably the greatest challenge, but several other issues need to be carefully studied, for example the selection of peaks to be included in the analysis, the meaning and implications of independence, seasonality etc. The POT method has rarely been used in a regional context, and there is a need for regional procedures based on the POT model, for instance of the index flood type.

REFERENCES

Ashkar, F. and Rousselle, J. (1981) "Design discharge as a random variable: A risk study", Water Resources Research 17, 577-591.

Ashkar, F. and Rousselle, J. (1983a) "Some remarks on the truncation used in partial flood series models", Water Resources Research 19, 477-480.

Ashkar, F. and Rousselle, J. (1983b) "The effect of certain restrictions imposed on the interarrival times of flood events on the Poisson distribution used for modeling flood counts", Water Resources Research 19, 481-485.

Ashkar, F. and Rousselle, J. (1987) "Partial duration series modeling under the assumption of a Poissonian flood count", Journal of Hydrology 90, 135-144.

Askhar, F., El-Jabi, N., and Bobée, B. (1987) "On the choice between annual flood series and peaks over threshold series in flood frequency analysis", in A.D. Feldman (ed.), Proc. Eng. Hydrol., Virginia, August 3-7, 1987, ASCE, New York, pp. 276-280.

Ben-Zvi, A. (1991) "Observed advantage for negative binomial over Poisson distribution in partial duration series", Stochastic Hydrology and Hydraulics 5, 135-146.

Bernier, J., (1967) "Sur la théorie de renouvellement et son application en hydrologie", HYD. 67(10), Elec. France.

Bobée, B., Cavadias, G., Ashkar, F., Bernier, J., and Rasmussen, P.F. (1993) "Towards a systematic approach to comparing distributions used in flood frequency analysis", Journal of Hydrology 142, 121-136.

Borgman, L.E. (1963) "Risk criteria", Journal of Waterways and Harbors Division 80, 1-35.

Buishand, T.A. (1989) "The partial duration series with a fixed number of peaks", Journal of Hydrology 109, 1-9.

Buishand, T.A. (1990) "Bias and variance of quantile estimates from a partial duration series", Journal of Hydrology 109, 35-49.

Cruise, J.F. and Arora, K. (1990) "A hydroclimatic aplication strategy for the Poisson partial duration model", Water Resources Bulletin 26, 431-442.

Cunnane, C. (1973) "A particular comparison of annual maxima and partial duration series methods of flood frequency prediction", Journal of Hydrology 18, 257-271.

Cunnane, C. (1979) "A note on the Poisson assumption in partial duration series models", Water Resources Research 15, 489-494.

Fitzgerald, D.L. (1989) "Single station and regional analysis of daily rainfall extremes", Stochastic Hydrology and Hydraulics 3, 281-292.

Gupta, V.K., Duckstein, L., and Peebles, R.W. (1976) "On the joint distribution of the largest flood and its time of occurrence", Water Resources Research 12, 295-304.

Hosking, J.R.M. and Wallis, J.R. (1987) "Parameter and quantile estimation for the generalized Pareto distribution", Technometrics 29, 39-349.

Jin, M. and Stedinger, J.R. (1989) "Partial duration series analysis for a GEV annual flood distribution with systematic and historical flood information", Dept. Civ. Eng., Penn. State Univ., PA. (unpublished paper).

Jenkinson, A.F. (1955) "The frequency distribution of the annual maximum (or minimum) of meteorological elements", Quarterly Journal of Royal Meteorological Society 81, 158-171.

Karr, A.F. (1976) "Two extreme value processes arising in hydrology", Journal of Applied Probability 13, 190-194.

Kavvas, M.L. (1982a) "Stochastic trigger model for flood peaks. 1. Development of the model", Water Resources Research 18, 383-398.

Kavvas, M.L. (1982b) "Stochastic trigger model for flood peaks. 2. Application of the model to the flood peaks of Goksu-Karahacili", Water Resources Research 18, 399-411.

Kirby, W. (1969) "On the random occurrence of major floods", Water Resources Research 5, 778-784.

Kirby, W. and Moss, M. (1987) "Summary of flood frequency analysis in the United States", Journal of Hydrology 96, 5-14.

Konecny, F. and Nachtnebel, H.P. (1985) "Extreme value processes and the evaluation of risk in flood analysis", Applied Mathematical Modelling 9, 11-15.

Langbein, W.B. (1949) "Annual floods and the partial duration flood series", Transaction (AGU) 30, 879-881.

Miquel, J. (1984) "Guide Practique d'Estimation des Probabilités de Crues", Editions Eyrolles, Paris, 160 pp.

Nachtnebel, H.P. and Konecny, F. (1987) "Risk analysis and time-dependent flood models", Journal of Hydrology 91, 295-318.

NERC (Natural Environment Research Council) (1975) "Flood Studies Report", Vols. 1-5, NERC, London, 1100 pp.

North, M. (1980) "Time-dependent stochastic model of floods", Journal of the Hydraulics Division 106, 649-665.

Pickands, J. (1975) "Statistical inference using extreme order statistics", Annals of Statistics 3, 119-131.

Rasmussen, P.F. and Rosbjerg, D. (1989) "Risk estimation in partial duration series", Water Resources Research 25, 2319-2330.
Rasmussen, P.F. and Rosbjerg, D. (1991a) "Evaluation of risk concepts in partial duration series", Stochastic Hydrology and Hydraulics 5, 1-16.
Rasmussen, P.F. and Rosbjerg, D. (1991b) "Prediction uncertainty in seasonal partial duration series", Water Resources Research 27, 2875-2883.
Rasmussen, P.F. and Rosbjerg, D. (1991c) "Application of Bayesian principles in regional flood frequency analysis", in G. Tsakiris (ed.), Advances in Water Resources Technology, Balkema, pp. 65-75.
Rosbjerg, D. (1985) "Estimation in partial duration series with independent and dependent peak values", Journal of Hydrology 76, 183-195.
Rosbjerg, D. (1987a) "On the annual maximum distribution in dependent partial duration series", Stochastic Hydrology and Hydraulics 1, 3-16.
Rosbjerg, D. (1987b) "Partial duration series with log-normal distributed peak values", in V.P. Singh (ed.), Hydrologic Frequency Modeling, Hingham, Mass., pp. 117-129.
Rosbjerg, D., Rasmussen, P.F., and Madsen, H. (1991) "Modelling of exceedances in partial duration series", Proc. of the International Hydrology and Water Resources Symp., Perth, pp. 755-760.
Rosbjerg, D. and Madsen, H. (1992) "On the choice of threshold level in partial duration series", Proc. of the Nordic Hydrological Conference 1992, NHP-Report No 30, pp. 604-615.
Rosbjerg, D., Madsen, H., and Rasmussen, P.F. (1992) "Prediction in partial duration series with generalized Pareto-distributed exceedances", Water Resources Research 28, 3001-3010.
Rousselle, J. and Hindie, F. (1976) "Incertitude dans les débits de crues: Approche bayesienne", Journal of Hydrology 30, 341-349.
Shane, R.M. and Lynn, W.R. (1964) "Mathematical model for flood risk evaluation", Journal of the Hydraulics Division, 90: 1-20.
Smith, J.A. (1987) "Estimating the upper tail of flood distribution", Water Resources Research 23, 1657-1666.
Smith, R.L. (1984) "Threshold methods for sample extremes", in J. Tiago de Oliveira (ed.), Statistical Extremes and Applications, Reidel, pp. 621-638.
Taesombut, V. and Yevjevich, Y. (1978) "Use of partial flood series for estimating distribution of maximum annual flood peak", Hydrol. Paper 97, Colorado State University, Fort Collins, CO.
Tavares, L.V. and Da Silva, J.E. (1983) "Partial duration series method revisited", Journal of Hydrology 64, 1-14.
Todorovic, P. (1970) "On some problems involving random number of random variables", Annals of Mathematical Statistics 41, 1059-1063.
Todorovic, P. (1978) "Stochastic models of floods", Water Resources Research 14, 345-356.
Todorovic, P. and Rousselle, J. (1971) "Some problems of flood analysis", Water Resources Research 7, 1144-1150.

Todorovic, P. and Woolhiser, D.A. (1972) "On the time when the extreme flood occurs", Water Resources Research 8, 1433-1438.
Todorovic, P. and Zelenhasic, E. (1970), "A stochastic model for flood analysis", Water Resources Research 6, 1641-1648.
USWRC (United States Water Resources Council) (1976) "Guidelines for Determining Flood Flow Frequency", Bulletin 17, Hydrol. Comm., Washington D.C., 73 pp.
Van Monfort, M.A.J. and Witter, J.V. (1985) "Testing exponentiality against generalized Pareto distribution", Journal of Hydrology 75, 305-315.
Van Monfort, M.A.J. and Witter, J.V. (1986) "The generalized Pareto distribution applied to rainfall depths", Hydrological Sciences Journal, 31:151-162.
Wang, Q.J. (1991) "The POT model described by the generalized Pareto distribution with Poisson arrival rate", Journal of Hydrology 129, 263-280.
Waylen, P. and Woo, M.K. (1982) "Prediction of annual floods generated by mixed processes", Water Resources Research 18, 1283-1286.
Yevjevich, V. and Taesombut, V. (1979) "Information on flood peaks in daily flow series", in E.A. McBean, K.W. Hipel, and T.E. Unny (eds.), Inputs for Risk Analysis in Water Systems, Water Resources Publications, Fort Collins, CO, pp. 171-192.
Zelenhasic, E. (1970) "Theoretical probability distributions for flood peaks", Hydrology Paper 42, Colorado State University, Fort Collins, CO.

A REGIONAL APPROACH IN THE IDENTIFICATION, ESTIMATION, AND HYPOTHESIS TESTING OF PROBABILITY DISTRIBUTIONS OF ANNUAL MAXIMUM FLOWS

Y. ALILA, K. ADAMOWSKI [1], P.J. PILON [2], and M.Z. KOWALCHUK[3]

Greater Vancouver Regional District
Liquid Waste Management Planning Division
4330 Kingsway, Burnaby, V5H 4G8
British Columbia, Canada

Statistical analyses of hydrologic data involve the identification of the probability distribution from which the sample was drawn, and the estimation of the parameters of the distribution. Standard approaches involve the selection of a distribution by fit or by reference to one or more standard at-site goodness-of-fit tests. The parameters of the distribution are estimated by procedures such as the method of moments or maximum likelihood. It has, however, long been recognized that the statistical distributions cannot be reliably delineated from small data sample sizes commonly available in hydrology using this at-site approach.

Recent studies (Hosking, 1990; Hosking and Wallis, 1993) have shown that the method of L-moments (data analysis using linear combination of an ordered sample) provides a unified, efficient, and convenient approach to all the stages of identification, estimation, and hypothesis testing of probability distributions. The apparent advantages of the L-moments is that they suffer less from the effects of sampling variability than does the conventional moments, they are more robust to outliers in the data, and by providing information on the regional variability of statistics, they enable more secure inferences to be made from samples about the underlying probability distribution.

The purpose of this study is to investigate the statistical distribution(s) associated with annual maximum flows in a region of the Saskatchewan-Nelson River Basin utilizing the method of L-moments. The results of the study indicate that the study area can be delineated into three hydroclimatically distinct homogeneous zones. The success of the approach may have significant implications on the current understanding of hydrologic network data, as well as on the broader aspects of water planning and management that rely upon statistical hydrologic information.

INTRODUCTION

In Canada, large expenditures are invested each year on the design and construction of various water structures to convey runoff from small catchments. The planning and design of these structures requires the estimation of appropriate design values of runoff discharge. Often, practicing engineers resort to the frequency analysis of

[1]University of Ottawa, Civil Engineering Department, Ottawa, K1N 6N5; Canada
[2]Environment Canada, Surveys and Information Systems Branch, Ottawa, K1A OH3, Canada
[3]Environment Canada, Water Resources Branch, Winnipeg, Manitoba, R3C 1B2, Canada

K. W. Hipel (ed.),
Stochastic and Statistical Methods in Hydrology and Environmental Engineering, Vol. 1, 27–40.

at-site streamflow records for the computation of design flood. When streamflow records are not available, however, regionalization procedures are used to transfer streamflow characteristics from gauged to ungauged locations.

A common problem to both at-site and regional frequency analyses is the fact that the form of the underlying flood distribution is not known, nor can it necessarily be assumed on theoretical grounds (Potter and Lettenmaier, 1990). Conventionally, more than one distribution is fitted to the same data and the one providing the best fit is chosen. The selection of the best-fit distribution can be based on graphical techniques (Dalrymple, 1960), through the assignment of plotting positions and the use of probability papers. The comparison of the plotted data on appropriate probability papers to the theoretical forms can assist in the selection. However, such procedures can be subjective and limit the assessment of fit to the range of the observed data. There are other statistical at-site goodness-of-fit tests, such as Chi-Square and Kolmogorov Smirnov (Keeping, 1966), that can be used to discriminate between different distributions. Reservations against the use of such tests have been raised (Moran, 1957), because of the relative degree of subjectivity of the index of fit and the plotting position. In other studies, the choice of a suitable distribution was based on comparison of estimated and theoretical values of the coefficients of skewness and kurtosis. This approach lacks credibility because of the inherent sampling errors in estimating higher order moments associated with existing range of record lengths (Hazen, 1924). Other, not fully developed, techniques include the Bayesian decision theory and nonparametric statistics (Bodo and Unny, 1976). In many parts of the world, the selection of a probability distribution is based on procedural policies and standardization of practices (NERC, 1975; USWRC, 1982). Some other studies avoided the parametric distribution assumption by adopting the nonparametric estimators (Adamowski, 1985; Wu and Woo, 1989). Recently, however, Hosking and Wallis (1993) introduced a new goodness-of-fit test based on the virtually unbiased linear moment statistics (L-moments). This test is different from conventional tests as it is based on regional as opposed to at-site data. It assumes, however, that the data at all sites of a pre-specified network of hydrometric stations are homogeneous, i.e., drawn from the same parent probability distribution.

Over the years, two approaches have evolved in the domain of identification of homogeneous regions. In some studies the geographical approach to homogeneity has been adopted whenever factors that control the flood generating mechanisms are thought of being conducive to spatial mapping. Gingras and Adamowski (1992), for instance, delineated hydroclimatically homogeneous regions based on the flood generating mechanism, using the nonparametric density function in the Province of New Brunswick. In other studies, groups of gauged basins that are not necessarily contiguous in space, have been categorized as being homogeneous using different classification philosophies such as cluster, discriminant, variance, factor, and canonical correlation analyses (Wiltshire, 1985; Cavadias, 1989; Burn, 1990). However, most of these techniques remain subjective as they are not based on a clear cut statistical test of homogeneity.

In this study, two methodologies are combined to assess the regional characteristics of floods. The nonparametric density function is first used to delineate hydroclimatically defined regions based on the flood generating mechanisms. To avoid subjectivity in the analysis, the homogeneity is verified using appropriate statistical testing. Regional goodness-of-fit testing is then applied for each hydroclimatic region to identify the underlying regional probability distribution. The composite technique is demonstrated using annual maximum flows in a large region of the Saskatchewan-Nelson River Basin. The analysis presented herein is part of a project prepared for Environment Canada, Winnipeg, Manitoba (Alila and Adamowski, 1992).

CLIMATE AND VEGETATION OF THE STUDY REGION

The study region comprises a large portion of the Saskatchewan-Nelson River Basin located in Saskatchewan, Manitoba, and Ontario as shown in Figure (1). Referring the Water Survey of Canada system of basin subdivision, this area is formed by the drainage basins 05P, 05R, 05N, 05M, 05S, 05O, 05K, 05J, and 05L. The major rivers in the study area include the Assiniboine, Red, Souris, Qu'Appelle, Saskatchewan, and Winnipeg River. Annual daily maximum discharges from 280 hydrometric stations, with a minimum record length of 10 years, within the area were used in the analysis.

In macro-climatic terms, the northern extreme of the study area borders upon the Boreal climatic region, while the south and southwestern portion includes parts of the Prairie climatic region. The Prairie portion, composed of the Manitoba lowland and the Saskatchewan plains, experiences very cold winters and a light precipitation regime. The summer period occasionally experiences influxes of moist air from the south. This area is classified as sub-humid, has no major relief, and is suitable for agriculture. It is also covered by natural grass to the south and by wooded parklands to the east and north. The latter land cover serves to separate the semi-arid from the Boreal climatic zones.

The Boreal portion of the study region is part of the largest climatic zone of Canada stretching from Alaska to Newfoundland. The southern limit of this zone lies to the north of the study region. It is characterized by its somber forest of spruce, fur, pine, larch, and other hardy broad leafed trees. This land cover is in a dynamic state of equilibrium with the physical climate of the region. The amount of net radiation available at the earth, and hence the temperature, have been shown to be closely affected by the land cover (Hare and Ritchie, 1972). This, along with the strongly contrasted seasons in the north, affects greatly the hydrologic characteristics of the region.

Temperatures in the study area vary from a minimum of negative 40 to a maximum of 38 degrees Celsius. The average January temperature is negative 17 degrees while a mid-summer average temperature is 20 degrees. Annual precipitation in the Prairie zone of the study area increases northward to over 410 mm in the wooded parkland belt and eastward in Manitoba to more than 510 mm. Precipitation decreases westward with an average of less than 300 mm in the heart of the Prairie dry zone. Precipitation throughout the year varies in a continental fashion from moderately heavy falls in summer to very light falls in winter. The region experiences severe snowmelt rather than heavy rainfall (Hare and Thomas, 1979).

DELINEATION OF HYDROCLIMATIC HOMOGENEOUS REGIONS

Frequency analysis is usually based on the assumption that annual flood series is a random sample from a single population, which may not be the case as annual floods could be the result of at least two types of climatic factors. Common examples are spring floods caused by snowmelt and floods caused by rainfall. Diehl and Potter (1987) showed that these two groups of floods are hydrologically distinct. The explicit recognition that floods come from mixed distributions would, however, affect the quantile estimation procedure when the two flood components are shown to be statistically different.

To investigate mixtures in flood data caused by climatic factors, the time of occurrence can be used as a somewhat objective criterion for the classification of annual extreme floods. This can further be supported by the use of the Mann-Whitney split sample test for homogeneity (Pilon and Harvey, 1993). Since there is a time overlap between the rain and snowmelt seasons in the study region, this method is found to be of limited use in this application.

A new approach for the study of mixtures in floods by the nonparametric density function is applied. The nonparametric density function is defined as (Adamowski, 1985)

$$f(x) = \frac{1}{nh}\sum_{i=1}^{n} K\left(\frac{x - x_i}{h}\right) \tag{1}$$

where x_i and x_n are the observations in an annual flood series, $K(\cdot)$ is a kernel function that itself is a probability density function, and h is a smoothing factor. It has been found that the choice of the kernel function is not crucial for the performance of the technique and as such the Gaussian kernel (Adamowski, 1989) is used. The shape of the density function is sensitive to the smoothing factor h, and an optimal algorithm that minimizes the integrated mean square errors by a cross-validation technique is adopted (Labatiuk and Adamowski, 1987).

Alila (1987) had shown that different modes in the nonparametric density function of eqn (1) correspond to different flood generating mechanisms (snowmelt and rain). More often hydrometeorological conditions follow a specific geographic pattern in space (Waylen and Woo, 1983). Hence, an attempt is made in this study to coordinate between the shape of the at-site nonparametric probability density function (PDF) and its geographic patterns in space to define hydroclimatically homogeneous sub-regions.

The nonparametric PDF plots for the hydrometric stations in the study region are classified on the basis of distribution shape. Sites with record lengths less than 15 years are not considered in the classification. It is believed that such sites do not have enough data to clearly define the PDF shape. Three classes of nonparametric PDFs are distinguished:

1. PDFs with a dominant unique mode resembling a symmetrical density function (UM group). A sample plot at station 05LD001 from the 90

unimodal PDFs is shown in Figure (2). Basins with UM PDFs are found to occupy the northwest-southeast part of the study region. In this area, winter temperature and spring rainfall are lower. The last frost in spring is quite late and, consequently, the thaw season is longer. Most annual floods, therefore, are related to spring melt or to some extent light rain on snowmelt. These two flood generating mechanisms seem to be statistically indiscernible from each other as they result in PDFs with a unique mode.

2. PDFs with a dominant unique mode having a heavy right-hand tail (HT group). A sample plot at station 05OA005 from the 50 heavy-tailed unimodal PDFs is shown in Figure (3). Basins with HT PDFs happen to occupy the agricultural part of the Prairie region, mainly consisting of the Manitoba lowland and the Saskatchewan plains. This area is sub-humid and experiences very cold winters. It is also known to be semi-arid with a short warm summer and a light precipitation regime. The near absence of significant precipitation events in this low relief area leads to rainfall induced floods being statistically indiscernible from snowmelt floods. While the latter flood type is predominant in this region (Hare and Thomas, 1979), occasionally the superposition of rain on snowmelt produces high extreme events explaining the heavy-tailed nature of the PDFs.

3. PDFs with at least two dominant modes which are referred to as bimodal density functions (BM group). A sample plot at station 05LL015 from the 90 bimodal PDFs is given in Figure (4). Basins with BM PDFs lie geographically in between the UM and HT regions. The area is a transition zone, where annual floods may be low in some years reflecting a dominance in snowmelt characteristics, but in other years larger rain-induced floods occurring in the summer might dominate. Early summer rains superimposed on late spring thaw can produce a distinct flood generating mechanism explaining the bimodal nature of the PDFs.

Thus, a geographical trend to the nonparametric PDF shape is evident as illustrated in Figure (1). Such an approach in relating climatic trends to flood generating mechanisms and, consequently, to the geographic trend of flood characteristics has also been adopted for southwestern British Columbia (Waylen and Woo, 1983), but using the time of occurrence for the classification of annual floods.

The geographic trend of the PDF shapes is also evidently related to the vegetation land cover. Trends in the vegetation land cover should not be viewed as distinct from climatic trends as both are intimately related (Hare and Ritchie, 1972). The study region is covered by three distinct types of vegetation, extending in northwest - southeast direction, that happen to be geographically delineated with no discontinuity. This geographic delineation of vegetation cover is found to be coherent, if not similar, to the nonparametric PDF shape geographic trend of Figure (1).

The geographic trend to the nonparametric PDF shape is also found to be in agreement with the geographic delineation of soil temperature and soil moisture regimes within the study region (refer to the National Atlas of Canada). Realizing that the vegetation cover affects the net radiation, it follows that the nature of the surface would affect the soil moisture and temperature. Of course, there is a complex feedback mechanism between the vegetation and climate.

It is therefore demonstrated that the extreme flood characteristics are governed by climatic as well as physiographic factors, both of which are closely linked. The conventional practice of assuming an a priori flood distribution is shown to be not only unnecessary but may not be valid or physically justifiable. The latter case arises when floods are induced by two or more significantly different generation mechanisms such as snowmelt and rain.

STATISTICAL TESTING OF HOMOGENEOUS REGIONS

The homogeneity of the defined regions within the study area is further verified using statistical tests derived based on L-moments. The r-th L-moments of a random variable have been defined by Hosking (1990). For brevity the equations are not repeated herein, but emphasis is placed on their potential meaning. The first L-moment, λ_1, is the arithmetic mean, while the second L-moment, λ_2, is a measure of dispersion analogous to the standard deviation. The coefficient of variation is defined in this application by the ratio $\tau_2 = \lambda_2/\lambda_1$. L-moments of order greater than two have been standardized by $\tau_r = \lambda_r/\lambda_2$, whereas τ_3 is a measure of symmetry and is referred to as L-skewness, while τ_4 is a measure of peakedness and is referred to as L-kurtosis. For the finite sample, the L-coefficient of variation, L-skewness, and L-kurtosis are denoted by t_2, t_3, and t_4, respectively. In the derivation of the L-moment estimates, plotting position estimators are used as suggested by Hosking (1990). While sample L-moments are considered virtually unbiased, this statement is more accurate for the L-skewness than for the L-kurtosis for short sample length ($n < 20$). To avoid this problem, a correction factor for the L-kurtosis as suggested by Hosking and Wallis (1993) is adopted.

The verification of the likeliness of sites in flood generating mechanism terms, or hydrolclimatic homogeneity, is performed by assessing the variability of the L-moments. The first homogeneity measure is given by (Hosking and Wallis 1993),

$$H = \frac{(V - \mu_v)}{\sigma_v} \tag{2}$$

where V is a measure of dispersion of an at-site L-moment around the weighted average value of the network. In t_2 terms, V is the weighted standard variation of the observed t_2. In t_3 terms, V is the weighted average distance from the site to the network's weighted mean on the observed t_2 vs. t_3 L-moment diagram. In t_4 terms, V is the weighted average from the site to the network's weighted mean on the observed t_3 vs. t_4 L-moment diagram. The μ_v and σ_v are the mean and standard deviation of the same parameter V derived from a large number of synthetic network replications from a Kappa distribution. This four-parameter Kappa distribution is suggested by Hosking and Wallis (1993) in order to avoid a early commitment to a particular two or three-parameter distribution as the assumed parent of the observed data. They suggest that the region under testing should be regarded as *acceptably homogeneous* if $H < 1$, *possibly heterogeneous* if $1 < H < 2$, and *definitely heterogeneous* if $H \geq 2$.

Pilon et al. (1991) suggested a variance ratio test that is somewhat similar to equation (2), namely:

$$S(\%) = \left[\frac{\sigma_{obs}{}^2 - \sigma_{sim}{}^2}{\sigma_{obs}{}^2}\right] 100 \tag{3}$$

where $\sigma_{obs}{}^2$ is the variance of the L-moment under scrutiny for the network and $\sigma_{sim}{}^2$ is its synthetic counterpart based on a large number of network replications. Pilon et al. (1991) indicate that $\sigma_{sim}{}^2$ represent the *noise* of the network due to sampling error for the particular L-moment. In comparison, $\sigma_{obs}{}^2$ represents a combination of sampling error and process *signal.* There is of course, no physical reason why the items that comprise signal should be independent. However, the assumption is made to simplify the analysis. Hence, S represents the percentage of signal that is evident in the network. The percentage of noise on the network would be equal to $100 - S$.

Both tests are applied to the study region as a whole and to each of the hydroclimatically delineated homogeneous regions, and the results are listed in Table (1). The H statistics for the entire area are 27.6 for the L-coefficient of variation, 13.6 for the L-skewness, and 5.8 for the L-kurtosis. Since standard H values for acceptably homogeneous regions lie between zero and one, the study region is viewed to be extremely heterogeneous as expected. Differences between observed and synthetic data variances, as measured by the S statistic, amount to 70 % for t_2, 43 % for t_3, and 45 % for t_4. This implies that the spatial variability of the various environmental factors affecting the flood generation mechanisms is responsible for 70 %, 43 %, and 45 % of the spatial variability of t_2, t_3, and t_4, respectively.

Applied to each of the three hydroclimatic regions, the S statistic gave values that are close to zero for both the L-skewness and L-kurtosis. This indicates that most of the spatial variability of t_3 and t_4 of observed data is due to random sampling errors. The H statistic, generally, leads to the same conclusion. The exception is in the L-skewness for which the H values are greater than unity. The H statistic is devised in a way that the three measures of homogeneity are not independent. Extreme heterogeneity detected by H in t_2, for instance, would also be reflected in the bidimensional L-moment space t_2 vs. t_3. The three indicators of heterogeneity measured by S, however, are independent of each other and should be more reliable. Thus, the three hydroclimatic regions, delineated based on the geographic trends of the nonparametric PDF shape, can generally be considered homogeneous in both the L-skewness and L-kurtosis.

In contrast, the regions could obviously not be considered homogeneous in the L-coefficient of variation as indicated in Table (1). The variation in t_2 exceeds that caused by random sampling. It is quantified by the statistic S and amounts to 56 %, 56 %, and 30 % for the unimodal, bimodal, and heavy-tailed regions, respectively. This is the amount of signal variability that is possibly due to physiographic and climatic causes. The formulation of a regional flood frequency model, applicable within each of the hydroclimatic regions, would involve a multivariate correlation between t_2 and variables such as basin area, percent forest cover, slope of channel, mean annual precipitation, etc. This, however, is beyond the scope of the study.

IDENTIFICATION OF REGIONAL PARENT DISTRIBUTIONS

Once a region is determined as being homogeneous, data at all sites within this region are assumed to be drawn from the same parent distribution. The shape indices, L-skewness and L-kurtosis, of such a parent are best represented by the regional average values based on data from all sites within the region.

The three identified hydroclimatic homogeneous regions are subjected to a regional goodness-of-fit test. The test discriminates between the most frequently used five distributions in the analysis of extreme hydrologic variables. These are the Generalized Extreme Value, Pearson Type III or Gamma, Generalized Log-Normal, Generalized Logistic, and the Generalized Pareto.

Since there is more confidence in estimating the regional L-skewness, the suitability of fit of a specific distribution (DIST) is judged by the difference between the theoretical L-kurtosis of the distribution and its observed regional counterpart. The significance of this difference is assessed through the statistic (Hosking and Wallis, 1993):

$$Z^{DIST} = \frac{\left\{\bar{t}_4 - \tau_4{}^{DIST}\right\}}{\sigma_{\bar{t}_4}} \tag{4}$$

where $\bar{t}_4$ is the regional average L-kurtosis of the observed network and $\tau_4{}^{DIST}$ is the theoretical τ_4 for the regional L-skewness of the same network. The $\sigma_{\bar{t}_4}$ is the standard deviation of $\bar{t}_4$ obtained by repeated simulations of a homogeneous region with the DIST frequency distribution as a parent. Since the test is applied to discriminate between different alternative three-parameter distributions with the same skewness and to avoid extensive simulations, $\sigma_{\bar{t}_4}$ is assumed to be the same for all distributions. As proposed by Hosking and Wallis (1993), the four-parameter Kappa distribution is used for the determination of $\sigma_{\bar{t}_4}$. In computing the regional L-kurtosis values for the synthetically generated homogeneous networks, the same bias correction factor mentioned earlier is applied.

The fit is declared satisfactory at the 90 % confidence level if $\mid Z \mid< 1.64$. The first assumption underlying the use of such a significance test is that the region is exactly homogeneous. The exact homogeneity, as defined by a constant shape parameter (equivalently L-skewness) across the region, allows the scale and location parameters to vary from one location to another. Such an assumption is more compelling than what have been implied in recently developed goodness-of-fit tests (Chowdhury et al., 1991), where only the scale parameter is allowed to vary within a homogeneous region. The second assumption is that data from different sites are independent. While this hypothesis is made to simplify the analysis, it has been demonstrated that the influence of the inter-station correlation decreases from the mean to the coefficient of skewness (Stedinger, 1983).

The results of the goodness-of-fit test indicate that at the 90 % confidence level the Generalized Log-Normal distribution is the only candidate that passed the test for the unimodal region; the Pearson Type III is the only candidate acceptable for the heavy-tailed region; whereas none of the tested distributions is accepted for the bimodal region. Since all the tested parametric distributions are exclusively unimodal, these results are thought to be in harmony with the findings from the nonparametric density function assessment.

CONCLUSIONS

The paper combines two techniques for studying the regional characteristics of annual flood maxima in the Saskatchewan-Nelson River Basin: (i) the nonparametric density function and (ii) the linear moment statistics. In an attempt to uncover the geographic nature of the heterogeneity of flood characteristics, the nonparametric density function of each site was located on a map of the study area. The shape of the nonparametric density function displayed a definite geographic trend. The entire area was then delineated into three geographically defined continuous regions based on three distinct shapes of the density function. Sites in the northwestern-southeastern part of the study area manifested bell-shaped unimodal density functions. In contrast, the southwestern part of the study area displayed unimodal densities having heavy right-handed tails. In a transitional zone between these two regions, all the sites displayed a bimodal density function.

It is believed that the three distinct density shapes are a result of the individual flood producing mechanisms - rain, snowmelt, and rain on snowmelt - that vary in a physically meaningful way within the study region. The bioclimatic zones of the study area and the three hydroclimatic regions have similar geographic tendencies. The effect of land cover on climate indices such as net and global radiation is hypothesized to be important in the regional characteristics of floods.

The unimodal, bimodal, and heavy-tailed regions are tested individually for statistical homogeneity using L-moments. Each region is found to be homogeneous in both L-skewness and L-kurtosis. Observed variability in these two L-statistics can be attributed to random space and time sampling errors (noise). Varying degrees of signal, however, have been detected for the L-coefficient of variation in all three regions. This signal is thought to be related to physiographic and climatic causes.

Finally, a statistical regional goodness-of-fit test is used to select the distribution, the one that gives the best fit to the data. At the ninety percent confidence level, only the Generalized Log-Normal is accepted for the unimodal region. The Pearson Type III distribution is cautiously acceptable for the heavy-tailed region. However, among five different parametric distributions tested, none are acceptable for the bimodal region. Thus, the use of nonparametric frequency distributions, combined probability analysis, or unclassified two component distributions might be potential candidates for this region.

REFERENCES

[1] Adamowski, K. (1985), "Nonparametric Kernel Estimation of Flood Frequencies", Water Resources Research, 21(11), 1585-1590.

[2] Adamowski, K. (1989), "A Monte Carlo Comparison of Parametric and Nonparametric Estimation of Flood Frequencies", Journal of Hydrology, 108, 1585-1590.

[3] Alila, Y. (1987), "Nonparametric Flood Frequency Analysis with Historic Information and Hydroclimatically Defined Mixed Distributions", M.A.Sc. Thesis, University of Ottawa, Ottawa, Ontario, Canada.

[4] Alila, Y. and Adamowski, K. (1992), " Comprehensive Analysis and Interpretation of Extreme Flood Data and Regionalization of Hydrologic Parameters in the Saskatchewan-Nelson River Basin", A Report Prepared for Environment Canada, Hydrology Division, Water Resources Branch, Manitoba-NW Ontario District, Inland Waters Directorate, Winnipeg, Manitoba.

[5] Bodo, B. and Unny, T. E. (1976), "Model Uncertainty in Flood Frequency Analysis and Frequency - Based Design", Water Resources Research, 12(6), 1109-1117.

[6] Burn, D. B. (1990), "An Appraisal of the 'Region of Influence' Approach to Flood Frequency Analysis", Hydrologic Sciences, 35(2), 149-165.

[7] Cavadias, G. S. (1989), "Regional Flood Estimation by Canonical Correlation", Proceeding of the Annual Conference of the Canadian Society for Civil Engineers, June 8-10, St. John's, Newfoundland.

[8] Chowdhury, J. U., Stedinger, J. R., and Lu, L. (1991), "Goodness-of-fit Tests for Regional Generalized Extreme Value Flood Distributions", Water Resources Research, 27(7), 1765-1776.

[9] Dalrymple, T. (1960), "Flood Frequency Analysis", U.S. Geological Survey Water Supply, Paper 1543-A, 80 pp.

[10] Diehl, T. and Potter, K. W. (1987), "Mixed Flood Distributions in Wisconsin", Paper Presented at the International Symposium on Flood Frequency and Risk Analysis, Baton Rouge, Louisiana.

[11] Gingras, D. and Adamowski, K. (1992), "Homogeneous Region Delineation Based on Flood Generation Mechanisms", Journal of Hydrological Sciences (in press).

[12] Hare, F. K. and Ritchie, J. C. (1972), "The Boreal Bioclimates", Geographical Review, 62, 333-398.

[13] Hare, F. K. and Thomas, M. K. (1979), "Climate Canada", 2nd Edition, John Wiley and Sons Canada Limited.

[14] Hazen, A. (1924), "Discussion on Theoretical Frequency Curves and their Application to Engineering Problems", Transactions of the American Society of Civil Engineers, 87, 143-173.

[15] Hosking, J. R. M. (1990), "L-moments: Analysis and Estimation of Distribution Using Linear Combination of Order Statistics", Journal of the Royal Statistical Society, Series B, 52(1), 105-124.

[16] Hosking, J. R. M. and Wallis, J. R. (1993), "Some Statistics Useful in Regional Frequency Analysis", Water Resources Research, 29(2), 271-281.

[17] Keeping, E. S. (1966), "Distribution-Free Methods in Statistics", Statistical Methods in Hydrology, Proceedings of Hydrology Symposium No.5, McGill University, Montreal, Canada.

[18] Labatiuk, C. and Adamowski, K. (1987), "Application of Nonparametric Density Estimation to Computation of Flood Magnitude/Frequency", In Stochastic Hydrology, I. B. MacNeil and G.J. Umphrey (Editors), 161-180.

[19] Moran, P. A. P. (1957), *The Statistical Treatment of Flood Flows*, Transactions American Geophysical Union, 38(4), 519-523.

[20] NERC - Natural Environment Research Council (1975), "Flood Studies Report Volume I - Hydrologic Studies", Water Resources Publications, Fort Collins, Colorado.

[21] Pilon, P. J., Alila, Y. and Adamowski, K. (1991), "Assessment of Risk of Flooding Based on Regional Information", Presented at the NATO Advanced Study Institute on Risk and Reliability in Water Resources and Environmental Engineering, 18-28 May, Porto Carras, Greece.

[22] Pilon, P. J. and Harvey, K. D. (1993), "Consolidated Frequency Analysis (CFA) Package - Version 3.1", Surveys and Information Systems Branch, Environment Canada, Ottawa.

[23] Potter, K. W. and Lettenmaier, D. P. (1990), "A Comparison of Regional Flood Frequency Estimation Methods Using a Resampling Method", Water Resources Research, 26(3), 451-424.

[24] Stedinger, J. R. (1983), "Estimating a Regional Flood Frequency Distribution", Water Resources Research, 19(2), 503-510.

[25] USWRC - U.S. Water Resources Council (1982), "Guidelines for Determining Flood Flow Frequency", Bulletin 17B, Hydrology Committee, Washington, D.C.

[26] Waylen, P. and Woo, M. K. (1983), "Annual Floods in Southwestern British Columbia, Canada", Journal of Hydrology, 62, 95-105.

[27] Wiltshire, S. E. (1985), "Grouping Basins for Regional Frequency Analysis", Hydrological Sciences Journal, 30(1), 151-159.

[28] Wu, K. and Woo, M. K. (1989), "Estimating Annual Flood Probability Using Fourier Series Method", Water Resources Bulletin, 24(4), 743-750.

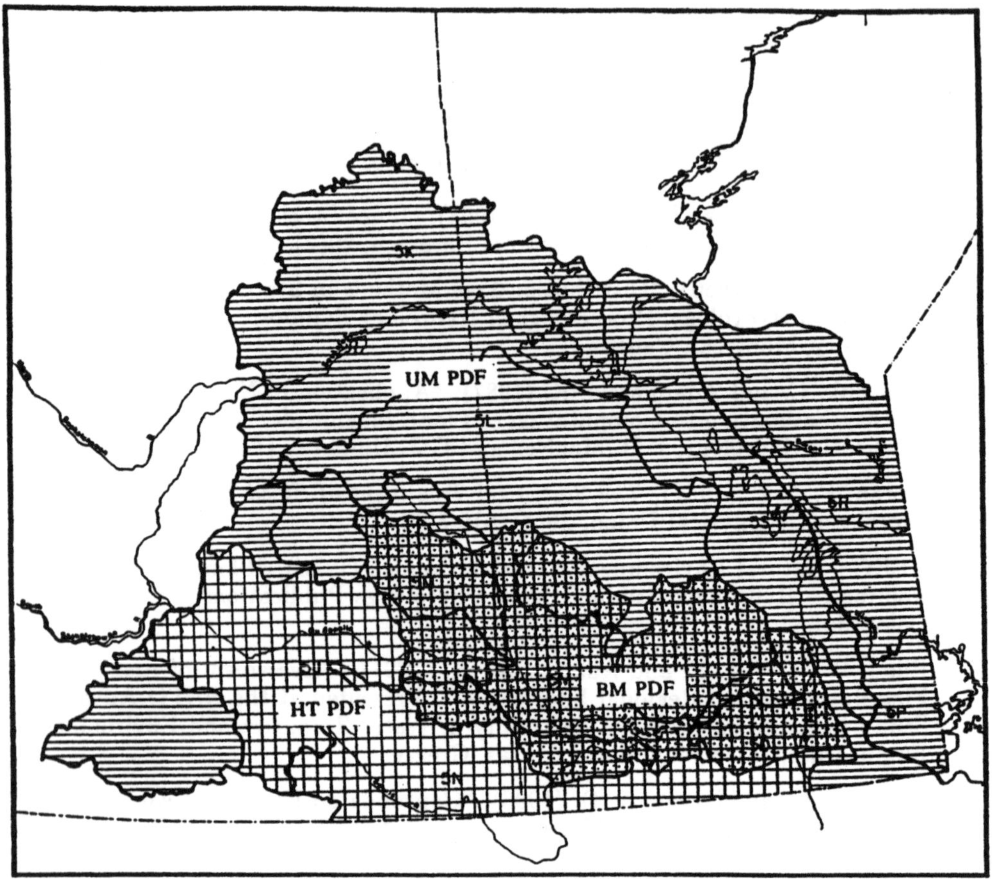

Figure 1: Location of the Study Region in the Saskatchewan- Nelson River Basin (UM, HT, and BM PDF refer to subregions with unimodal, heavy-tailed and bimodal probability density functions, respectively)

Table (1): Results of the Statistical Testing of Homogeneity for the Study Area and the Three Hydroclimatic regions

L-Moment	Sudy Area		UM Region		HT Region		BM Region	
	H	S(%)	H	S(%)	H	S(%)	H	S(%)
t_2	27.6	72	7.9	56	2.9	30	7.6	56
t_3	13.6	43	3.9	-11	1.7	-1	3.2	-1
t_4	5.8	45	0.9	2	0.4	-23	0.3	4

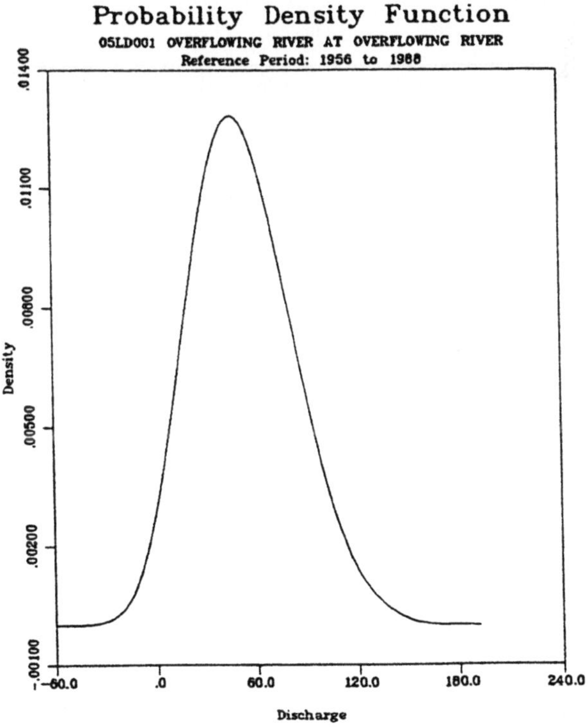

Figure 2: Sample Probability Density Function Plot from the Unimodal Hydoclimatically Homogeneous Subregion

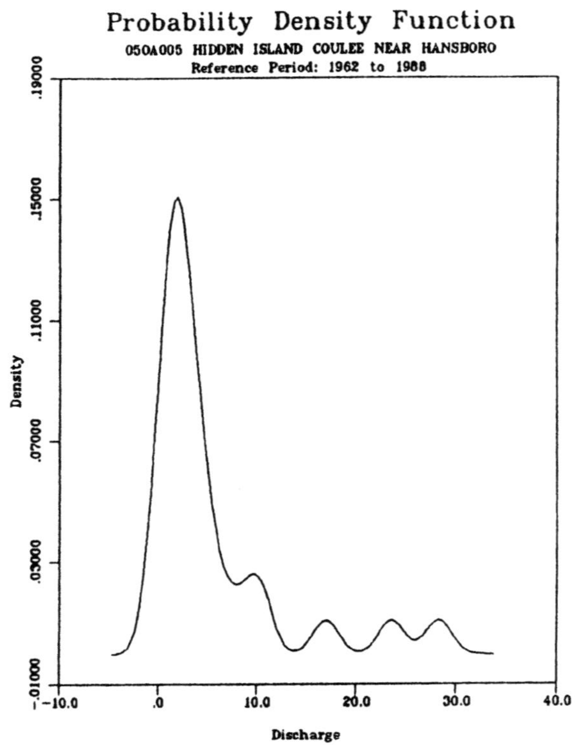

Figure 3: Sample PDF Plot from the Heavy-Tailed Hydoclimatically Homogeneous Subregion

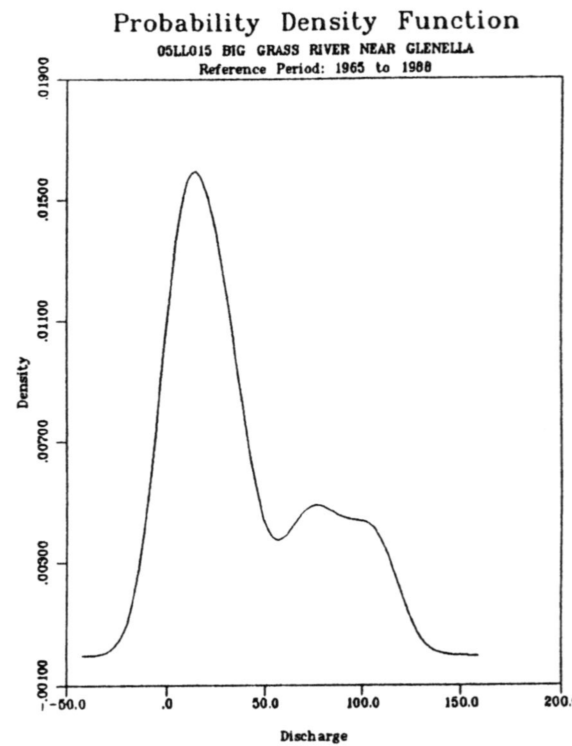

Figure 4: Sample PDF Plot from the Bimodal Hydoclimatically Homogeneous Subregion

EXPERIENCE WITH WEIGHTED $n\omega^2$-TEST-STATISTICS BY FITTING EXTREME VALUE DISTRIBUTIONS TO ANNUAL FLOOD FLOWS INCLUDING HISTORICAL FLOODS

CH. KLUGE[1], M. SCHRAMM[2], and P. SCHIEKEL[2]
[1]Institute of Hydrology and Meteorology, Technical University of Dresden, Mommsenstraße 13, 01069 Dresden, Federal Republic of Germany
[2]Company for Water Management and Systems Research Ltd., Branch Office Dresden, Goetheallee 21, 01309 Dresden, Federal Republic of Germany

The $n\omega^2$-test-statistic modified by weight functions is apt for better assessing the goodness of fit of flood frequency models for annual maximum series at gauging stations in the foreland of the Erzgebirge and the North German plains. Fitted models are the extreme value distribution type 1, the generalized extreme value distribution, and the two-component extreme value distribution; the parameters are estimated by the method of moments, of probability weighted moments and of maximum likelihood. All formulae (for plotting positions, parameter estimation, the test statistic and its moments) include historical flood information.

The variable weighting of the regions of the distribution function F is enabled by different power functions under the integral defining of the test statistic. A weight function recommendable for many applications is $F^5/(1-F)$. By this increased weighting of the upper region the assessment of fit of each distribution function and estimation method is remarkedly changed at all sites, and the discrepancies to visual assessment by graphical procedures can be removed or reduced.

Due to the lack of exact confidence levels for the weighted $n\omega^2$-statistic an approximation is used based on standardization with its interval of double the standard deviation.

INTRODUCTION

An important problem in statistical hydrology is to assess the goodness of fit of flood frequency models to observed or simulated annual maximum flood series. For this purpose the $n\omega^2$-statistic, i.e. the integrated square errors between model and sample distribution is frequently used. In the formulation by Kluge and Finke (1986) it is applicable also to a mixed sample of single and grouped values as it is often found e.g. in precipitation series.

It is well known in hydrologic practice that discrepancies often occur between the assessment of fit by the $n\omega^2$-statistic and by the visual inspection of sample plots,

K. W. Hipel (ed.),
Stochastic and Statistical Methods in Hydrology and Environmental Engineering, Vol. 1, 41–54.

especially in probability nets as that of Gumbel (1966). This is mainly based on the fact that there the reach of small exceedance probabilities is considerably extended, while the $n\omega^2$-formula is commonly used with equal weights in the whole range of values. In contrast the original definition by Anderson and Darling (1952) contains a function which weighs more heavily the ranges of great and also of small exceedance probabilities. One factor of this function emphasizing the range of small exceedance probabilities is used by Ahmad and Sinclair (1988) in the field of flood statistics.

Our work is mainly initiated by the occurrence of the discrepancies in analyzing a great number of flood series, especially by fitting the extreme value distribution type 1 (E1), the generalized extreme value distribution (GE or AE) given by Jenkinson (1955), and the two-component or mixture extreme value distribution (TCE or ME) given by Rossi et al. (1984). We found that the statements of the usual test statistic did not change when using the modified version by Ahmad and Sinclair (1988). Therefore, the intention of our work is to provide a set of functions which enable a stronger and graduated weighting of the upper range of flood values, and to test this method when it is applied on the three extreme value distributions.

DEFINITION OF THE TEST STATISTIC FOR FLOOD SERIES COMPRISING HISTORICAL FLOODS

The test statistic is defined as

$$\omega^2=\int_{x_A}^{x_E}[F(x)-p(x)]^2 * w[F(x)] * dF(x). \tag{1}$$

with F(x) as model distribution of flood flows x, defined in (x_A,x_E), where often $x_A= -\infty$, $x_E= \infty$, p(x) as the empirical distribution function of the sample, and w[F(x)] as the weight function. The most convenient assumption on the historical floods is, that all annual maxima above the level h are known in the long time period of n_h years. Besides the n maximum floods x_i of the currently observed series a number z_h' of the historical floods $x_{ih} > h$ is known, which are completed by n' values $x_i > h$ of the current series, giving the total number $z_h= z_h'+n'$ (Fig. 1).

The values arranged in ascending magnitude are x_j with $j= 1,2,...,n_r$ for $x<h$ within a sample of reduced size $n_r= n-n'$, and x_{jh} for $x\geq h$ with $j_h= 1,2,...,z_h$. The total number of numerically known values is $n_g= n+z_h'= n_r+z_h$.

To take into account the number n_h-n_g of numerically unknown values occurring in the time period n_h-n, each x_j gets the multiplicity factor $v= (n_h-z_h)/n_r$. Then a x_{jh} has the index $vn_r+ j_h$ within the series of all values, or better $j'+n_h-n_g$, if we use for computational convenience instead of j_h the index $j'= n_r+j_h$, with $j'= n_r+1,...,n_g$.

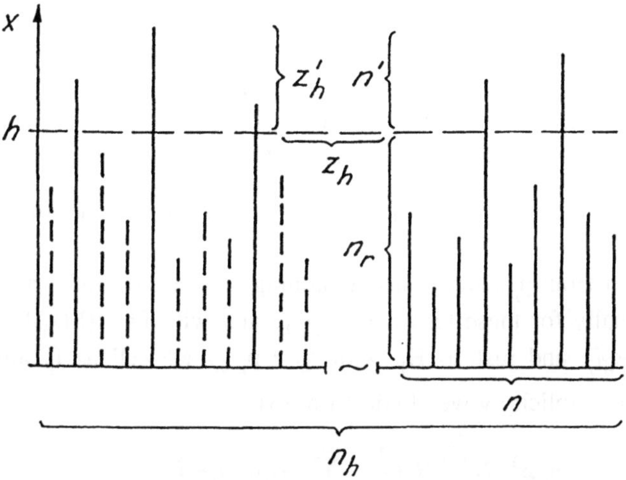

Fig. 1. Model of historical floods: numerically known annual maxima (ordinate ——), and numerically unknown maxima (- - - -), with its numbers n resp. z (with specific index) during the currently observed period of n years, and the total historical period of n_h years.

Thus the empirical distribution function of the sample is

$$p(x)=\begin{cases} p_j=\dfrac{j*v}{n_h} & \text{for} \quad x_j < x \le x_{j+1} < h, \ j=0,\ 1,\dots,\ n_r \\ p_{j'}=\dfrac{j'+n_h-n_g}{n_h} & \quad h < x_{j'} < x \le x_{j'+1},\ j'=n_r+1,\dots,\ n_g \end{cases} \tag{2}$$

with $x_0= x_A= -\infty$, $x_{ng+1}= x_E= \infty$ as an abbreviation. This step function has jumps at $x_j^{(+)}$ (i.e. "behind" x_j as middle of the v-fold filled class j) and also at $x_{j'}^{(+)}$. Then we obtain from (1)

$$\omega^2=\sum_{j=0}^{n_r}\int_{x_j^{(-)}}^{x_{j+1}}(F-p_j)^2*w(F)*dF+\sum_{j'=n_r+1}^{n_g}\int_{x_{j'}^{(-)}}^{x_{j'+1}}(F-p_{j'})^2*w(F)*dF. \tag{3}$$

In the case without historical floods and without weighting, the well-known formula of the $n\omega^2$-test-statistic can be obtained by integration and rearranging the remaining terms of (3), cf. Smirnow and Dunin-Barkowski (1969).

In the same way we find with existence of historical floods but still with $w(F)=1$

$$n_e\omega^2=\sum_{j=1}^{n_r}[F(x_j)-\frac{j-0{,}5}{n_e}]^2+\frac{1}{v}\sum_{j'=n_r+1}^{n_g}[F(x_{j'})-\frac{j'-0{,}5+n_h-n_g}{n_h}]^2+\frac{1}{12n_e}[1-\frac{z_h}{n_h}(1-\frac{1}{v^2})] \tag{4}$$

where $n_e= n_h/v$ may be called "effective sample size".

EVALUATION WITH WEIGHT FUNCTIONS

The evaluation of (1) is given by Anderson and Darling (1952) without historical floods for arbitrary functions w(F) if these are not negative in (0,1) and fulfil such conditions of integrability that the following auxiliary functions exist:

$$f(F)=\int_0^F w(t)*dt, \quad g(F)=\int_0^F t*w(t)*dt, \quad R=\int_0^1 (1-t^2)*w(t)*dt. \tag{5}$$

Integrating (3) we get from p_j some terms containing j-0.5 instead of j, as can be seen in (4). Writing formally for these terms $q_j = p_{j-0.5}$ and with $f_j = f[F(x_j)]$, $g_j = g[F(x_j)]$ we define $H_j = g_j - q_j f_j$, and just so $H_{j'}$ with $q_{j'} = p_{j'-0.5}$ as well as $f_{j'}$ and $g_{j'}$. If x_j possesses again the multiplicity v we obtain from (3)

$$n_e\omega^2 = 2\Big(\sum_{j=1}^{n_r} H_j + \frac{1}{v} * \sum_{j'=n_r+1}^{n_g} H_{j'}\Big) + n_e * R. \tag{6}$$

Expectation value $E[n_e\omega^2] = m$ and variance $Var[n_e\omega^2] = s^2$ are also given by Anderson and Darling (1952) as

$$m=\int_0^1 t*(1-t)*w(t)*dt, \quad s^2=4\int_0^1 (1-t^2)*w(t)\int_0^t u^2*w(u)*du*dt. \tag{7}$$

To emphasize the desired ranges of the variate and to allow relatively simple integrations of (5) and (7), we consider now product forms $w_{k,l}(F) = F^k(1-F)^l$. For $k = l = 0$ we get again the usual statistic $n_e\omega_{0,0}^2$, stronger weighting the range of great values F (near 1) is accomplished by $k>0$ and/or $l<0$, while the same result in the range of small F (near 0) is given by reversed signs of the exponents (drought analysis). In the case of flood analysis, l is restricted to values $l = 0$ and $l = 1$ to ensure the existence of the integrals in (5) and (7).

With help of formulas for the auxiliary functions in (5), (6) and (7) given in Table 1, the statistic $n_e\omega_{k,l}^2$ as well as its expectation values and variances can be evaluated for each admissible k and l.

Because in the whole region the differences between the model and sample distributions become smaller by F^k with $k>0$ and larger by $(1-F)^l$ with $l<0$, for better comparison the weight functions should be standardized. Therefore in Fig. 2 w(F) was related to w(0.5), i.e. $w_{k,l}^*(F) = 2^{k+l}F^k(1-F)^l$ is shown.

Alternative approach by integration of transformed variates

An alternative approach suggested by the visual assessment in the Gumbel net is to integrate the differences of the distributions in the doubly logarithmically transformed scale T, i.e. to consider

$$n_e\omega_T^2=n_e\int_{y_A}^{y_E}[y-T(p)]^2*dy=\frac{n_e}{3}\sum_{j=1}^{n_g-1}[(y_{j+1}-T_j)^3-(y_j-T_j)^3] \tag{8}$$

where T_j= -ln(-lnp_j) and y_j= -ln(-lnF_j); y_A and y_E are the borders of the transformed region of F. T_j follows from (2), and y is given in Appendix III for the model distributions. Corresponding to the usual graphical method the multiplicity of the x_j is not considered here, and the whole sample including historical floods is ordered with index j= 1,2,...,n_g.

Owing to the nonlinearity of the transformation we did not succeed in integration and an arrangement with quadratic terms analogous to (4), and also could not find formulas for the expectation value and the variance of the statistic (8). Therefore Table 2 shows only non-standardized values $n_e\omega_T^2$.

TABLE 1. Auxiliary functions H,R and moments m, s^2 of the test statistic $n_e\omega_{k,l}^2$ with weight functions $w_{k,l}$= F^k and $w_{k,l}$= $F^k/(1-F)$

$w_{k,l}$	F^k	$F^k/(1-F)$
H	$F^{k+1}*(\frac{F}{k+2}-\frac{q}{k+1})$	$(q-1)*[\ln(1-F)+\sum_{l=1}^{k}\frac{F^l}{l}]-\frac{F^{k+1}}{k+1}$
R	$\frac{2}{(k+1)*(k+2)*(k+3)}$	$\frac{1}{(k+1)*(k+2)}$
m	$\frac{1}{(k+2)*(k+3)}$	$\frac{1}{k+2}$
s^2	$\frac{2}{(k+2)*(k+3)^2*(2k+5)}$	$4\sum_{l=1}^{k+2}\frac{1}{l}[\frac{1}{(k+1)*(k+2)}-\frac{1}{(l+k+1)*(l+k+2)}]$ $-\frac{4}{(k+2)^2}[1+\frac{1}{2(2k+3)}]$

ESTIMATION OF CONFIDENCE INTERVALS

The methods for evaluation of the sample distribution of $n_e\omega_{k,l}^2$ given by Anderson and Darling (1952) are mathematically expensive. But the tables given there with k=l=0 and k=l=1 allow for the following conclusion: with expectation value $m_{k,l}$ and standard deviation $s_{k,l}$ of the test statistic the value $S_{k,l} = m_{k,l} + 2s_{k,l}$ can be defined, which agrees rather well with the significance point of $\alpha = 0.05$, for the probabilties are $P(n_e\omega_{0,0}^2 < S_{0,0}) = 0.9510$ and $P(n_e\omega_{-1,-1}^2 < S_{-1,-1}) = 0.9518$. To make the new statistic equivalent with the values without weight function with regard to assessment of significance, we will consider in the following the corrected values

$$n_e \omega_{k,l}^{*2} = n_e C_{k,l} * \omega_{k,l}^2 \tag{9}$$

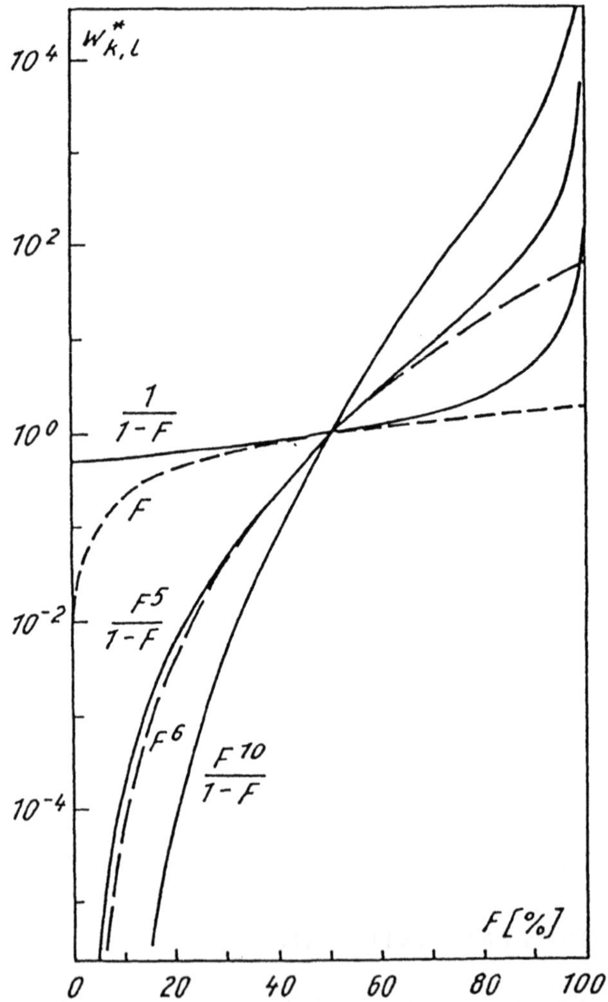

Fig. 2. Standardized power functions $w_{k,l}^* = 2^{k+l}F^k(1-F)^l$ for weighting the range of F.

with $C_{k,l} = S_{0,0}/S_{k,l}$. But it should be kept in mind that the sample distribution of the test statistic could have greater skew for $k>0$, $l=1$ than for symmetric exponents. In addition it must be noted that the methods given by Anderson and Darling (1952) and applied in our work are valid only for the simplest case, i.e. asymptotically for great samples and exactly known parameters of the model distribution. For estimation of the parameters from the sample the significance borders are arbitrarily diminished, namely in the case of extreme value distributions by the factor 1/5 until 1/4, as shown by Stephens (1977) from simulation experiments. Therefore, it must be concluded that the often used significance points of $n\omega_{0,0}^2$ are only a crude approximation, also for the corrected test statistic (9). But for a given sample the relative goodness of fit of different model distributions can be objectively compared, which is true also for the non-standardized statistic $n_e\omega_T^2$.

NUMERICAL EXAMPLES

We analyzed annual maximum series at 17 gauging stations in the foreland of the Erzgebirge in Saxony and in the north German plains. To use data as independently and as comparably as possible for the results given below these are based on the series without historical floods, and for simplicity we will write furtheron $n\omega_{k,l}^2$ for the corrected value (9).

The formulas of model distributions and of parameter estimation methods as well as abbreviations further used are given in the appendix.

Due to the randomness of data in each sample the test statistic exhibits a rather erratic behaviour for increasing weighting. Nevertheless the following general conclusions can be drawn:

- weighting by low powers of F ($0<k\leq3$) does not yield a significant change compared to $n\omega_{0,0}^2$, and the discrepancies against the visual assessment of the fit remain.
- weighting by high powers of F ($k>6$) prefers that model which meets best the highest flood or some few of the highest floods, while the fit of all other sample values has no influence. This choice is too extreme even for flood statistics.
- powers in the mean range give the most reasonable results. Therefore we recommend the weight function $w_{5,-1} = F^5/(1-F)$. It best removes the discrepancies against the visual inspection in the Gumbel net. In most cases the statements of $n\omega_{5,-1}^2$ agree with that of $n\omega_T^2$.

Some more inferences mainly based on this criterion are:

- a distinct favour given to a model distribution or parameter estimation method before or after weighting is not apparent. But the worst fit has the E1 without weighting in 11 cases and with weighting in 13 cases from all the 17 ones. A model distribution which has best fit without weighting keeps this property with weighting in only two cases of all (in both cases the ME)

- if the sample plots in the Gumbel net allow good fit by a straight line in the range of small and medium values that is continued in the upper range by an upward flexure of the plotting positions, then the best fit turns from the E1 to the AE or ME with increased weighting (Example: station Borstendorf, Table 2)
- if the sample plots in the Gumbel net show in the lower reach an upward flexure which does not continue in the upper reach, there is a reversed exchange of the model distributions, i.e. substitution of the AE or ME by the E1 for best fit with increased weighting
- an overall consequence of increased weighting is a moderate augmentation of expectation value and variance of the test statistic owing to the greater sample variability in the upper range of sample values.

TABLE 2. Estimated distribution parameters, and also $n\omega_T^2$ and $n\omega_{k,l}^2$ for the weight functions $w_{k,l} = F^k(1-F)^l$. Gauging station Borstendorf (river Flöha), period 1931-1981

type of distribution		E1			AE		ME
estimation method		Mom.	WG. Mom.	Max.Lik.	WG. Mom.	Max.Lik.	Max.Lik.
parameter		estimates					
c		66.546	66.946	67.053	66.239	63.149	52.871
d		39.970	39.277	37.342	37.786	34.187	28.838
a					-0.040	-0.152	
c_1							-1.482
d_1							61.041
		$n\omega_T^2$					
		27.319	32.183	54.438	21.316	8.756	9.340
k	l	$n\omega_{k,l}^2$					
0	0	0.057	0.059	0.067	0.063	0.108	0.086
1	0	0.044	0.044	0.045	0.045	0.095	0.067
6	0	0.033	0.032	0.033	0.026	0.035	0.025
0	-1	0.053	0.055	0.064	0.052	0.084	0.064
5	-1	0.056	0.060	0.082	0.048	0.042	0.037
10	-1	0.070	0.077	0.114	0.061	0.045	0.045

Some experience with fitting and estimation methods

The following general properties of the model distributions and parameter estimates could be detected:

E1: For an upward flexure of sample plotting positions in the Gumbel net, i.e. if the AE-parameter $a<0$, holds d(Max.Lik.) < d(WG Mom.) < d(Mom.). For a downward flexure in most cases the reversed inequality is valid, except for very small $|a|$.

AE: For a distinct upward flexure is a(Max.Lik.) < a(WG Mom.), while for slight or downward flexures often the reversed relation holds.

ME: For a marked upward flexure is $d_1>d$, while for minor flexures the parameters depend strongly on sample fluctuations and the interpretation is difficult.

Another experience was that the parameter estimates by the the methods of moments and by maximum likelihood are particularly different in samples with historical floods. It is possible that this is due to the differences in taking into account the numerically unknown values $x_i<h$ in the estimation formulas. It seems useful to realize the v-fold multiplicity of the known x_i within the likelihood function in the same way as in the formulas of moment estimates.

Special features of the ME

Figures 3 and 4 are examples for the fit of the model distributions for a distinct flexure in the upper reach. An idenfication of the greatest sample values as outliers is not considered there.

In Fig.3 the advantage of the ME is apparent in that its curvature turns to a straight line in the upper reach if one of the doubly exponential factors gets a value near 1. Thus the risk of overdesign for floods of high recurrence intervals diminishes. On the other hand Fig.4 shows that the ME does not fit a downward flexure of the sample plottings, but proves practically a straight line over the whole reach.

To reveal this special property two lines E1' and E1" were visually fitted to the two tails of the plotting positions, the parameters of the belonging doubly exponential factors were graphically estimated, and by its products the distribution ME' was evaluated. Obviously ME' would fit sample plottings reflected at the diagonal between E1' and E1", but not the observed ones. We suppose that this disadvantage is owing to the fact that according to the definition by Rossi et al. (1984) the mixture of distributions does not take place within the the annual maxima but in the underlying partial duration series. For a genuine mixture of annual maxima an additional parameter to be estimated would be necessary.

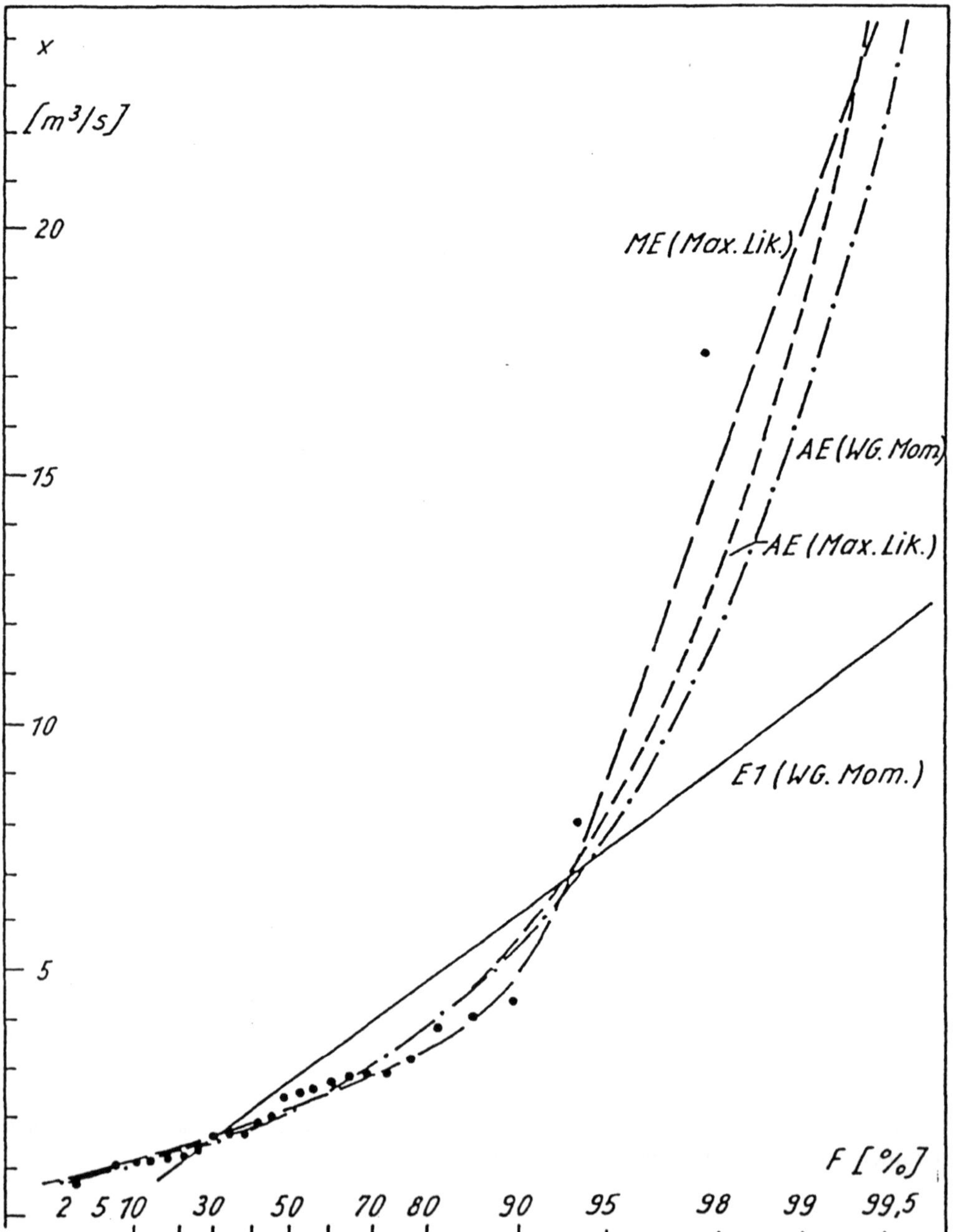

Fig.3. Model fitting to annual maxima x of gauging station Sosa (river Kleine Pockau), period 1913-1937.

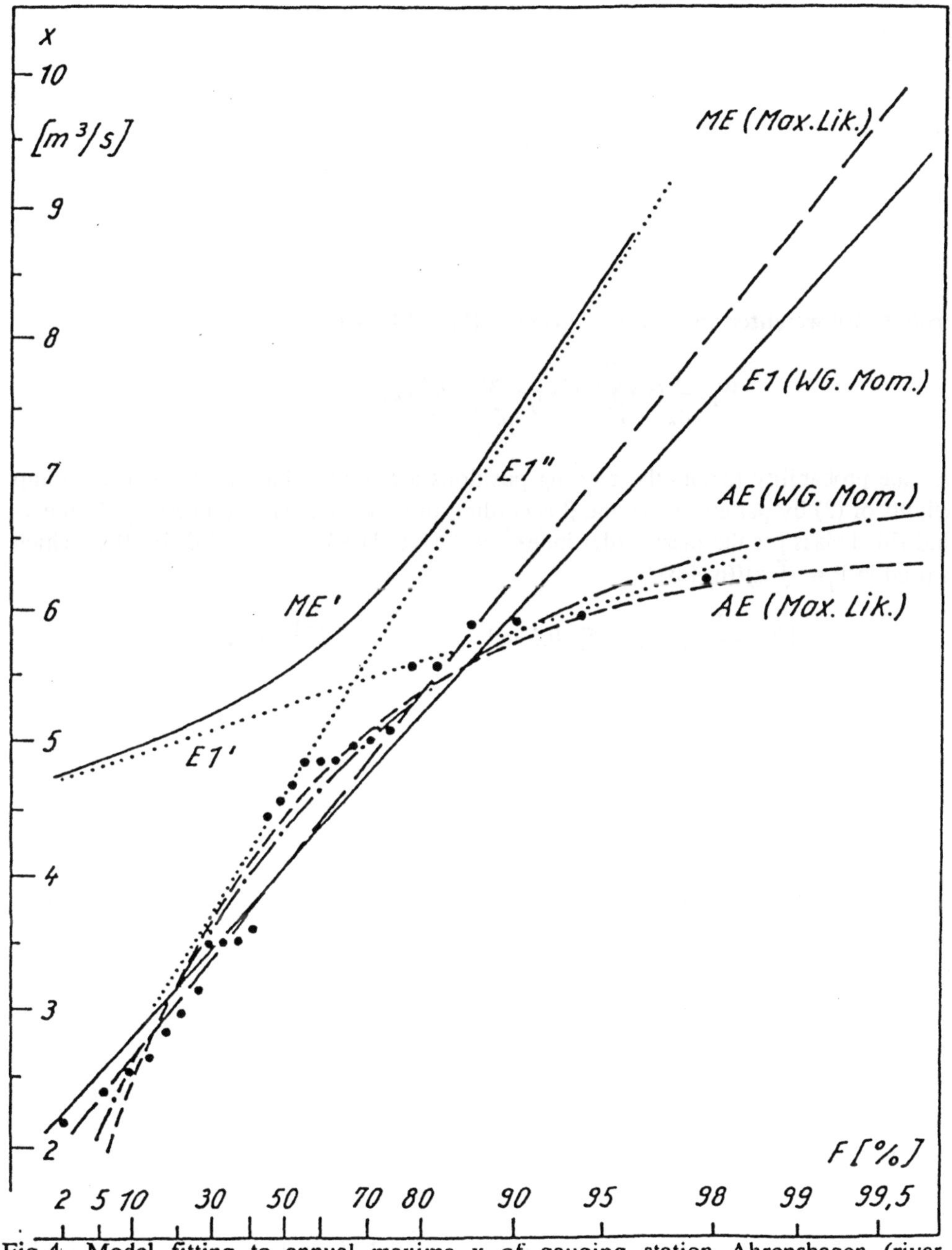

Fig.4. Model fitting to annual maxima x of gauging station Ahrenshagen (river Nebel), period 1955-1980.

APPENDIX: ESTIMATION FORMULAS FOR MOMENTS AND FOR PARAMETERS OF THE MODEL DISTRIBUTIONS

Moments

Expectation value $m=\frac{1}{n_h}(\nu*\sum_{j=1}^{n_r} x_j+\sum_{j'=n_r+1}^{n_g} x_{j'})$

Variance $s^2=\frac{1}{n_h-1}[\nu*\sum_{j=1}^{n_r}(x_j-m)^2+\sum_{j'=n_r+1}^{n_g}(x_{j'}-m)^2]$

Probability weighted moments of order k (WG.Mom.)

$$W_k=\frac{1}{n_h}(\nu*\sum_{j=1}^{n_r} P_j^k*x_j+\sum_{j'=n_r+1}^{n_g} P_{j'}^k*x_{j'})$$

For the probability factors the plotting positions are used, which interpolate the jump heights of (2) by parameters α and β according to some criteria of optimality. In the v-fold filled class j is the mean order index $\bar{j}= 0.5[v(j-1)+1+vj]= v(j-0.5)+0.5$, which is used in $P_j= (\bar{j}-\alpha)/(n_h+\beta)$:

$$P=\begin{cases} P_j=\frac{1}{n_h+\beta}[\nu*(j-0,5)+0,5-\alpha] & j=1,\ldots, n_r \\ & \text{for} \\ P_{j'}=\frac{1}{n_h+\beta}[j'+n_h-n_g-\alpha] & j'=n_r+1,\ldots, n_g \end{cases}$$

For the extreme value distributions was set $\alpha= 0.44$ and $\beta= 1-2\alpha= 0.12$.

Model functions

For all numerically known x_i successively arriving in time and not ordered in magnitude the index i will be written.

Extreme value distribution type 1:

$F(x)=\exp[-\exp(-\frac{x-c}{d})]$ with $-\infty \le x \le \infty,\ d>0$ $\qquad y=\frac{x-c}{d}$

Estimation by moments (Mom.): $d=\frac{\sqrt{6}}{\pi}*s,\quad c=m-\gamma d,$ with $\gamma=0,5772$

Estimation by probability weighted moments (WG.Mom.): $d=\frac{2\ W_1-W_0}{ln2},\quad c=W_0-\gamma d$

Likelihood function: $-lnL=(n_h-n_g)*\ \exp\ (-\frac{h-c}{d})+\sum_{i=1}^{n_g}[\exp\ (-\frac{x_i-c}{d})+\frac{x_i-c}{d}]+n_g*lnd$

Equations for maximum likelihood estimation by iteration:

$$d=[\frac{1}{n_g}\sum_{i=1}^{n_g} x_i]*[1+\frac{(n_h-n_g)\frac{h}{d}\ \exp(-\frac{h}{d})+\sum_{i=1}^{n_g}\frac{x_i}{d}\ \exp(-\frac{x_i}{d})}{(n_h-n_g)*\ \exp(-\frac{h}{d})+\sum_{i=1}^{n_g}\ \exp(-\frac{x_i}{d})}]^{-1}$$

$$c=d*\ln\frac{n_g}{(n_h-n_g)*\ \exp(-\frac{h}{d})+\sum_{i=1}^{n_g}\ \exp(-\frac{x_i}{d})}$$

Generalized extreme value distribution (AE):

$$F(x)=\exp\ [-(1-a\frac{x-c}{d})^{1/a}] \quad \text{with } \frac{x-c}{d}<\frac{1}{a},\ a>-1,\ d>0 \qquad y=-\frac{1}{a}*\ln(1-a\frac{x-c}{d})$$

Estimation by probability weighted moments:
Hosking et al. (1985) have shown that

$$a=7{,}8590*U+2{,}9554*U^2, \quad d=\frac{(2W_1-W_0)*a}{(1-1/2^a)*\Gamma(1+a)}, \quad c=W_0+\frac{d}{a}[\Gamma(1+a)-1]$$

with the auxiliary function $U=\frac{2W_1-W_0}{3W_2-W_0}-\frac{ln2}{ln3}$ and the gamma function Γ.

Likelihood function:

$$-lnL=(n_h-n_g)*(1-a\frac{h-c}{d})^{1/a}+n_g*lnd+(1-1/a)*\sum_{i=1}^{n_g}\ln(1-a\frac{x_i-c}{d})+\sum_{i=1}^{n_g}(1-a\frac{x_i-c}{d})^{1/a}$$

Initial values for evaluation of extremes: a,c,d from WG.Mom. estimation.

Two-component extreme value distribution (ME):

$$F(x)=\exp\ [-\exp\ (-\frac{x-c}{d})]*\ \exp\ [-\exp\ (-\frac{x-c_1}{d_1})] \text{ with } -\infty\le x\le\infty,\ d>0,\ d_1>0$$

$$y=-\ln[\exp\ (-\frac{x-c}{d})+\ \exp\ (-\frac{x-c_1}{d_1})]$$

Likelihood function:

$$-lnL=(n_h-n_g)[\exp\ (-\frac{h-c}{d})+\ \exp\ (-\frac{h-c_1}{d_1})]+\sum_{i=1}^{n_g}[\exp\ (-\frac{x_i-c}{d})+\ \exp\ (-\frac{x_i-c_1}{d_1})-lne_i]$$

with $e_i = \frac{1}{d} \exp\left(-\frac{x_i - c}{d}\right) + \frac{1}{d_1} \exp\left(-\frac{x_i - c_1}{d_1}\right)$

Initial values for evaluation of extremes: c,d from WG.Mom. estimation of E1; $c_1 = 0$, $d_1 = 2.3d$ (deducted from Rossi et al. (1984)).

REFERENCES

Ahmad,M.J., Sinclair,C.D. and Spurr,B.D. (1988) "Assessment of flood frequency models using empirical distribution function statistics", Wat.Resour.Res. 24, 1323-1328.

Anderson,T.W. and Darling,A.D. (1952) "Asymptotic theory of certain 'goodness of fit' criteria based on stochastic process", Ann.Math.Stat. 23, 193-212.

Gumbel,E.J. (1966) Statistics of Extremes, Columbia University Press, New York.

Hosking,J.R.M.,Wallis,J.R. and Wood,E.F. (1985) "Estimation of the generalized extreme value distribution by the method of probabilty-weighted moments", Technometr. 27, 251-261.

Jenkinson,A.F. (1955) "The frequency distribution of the annual maximum (or minimum) of meteorological elements", Quart.J.Roy.Met.Soc. 81, 158-171.

Kluge,Ch. and Finke,W. (1986) "Anpassung stetiger Verteilungsfunktionen bei spezieller Klasseneinteilung der Beobachtungswerte, insbesondere des Niederschlages", Acta Hydrophys. 30, 19-40.

Rossi,F., Fiorentino,M. and Versace,P. (1984) "Two-component extreme value distribution for flood-frequency analysis", Wat.Resour.Res. 20, 847-856.

Smirnow,N.W. and Dunin-Barkowski,J.W. (1969) Mathematische Statistik in der Technik, Deutscher Verlag der Wissenschaften, Berlin.

Stephens,M.A. (1977) "Goodness of fit for the extreme value distribution", Biometrika 64, 585-588.

PRACTICAL USE OF THE L-MOMENTS

K.C. ANDER CHOW
Atria Engineering Hydraulics Inc.
8 Stavebank Road North, Suite 401
Mississauga, Ontario L5G 2T4

W. EDGAR WATT
Department of Civil Engineering,
Queen's University
Kingston, Ontario K7L 3N6

ABSTRACT

Sample moments are often used by practising engineers to select and fit an appropriate statistical distribution to geophysical data. The first and second moments are commonly used for parameter estimation, and if the sample is sufficiently large, third or higher order moments could provide information on the distribution form. However, it is known that moments higher than the first order (the mean) are biased and subject to sampling error. Rather than using the sample moments, Hosking (1990) proposed the use of L-moments for parameter estimation and hypothesis testing about distributional forms. A Monte Carlo simulation was performed to evaluate the sampling errors of the L-moments for the EV1 distribution, and the method of L-moments was applied to annual maximum flow data for 42 long-term hydrometric stations across Canada. This study indicates that the L-moments also suffer from the sampling error due to the small samples commonly encountered in flood frequency analysis. Use of the L-moment ratio diagram alone is not sufficient for selecting a distribution. Further research is required for the parameter estimators for the method of L-moments.

K. W. Hipel (ed.),
Stochastic and Statistical Methods in Hydrology and Environmental Engineering, Vol. 1, 55–69.

INTRODUCTION

Sample moments are often used by practising engineers to select and fit an appropriate statistical distribution to geophysical data. The first and second moments are commonly used for parameter estimation, and if the sample is sufficiently large, third or higher order moments could provide information on the distribution form. However, it is known that moments higher than the first order (the mean) are biased. Correction for bias of the higher moments has been observed to depend not only on the sample size but also on the population skewness and the underlying distribution (Wallis et al. 1974; Bobée and Robitaille 1975; Yevjevich and Obeysekera 1984).

Very often, in the case of flood frequency analysis sample sizes are small and hence the higher order moments are subject to such large sampling errors that they convey little information on the shape of the distribution, and may lead to wrong conclusions when outliers are present.

Many formula have been developed to correct the problem of bias, but the problem of sampling error is still unresolved due to the fact that the record length of the hydrological or meteorological data is relatively short in the statistical sense.

Hosking (1990) formalized the use of quantities called L-moments for statistical analysis of probability distributions. These L-moments are linear combinations of the ordered statistics and are considered an alternative approach to quantify the sample statistics. The objectives of the work reported in this paper are: (1) to evaluate the usefulness of L-moments for single station flood frequency analysis, (2) to investigate the sampling error of L-moments, and (3) to determine if L-moments are superior to conventional measures for identification of distributions for single station frequency analysis.

THE METHOD OF L-MOMENTS

The conventional approach to describe a sample is to compute the sample moments, i.e. sample mean, standard deviation, the standardized third and fourth moments (the coefficients of skewness and kurtosis). The moments of higher order are expressed as sums of powers of the observations about the sample mean. These conventional sample moments have been criticized as being biased and have serious practical drawbacks (Royston 1991). An alternative approach is to use linear combinations of the ordered sample values to obtain quantities so called L-moments. A description of this approach can be found in Kaigh and Driscoll (1985). Hosking (1990) formalized this approach for identification and parameter estimation of probability distributions, and hypothesis testing about distributional forms. He found that the L-moments are able to characterize a wider range of distributions and are more robust to outliers. Parameter estimators based on L-moments are less biased, and for small samples, parameter estimates obtained from L-moments are sometimes more accurate than even maximum likelihood estimates. The procedure to compute the sample L-moments based on Royston (1991) is presented below, interested readers are referred to the original paper of Hosking (1990) for a detailed description.

Suppose that $x_1, ..., x_n$ is a random sample of size n and $x_{1:n} \leq ... \leq x_{n:n}$ are the ordered sample; the L-moments and their ratios can be estimated by

$$\begin{aligned} l_1 &= \bar{x}, \\ l_2 &= 2\,w_2 - l_1, \\ l_3 &= 6\,w_3 - 6\,w_2 + l_1, \\ l_4 &= 20\,w_4 - 30\,w_3 + 12\,w_2 - l_1, \\ t_3 &= l_3 / l_2, \\ t_4 &= l_4 / l_2, \end{aligned} \tag{1}$$

where l_r is the r-th sample L-moment and t_r is the r-th sample L-moment ratio. The first L-moment, l_1, is the sample mean $\bar{x} = \sum x_i / n$, the second L-moment, l_2, is a measure of the dispersion of the sample, t_3 and t_4 are known as the sample L-skewness and L-kurtosis which can be considered as measure of asymmetry and tail weight respectively. The coefficients w_2, w_3 and w_4 can be computed as

$$\begin{aligned} w_2 &= \frac{1}{n(n-1)} \sum_{i=2}^{n} (i-1)\, x_{i:n}, \\ w_3 &= \frac{1}{n(n-1)(n-2)} \sum_{i=3}^{n} (i-1)(i-2)\, x_{i:n}, \\ w_4 &= \frac{1}{n(n-1)(n-2)(n-3)} \sum_{i=4}^{n} (i-1)(i-2)(i-3)\, x_{i:n}. \end{aligned} \tag{2}$$

Hosking (1990) derived the L-moments for 11 statistical distributions and their parameter estimators. The distribution form and the parameter estimators for three of these distribution - the normal, Gumbel (EV1) and general extreme value (GEV) are demonstrated in Table 1. It should be noted that the values of the L-skewness and L-kurtosis are 0.0 and 0.1226 respectively for the normal distribution and 0.1699 and 0.1504 for the EV1 distribution.

For the purpose of testing hypotheses about distributional form, Hosking (1990) proposed that the test statistic

$$N_n = (0.1866\, n^{-1} + 0.8\, n^{-2})^{-1/2}\, t_3 \tag{3}$$

be used to test whether the sample L-skewness t_3 is significantly different from zero.

For testing whether the shape parameter k of the GEV distribution is significantly different from zero, he proposed the use of the statistic

$$Z_n = (n/0.5633)^{1/2}\,\hat{k} \tag{4}$$

where $\hat{k}$ is the L-moment estimator of the GEV shape parameter k.

IDENTIFICATION OF STATISTICAL DISTRIBUTIONS

To select an appropriate distribution for a single station flood frequency analysis, many analysts may rely on a comparison of the sample moments with the theoretical distribution values. Rather than using the sample moments, Hosking (1990) recommended the use of L-moment ratios. He provided a L-moment ratio diagram (MRD) for 11 distributions in which the fourth L-moment ratio (L-kurtosis) of the theoretical distribution is plotted against its third L-moment ratio (L-skewness). An example of L-moment ratio diagram for the normal, EV1 and GEV distributions is displayed in Figure 1. Since distributions such as EV1 and normal have only one value of L-skewness and L-kurtosis, they are plotted as a single point in the diagram (i.e., (0.0, 0.1226) and (0.1699, 0.1504) respectively). Other distributions such as GEV have a set of possible combination of L-skewness and L-kurtosis and are represented by a curve. From this figure it can be seen that a distribution such as GEV can have a wider range of combined values of the L-moment ratios, and thus is more flexible to fit the sample data. The procedure to select an appropriate distribution for a sample is to compute the sample L-moment ratios and plot them on the diagram; the distribution that seems to better represent the location of the sample values is then selected.

Use of a MRD as an aid in the selection of distribution is not new. The widely used MRD for conventional moments is the square of the coefficient of skewness of the distribution against its coefficient of kurtosis. However, one major difficulty with this approach is the large sampling errors associated with the two coefficients due to small sample size encountered in hydrology (Bobée and Ashkar 1991). Confidence limits for these coefficients for various sample size should be added to the diagram, but analytical solution of the confidence limits is difficult to obtain.

Recently, the authors conducted a Monte Carlo simulation to estimate the sampling errors of the skewness coefficient for the EV1 distribution. The Monte Carlo simulation is repeated here for the L-skewness and L-kurtosis for the EV1 distribution. In the experiment, 1000 samples were drawn from an EV1 population for sample sizes of 10, 20, 25, 50 and 100. The population value, the expected value and one standard error of the sample L-skewness are plotted in Figure 2(a), while the simulation results of the previous study are plotted in Figure 2(b) for visual comparison. It can be seen that the bias of the L-skewness is smaller and that the sampling errors of the L-skewness are reducing significantly when the sample size increases.

The simulation results are also plotted on the L-moment ratio diagram. An elliptical form is fitted to the results to obtain the approximate 95% confidence limits. The approximate 95% confidence limits for sample sizes of 25, 50 and 100 are then plotted in Figure 1. If the L-moment ratios of a sample is plotted within the region enclosed by the confidence limits of the corresponding sample size, it may imply that the sample is drawn from a EV1 distribution. Considering this, the appearance of the EV1 distribution on the L-moment ratio diagram is not a point but rather a big bullet centred around the point (0.1699, 0.1504) on the diagram.

USE OF L-MOMENTS FOR SINGLE STATION FLOOD FREQUENCY ANALYSIS

The sample L-moments were computed for the annual maximum flow data collected at 42 long-term hydrometric stations across Canada which have been used in previous studies (Chow and Watt 1990; 1992). The sample L-skewness, L-kurtosis, and the test statistics computed by equations (3) and (4) are summarized for all stations in Table 2. The results are plotted on the L-moment ratio diagram in Figure 3. For each station, the test statistics from (3) and (4) were compared with the 95% significance levels. If the Z_n is greater than the 95% significance level (i.e. the estimated k value of the GEV distribution cannot be considered as zero), the GEV distribution is appropriate for the station and is shown as hollow dot in Figure 3. If the N_n is less than the 95% significance level (i.e. the value of L-skewness can be considered as zero), the normal distribution may be appropriate for the station and is shown as shaded dot in Figure 3.

The trend of all station data points appears to follow the GEV curve on the L-moment ratio diagram. However, considering the sampling errors of the L-moments, the EV1 distribution is still a potential candidate distribution for these study stations. With the aids of the test statistics given by equations (3) and (4), for only five stations the estimated shape parameter of the GEV distribution is significantly different than zero, and for another eight stations, the normal distribution can be considered appropriate. Even though the L-moments are less biased, with the consideration of their sampling errors, the L-moment ratio diagram alone cannot be used to identify the distributions. Visual examination of probability plots for all possible distributions or other measures of goodness-of-fit are still required in the decision process of selecting an appropriate distribution.

USE OF L-MOMENTS FOR REGIONAL FREQUENCY ANALYSIS

Adamowski, Pilon and Alila (1990) employed the method of L-moments for regional frequency analysis of rainfall for Ontario. They claimed that the classical Gumbel (EV1) distribution, which is used by the Atmospheric Environment Service for rainfall frequency analyses, is generally not appropriate, and the general extreme value distribution (GEV) should be used. The same recommendation was made in a regional flood frequency analysis for Nova Scotia by Pilon and Adamowski (1992). In both papers, the variability of the L-skewness was attributed largely to sampling error. No further explanations were given to explain the variability of L-skewness or to relate the magnitude of the sampling error to sample size.

The sample L-moments of the flood data given in Pilon and Adamowski (1992) are plotted on the L-moment ratio diagram shown in Figure 4. Most of the data points are within the 95% confidence limits for the EV1 distribution. Using the test statistics given by equation (4), for only three stations the GEV distribution is considered appropriate. The average L-skewness of 0.226 and the average L-kurtosis of 0.169 were also computed for the study sites by Pilon and Adamowski (1992), and these site values are also plotted in Figure 4. They argued that this data point suggested a GEV distribution. However, considering the sampling errors for the EV1 distribution this site average values lies well within the 95% confidence interval for the EV1 distribution.

FITTING THE DISTRIBUTIONS

The parameter estimation procedure using the method of L-moments is demonstrated for La Have River at West Northfield, Nova Scotia (01EF001). The reason for choosing this station as example is the significant extreme event (outlier) in the sample. Removing the high outlier for this station, the sample skewness changes from 4.2 to 0.7; but the L-skewness changes only from 0.28 to 0.12. It seems that the L-moments are more robust to outliers.

The fitted normal, EV1 and GEV distributions using the method of moments, L-moments and Maximum Likelihood for this station are plotted in Figure 5. It can been seen that for the normal distributions, the method of moments gave a steeper fitted line due to the presence of high outlier. For the GEV distributions the method of L-moments is as efficient as the method of Maximum Likelihood. However, for the EV1 distribution the method of L-moments seemed unable to provide an adequate fitted line to the sample. More research may be required for the method of L-moments to derive the robust parameter estimators.

SUMMARY

1. The L-moments are less biased and more robust to the presence of outliers.

2. The L-moments suffer from the sampling errors due to small samples commonly encountered in flood frequency analysis. Use of the L-moment ratio diagram alone for identification of statistical distribution is not sufficient.

3. Further research is required in the parameter estimators for the method of L-moments.

REFERENCES

Adamowski, K., Pilon, P.J. and Alila, Y. (1990) "Regional analysis of annual maximum precipitation", Paper presented at International Workshop on Urban Rainfall and Meteorology, St. Moritz, Switzerland.

Bobée, B. and Ashkar, F. (1991) "Choice of a Statistical Distribution in Practice" in The Gamma Family and Derived Distributions Applied in Hydrology, Water Resources Publications, 162-187.

Bobée, B. and Robitaille, R. (1975) "Correction of bias in the estimation of the coefficient of skewness", Water Resources Research, 11(6), 851-654.

Chow, K.C.A. and Watt, W.E. (1990) "A knowledge-based expert system for flood frequency analysis", Canadian Journal of Civil Engineering, 17(4), 597-609.

Chow, K.C.A. and Watt, W.E. (1992) "Use of Akaike information criterion for selection of flood frequency distribution", Canadian Journal of Civil Engineering, 19(4), 616-626.

Hosking, J.R.M. (1990) "L-moments: Analysis and estimation of distributions using linear combinations of order statistics", Journal of the Royal Statistical Society, Series B, 52(1), 105-124.

Kaigh, W.D. and Driscoll, M.F. (1987) "Numerical and graphical data summary using O-statistics", American Statistician, 41, 25-32.

Pilon, P.J. and Adamowski, K. (1990) "The value of regional information to flood frequency analysis using the method of L-moments", Canadian Journal of Civil Engineering, 19(1), 137-147.

Royston, P. (1991) "Which Measures of Skewness and Kurtosis are Best", Statistics in Medicine, Vol. 10, 1-11.

Wallis, J.R., Matalas, N.C. and Slack, J.R. (1974) "Just a moment", Water Resources Research, 10(2), 211-219.

Yevjevich, Y. and Obeysekera, T.B. (1984) "Estimation of skewness of hydrologic variables", Water Resources Research, 20(7), 935-943.

Table 1. L-moments Parameter Estimators for Normal, Gumbel and GEV Distributions (after Hosking 1990)

Distribution	Cumulative Function	Parameter Estimators
Normal	$F(x) = \int_{-\infty}^{\infty} \frac{1}{\sqrt{2\pi}\,\sigma} \exp\left[-\frac{1}{2}\left(\frac{x-\mu}{\sigma}\right)^2\right] dx$	$\hat{\sigma} = \sqrt{\pi}\, l_2,$ $\hat{\mu} = l_1$
Gumbel	$F(x) = \exp\left[-\exp\left[-\alpha(x-u)\right]\right]$	$\hat{\alpha} = \ln 2 / l_2,$ $\hat{u} = l_1 - \gamma / \hat{\alpha}$
GEV	$F(x) = \exp\left[-\left(1-\frac{(x-u)k}{\alpha}\right)^{\frac{1}{k}}\right]$	$\hat{\alpha} = l_2 \frac{\hat{k}}{(1-2^{-\hat{k}})\Gamma(1+\hat{k})},$ $\hat{u} = l_1 + \frac{\hat{\alpha}}{\hat{k}}(\Gamma(1+\hat{k})-1),$ $\hat{k} \approx 7.8590z + 2.9554z^2,$ $z = 2/(3+t_3) - \ln 2/\ln 3,$

Note: γ is Euler's constant

Table 2. Sample Statistics and Test Statistics

WSC No.	L-skewness	L-kurtosis	n	N_n	Z_n
09AB001	0.0072	0.0764	43	0.1039	2.3747
08EC001	0.0950	0.0379	41	1.3404	1.0294
08EE004	0.0642	0.1443	56	1.0719	1.7198
08LE031	0.0526	0.1295	64	0.9430	2.0500
08ME001	0.2867	0.2699	36	3.7638	-1.3960
08MG005	0.2668	0.2609	68	4.9407	-1.6050
08NH006	0.0213	0.1834	60	0.3685	2.5473
08NH032	0.1270	0.2172	59	2.1801	0.6992
08NN012	-0.0237	0.1273	57	-0.3989	3.2944
08NN013	0.0552	0.1260	58	0.9391	1.9062
05AA022	0.3923	0.3182	42	5.6066	-2.7633
05AD003	0.0230	0.1270	39	0.3159	2.0281
05BB001	0.1158	0.0985	78	2.3053	1.0176
05BJ005	0.3160	0.1444	44	4.6320	-1.9103
05CC002	0.3058	0.1302	72	5.8360	-2.2820
05HD036	0.2166	0.1163	32	2.6641	-0.5409
05PA006	0.0405	0.1445	64	0.7268	2.2718
05PB014	0.1403	0.0875	70	2.6386	0.5219
05QA001	0.2123	0.1656	60	3.6771	-0.6723
05QA002	0.2348	0.1602	65	4.2436	-1.0623
04LJ001	0.0653	0.1540	67	1.2004	1.8598
02AA001	0.1938	0.1402	63	3.4465	-0.3921
02BD003	0.1982	0.2319	47	3.0114	-0.3997
02EA005	0.1443	0.1525	71	2.7334	0.4546
02EC002	0.0054	0.1730	71	0.1019	3.0874
02FC001	0.0587	0.1516	72	1.1209	2.0553
02FC002	0.1755	0.1803	72	3.3487	-0.0980
02GA010	0.0755	0.0759	48	1.1601	1.4157
02HB001	0.2340	0.2052	67	4.2988	-1.0666
02OJ007	-0.1417	0.0881	49	-2.2011	5.1715
02ZK001	0.3434	0.2618	37	4.5775	-2.0593
01AD002	0.0713	0.1111	60	1.2351	1.6558
01AK001	0.1810	0.2357	66	3.2984	-0.1865
01AQ002	0.3417	0.3084	60	5.9187	-2.5979
01AR003	0.2660	0.1496	64	4.7698	-1.5445
01BE001	0.1194	0.0962	58	2.0305	0.8189
01DG003	0.1481	0.1087	65	2.6771	0.3700
01EC001	0.2301	0.1647	67	4.2279	-1.0038
01EF001	0.2776	0.2743	71	5.2581	-1.8142
01EO001	0.1918	0.1606	71	3.6326	-0.3804
01FB001	0.2630	0.1992	70	4.9453	-1.5665
01FB003	0.1135	0.1557	67	2.0845	0.9848

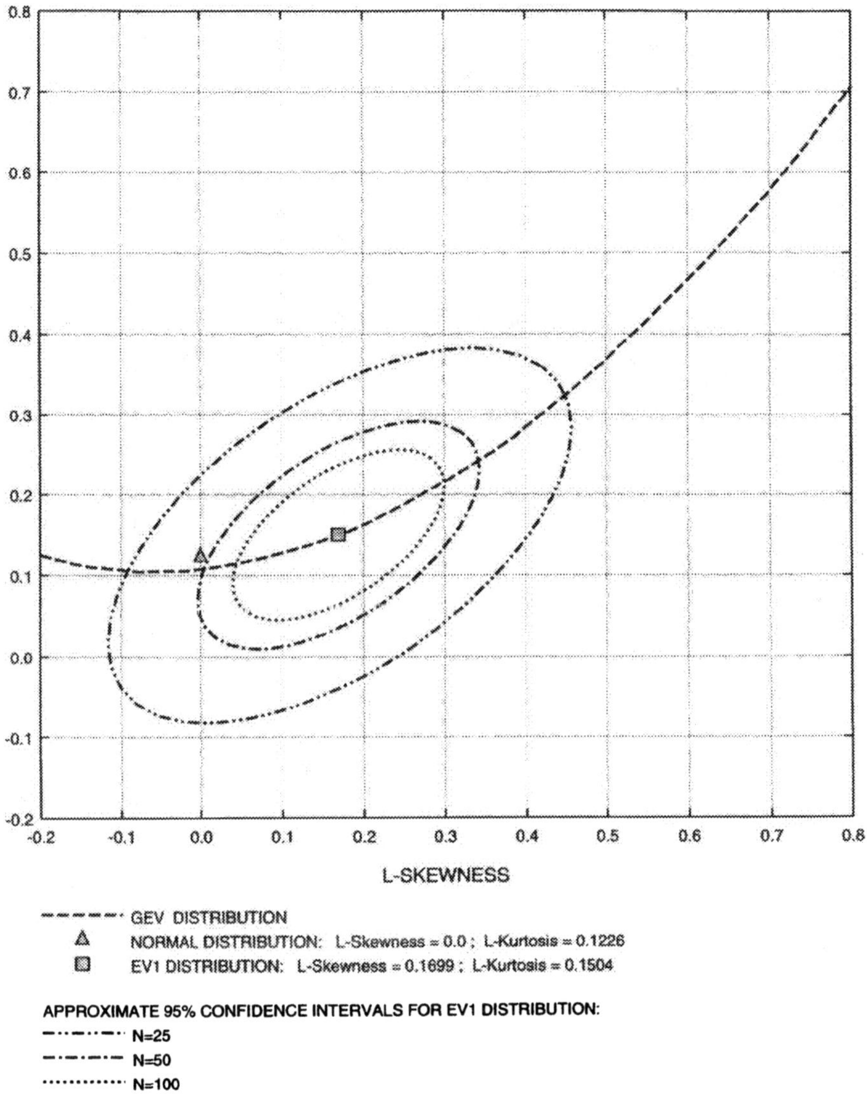

Figure 1. L-moment Ratio Disgram

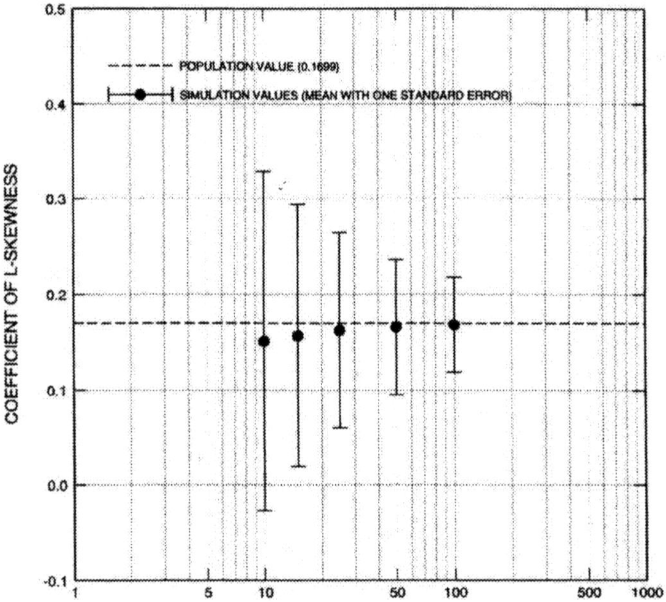

2 (a)

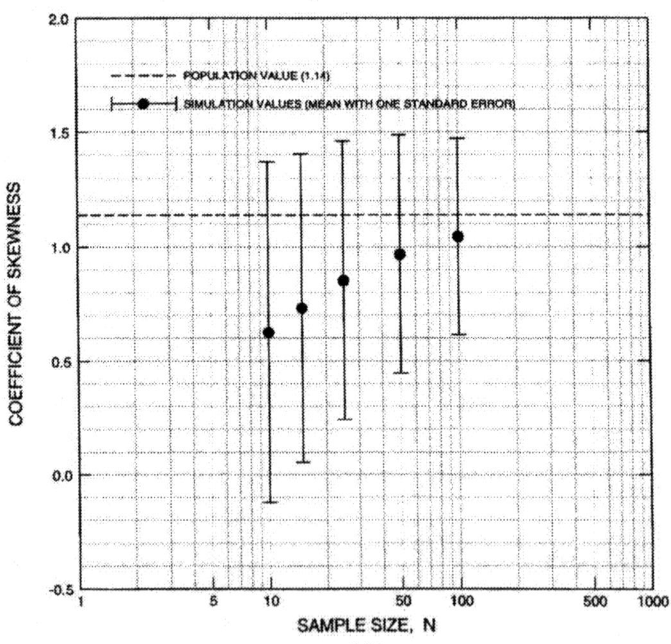

2 (b)

Figure 2. L-skewness and Skewness Estimates for a Sample drawn from an EV1 Distribution

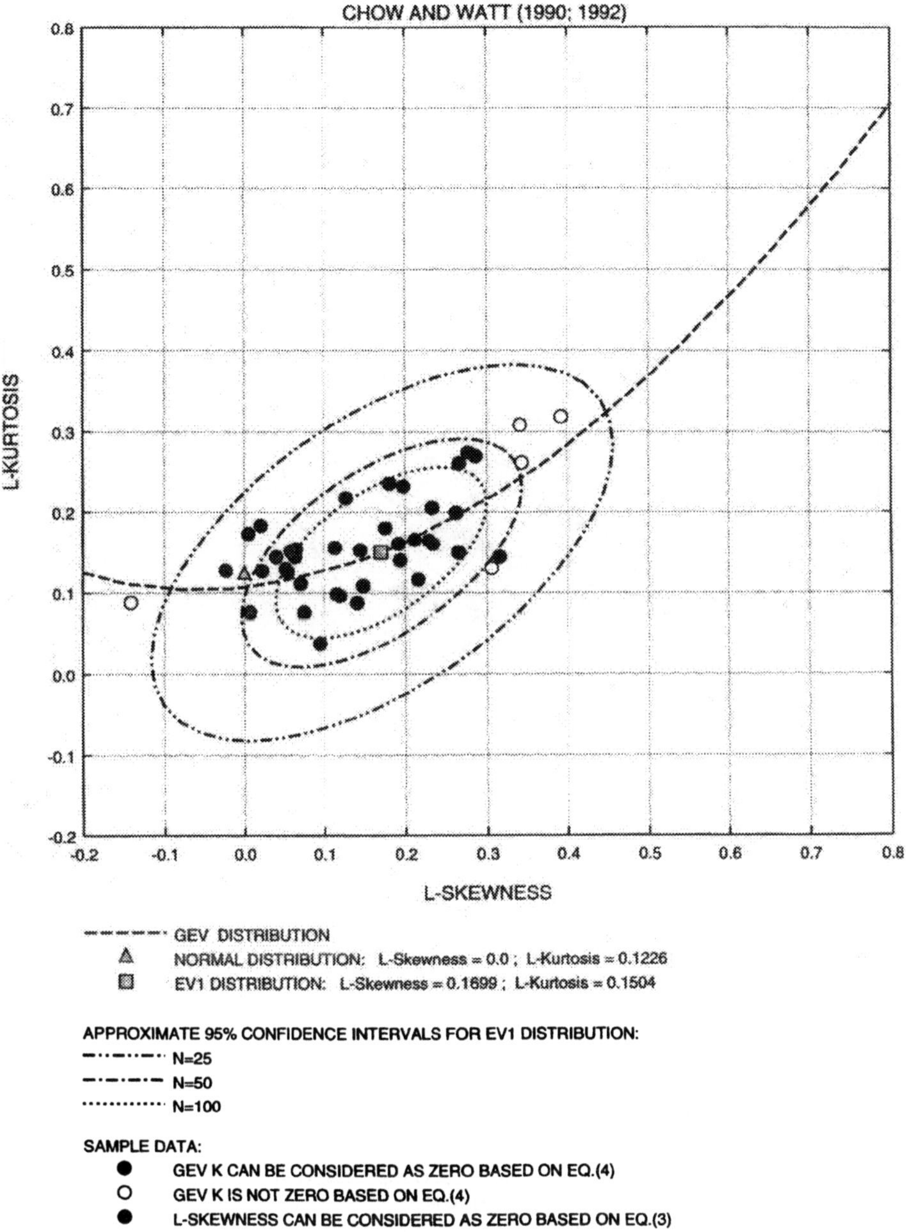

Figure 3. L-moment Ratio Values for Chow and Watt's Data

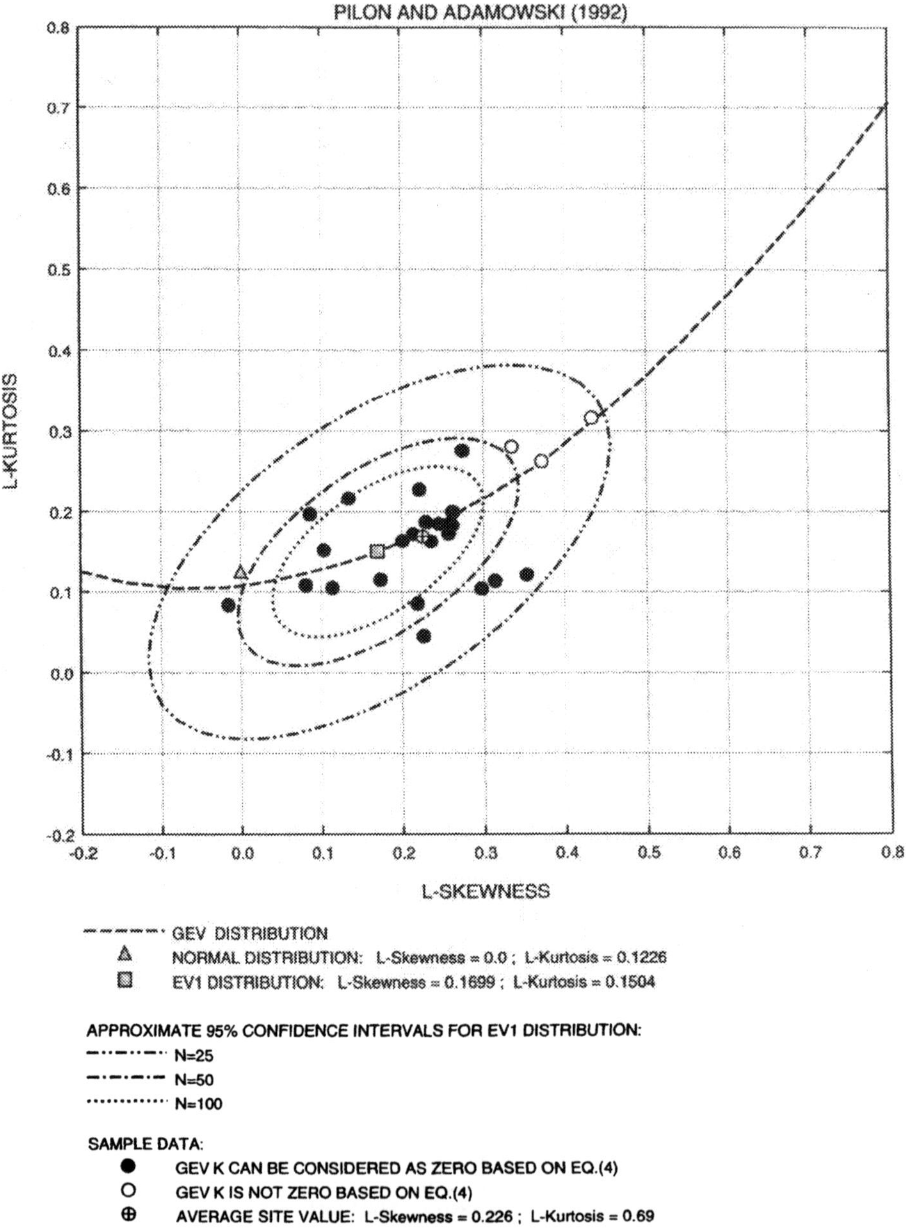

Figure 4. L-moment Ratio Values for Pilon and Adamowski's Data

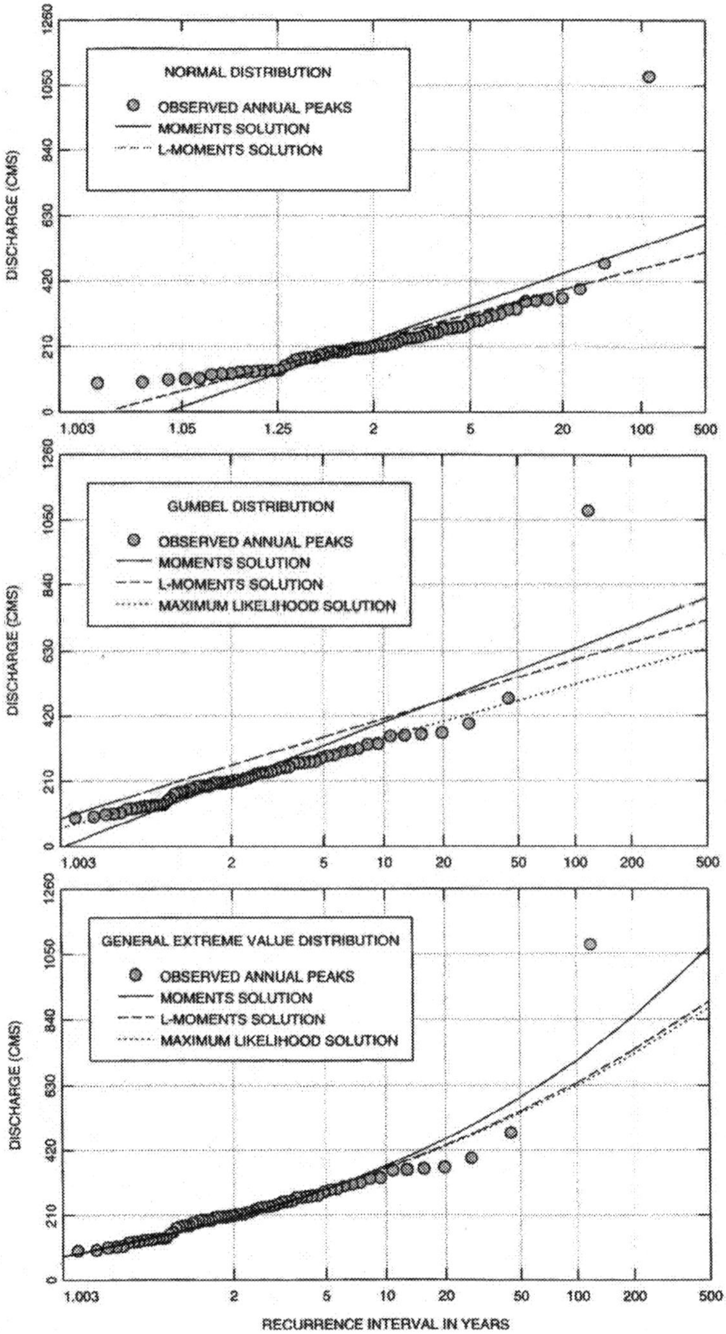

Figure 5. Fitted Distributions by the Method of Moments, L-Moments ans Maximum Likelihood

PART II

PARTIAL DURATION SERIES

RISK ESTIMATION IN PARTIAL DURATION SERIES OF SEASONAL FLOODS

H. P. NACHTNEBEL(1) and F. KONECNY(2)
(1) Institute for Water Resources Management, Hydrology and Hydraulic Engineering
(2) Institute of Mathematics and Applied Statistics
both University for Bodenkultur
Gregor Mendel Str. 33, A-1180 Vienna, Austria

In many river basins a distinct annual cycle is observed in the hydrograph which suggests also a seasonality in the magnitude and in the frequency of flood peaks. A stochastic flood model based on a partial duration series is proposed to consider seasonal phenomena and possible longterm trends. The flood series are described by a marked inhomogeneous Poisson process with time varying exponentially distributed marks. The design quantities and the parametric risk are derived from an extreme value process. Results are compared with those from models based on stationary distributions.

INTRODUCTION

Many hydrological variables exhibit both a pronounced annual cycle and frequently a longterm trend caused by human impacts on the water cycle. Also the discharge series in many river basins show a seasonality in the flood peaks and in the occurrence of floods. Thus, for a detailed description of flood events a model is required which explicitely consideres the seasonal characteristics.

A stochastic point process model based on a partial duration series approach (PDS) was developed by Todorovic and Zelenhasic (1970), and was subsequently generalized and extended by Todorovic and Rouselle (1971). North (1980) proposed a seasonal model and Ashkar and Rousselle (1981) evaluated the hydrological risk associated with this model class although the procedure was based on the maximum annual discharges. Rasmussen and Rosbjerg (1991) estimated design floods corresponding to a given return period for identically and exponentially distributed exceedances with Poissonian occurrence times. In' the work of Rasmussen and Rosbjerg (1991 a,b) the year was devided into two or four seasons with stationary exceedances. To avoid the somewhat arbitrarely division into seasons, which introduces besides an increased number of parameters also discontinuities at the interval boundaries Nachtnebel and Konecny

K. W. Hipel (ed.),
Stochastic and Statistical Methods in Hydrology and Environmental Engineering, Vol. 1, 73–83.

(1987) proposed a marked inhomogeneous point process with continuously varying parameters in the intensity function and in the distribution of the flood peaks.

The objective of this paper is the provision of design quantities with limited flood risk. This means, the design quantity will be exceeded over a given horizon with a small prespecified failure probability only. In this paper both, seasonal and nonseasonal models are used to estimate design quantities. The probabilities of large exceedances are estimated from the extreme value process which is directly derived from an inhomogeneous marked Poisson process (Konecny and Nachtnebel, 1985). The time dependent distribution function of the non decreasing step functions of the extreme value process was derived after Karr (1976). The procedure consideres both the occurrence of flood events and their exceedances.

The paper is organized in the following way. After a description of the model, the design exceedance with limited risk is defined. In the following section, the limit distribution of the design exceedance is identified. The case of independently and identically distributed (i.i.d.) exceedances is considered as well as the case of time-dependent distributed exceedances. Numerical results are given for a 86-year discharge series.

MODEL DESCRIPTION

In a hydrograph Q(t), those flood peaks are considered, which exceed a certain base level Q_o. From multiple-peaked hydrographs, only these maximum peaks were selected. The exceedance ξ_n denoting the occurrence time of the n-th flood peak over Q_o is observed at occurrence time τ_n are obtained according to

$$\xi_n = Q(\tau_n) - Q_o.$$

The following assumptions are made (cf. Karr (1976) and Todorovic (1978)):

(A.1) For n = 1, 2, ... τ_n is the occurrence time of the n-th event of an inhomogeneous Poisson process with the periodic intensity function

$$\lambda(t) = \theta_o \exp(\sum_{j=1}^{q} \theta_j \sin(\omega j t + \theta_{q+j})), \quad \begin{array}{l} \omega = 2\pi / t^* \\ t^* = 365 \end{array}.$$

The intensity function of a Poisson process expresses the expected number of events per time unit.

(A.2) The exceedances (ξ_n) are conditionally independent, given their occurrence times (τ_n).

(A.3) For k = 1,2, ... ξ_k is conditionally independent of (τ_n), given its occurrence time τ_k. The conditional distribution of ξ_n given τ_n is identical for all n.

For a fixed $x \geq 0$ the number N(t,x) of exceedances greater than x in the time interval (0,t), $t \geq 0$, is a Poisson process with a cyclic trend, which is described by the corresponding intensity function

$$\lambda(t,x) = R(x|t)\lambda(t).$$

with $$R(x|t) = P(\xi_n > x|\tau_n = t)$$

Obviously, it holds that $\lambda(t, 0) = \lambda(t)$, and the average number of events greater than x and observed in the time interval (0,t) is directly obtained from

$$\pi(t,x) = E(N(t,x)) = \int_0^t \lambda(s,x)ds$$

The probability to observe just n events is given by

$$P(N(t,x) = n) = \frac{1}{n!}\pi(t,x)^n \exp(-\pi(t,x))$$

DESIGN EXCEEDANCE UNDER SPECIFIED RISK

On the basis of the Poisson-model for partial duration series two different characteristic design quantities were proposed both related to a given design period T>0. Ashkar and Rousselle (1981) defined the design discharge as the (T-1)/T-quantile of the the distribution function for the maximal annual exceedance. In Nachtnebel and Konecny (1987), the design exceedance was derived from the extreme value process. This process is defined according to

$$\chi(t) = \max\{\xi_n : \tau_n \leq t\}.$$

The marginal distribution functions of this process can be expressed by

$$P(\chi(t)<x) = P(N(t, x) = 0) = \exp(-\pi(t, x)),$$

(cf. Karr, 1976). For a given design period T > 0, the design exceedance x_T is defined as the smallest x such that the expectation of its upcrossing time

$$\sigma_x = inf\{t: \chi(t) > x\}$$

is T, that is, $\bar{\sigma}_{X_T} = E(\sigma_{X_T}) = T$.

The reliability with respect to the horizon [0,T] is given by

$$r_T = P(\sigma_{X_T} > T),$$

and the probability of the complementary event

$$\bar{r}_T = P(\sigma_{X_T} \leq T)$$

is called flood risk in our context.

In order to specify explicitely the flood risk, a different design quantity will be introduced. For a design period D > 0 and 0 < ε < 1 the design exceedance $x_{D,\varepsilon}$ is defined as follows:

$$x_{D,\varepsilon} = inf\{x: P(\sigma_x > D) > 1 - \varepsilon\}$$

Since $\sigma_X > D$ iff N(D, x) = 0, we obtain

$$x_{D,\varepsilon} = inf\{x: \pi(D, x) = -\ln(1 - \varepsilon)\}$$

In applications π(D,.) is strictly monotonuously decreasing. Consequently $x_{D,\varepsilon}$ is the solution of

$$\pi(D,x) = -\ln(1-\varepsilon).$$

Next, the uncertainty in the parameter estimation from a limited series of observations will be considered.

The estimates $\hat{\theta}_p$ and $\hat{\gamma}_p$ referring to the intensity λ (t) and to the distribution of the exceedances R(x t) are based on p years of observation. Since the quantity $x_{D,\varepsilon}$ is a function of θ and γ the corresponding random variable $X_{D,\varepsilon,p}$ is defined implicitely by the equation

$$\pi(D, X_{D,\varepsilon,p}, \hat{\theta}_p, \hat{\gamma}_p) = -\ln(1 - \varepsilon),$$

and depends on the design period D, the flood risk ε and the sample size p. Subsequently $X_{D,\varepsilon,p}$ stands for the design exceedance with the flood risk ε.

THE DISTRIBUTION OF THE DESIGN EXCEEDANCE

In Todorovic and Zelenhasic (1970) and other papers, the occurrences of floods follow an inhomogeneous Poisson process with a cyclic trend but the distribution of the exceedances is independent of the occurrence time and thus i.i.d.

Since the upper tail of many distributions may be represented as exponential, it is assumed

$$R(x;\gamma) = e^{-\gamma x},$$

where γ is the scale parameter to be estimated. When T is given in years

$$\pi(T, x; \mu, \gamma) = \mu T e^{-\gamma x}$$

and with $\mu = E(N(t^*,0)$ the design exceedance $X_{D,\varepsilon,p}$ can be expressed as follows:

$$X_{D,\varepsilon,p} = \frac{\ln(\hat{\mu}_p D / \ln(1/(1-\varepsilon))}{\hat{\gamma}_p}$$

with

$$\hat{\mu}_p = N(pt^*)/p \text{ and } \hat{\gamma}_p = \left[\frac{1}{N(pt^*)}\sum_{i=1}^{N(pt^*)}\xi_i\right]^{-1}.$$

The determination of the distribution function of $X_{D,\varepsilon,p}$ is carried out similarly to Ashkar and Rousselle (1981) and Nachtnebel and Konecny (1987)

$$P(X_{D,\varepsilon,p} \leq x) = \sum_{n=o}^{\infty} P\left(X_{D,\varepsilon,p} \leq x \middle| N(pt^*) = n\right) P(N(pt^*) = n)$$

$$= e^{-\mu p}\left(1 + \sum_{n=1}^{\infty} \frac{(\mu p)^n}{n!} p\left(\frac{1}{n}\sum_{i=1}^{n} \xi_i \leq \frac{x}{\ln(nD/p\ln(1/1-\varepsilon))}\right)\right)$$

$$= e^{-\mu p}\left(1 + \sum_{n=1}^{\infty} \frac{(\mu p)^n}{n!} G_{n,\gamma}(nx/\ln(nD/p\ln(1/1-\varepsilon)))\right),$$

where $G_{n,\gamma}$ (.) denotes the distribution function of a Gamma variable with parameters n and γ.

Next, the more general case with not identically distributed exceedances following a time-dependent exponential distribution is analysed.

With a seasonally dependent scaling parameter $\beta(t;\gamma)$

$$\beta(t;\gamma) = \gamma_o \exp\left(\sum_{k=1}^{r} \gamma_k \sin(\omega k t + \gamma_{r+k})\right).$$

a survival function is obtained according to

$$R(x|t;\gamma) = e^{-\beta(t;\gamma)x}$$

and the integrated intensity π is given by

$$\pi(D,x;\theta,\gamma) = \int_{o}^{D} e^{-\beta(t;\gamma)x} \lambda(t;\theta)\,dt.$$

The design exceedance $X_{D,\varepsilon,p}$ cannot be expressed in a closed formula, but the asymptotic distribution of $X_{D,\varepsilon,p}$ for $p \to \infty$ can be derived.

The asymptotic joint distribution of the maximum likelihood estimators $\hat{\theta}_p$ and $\hat{\gamma}_p$ for $p \to \infty$ was identified in Konecny (1987). It was established that the limiting distribution of $(\hat{\theta}_p, \hat{\gamma}_p)$ is the normal distribution $N(a, M^{-1}/p)$, with the mean vector

$$a = \begin{pmatrix} \theta * \\ \gamma * \end{pmatrix}$$

and the covariance matrix

$$M = \begin{pmatrix} I(\theta *) & O \\ O & J(\theta *, \gamma *) \end{pmatrix}$$

with the elements

$$I(\theta)_{i,j} = \int_o^{t*} \frac{1}{\lambda(t;\theta)} \frac{\partial \lambda(t;\theta)}{\partial \theta_i} \frac{\partial \lambda(t;\theta)}{\partial \theta_j} dt$$

$$J(\theta,\gamma)_{k,l} = \int_o^{t*} \frac{\partial \log \beta(t,\gamma)}{\partial \gamma_k} \frac{\partial \log \beta(t;\gamma)}{\partial \gamma_1} \lambda(t;\theta) dt.$$

It was proven that $X_{D,\varepsilon,p}$ is normal distributed with mean $x^*_{D,\varepsilon}$ given by $\pi(T, x^*_{D,\varepsilon}, \theta *, \gamma *) = -\ln(1-\varepsilon)$, and variance $\sigma^2{}_{D,\varepsilon} / p$, where

$$\sigma^2_{D,\varepsilon} = \left\{ \sum_{i,j} \frac{\partial \pi}{\partial \theta_i} \cdot \frac{\partial \pi}{\partial \theta_j} (D, x^*_{D,\varepsilon}, \theta *, \gamma *) I^{ij}(\theta *) \right.$$
$$\left. + \sum_{k,e} \frac{\partial \pi}{\partial \gamma_k} \frac{\partial \pi}{\partial \gamma_e} (D, x^*_{D,\varepsilon}, \theta *, \gamma *) J^{kl}(\theta *, \gamma *) \right\} / \frac{\partial \pi}{\partial x} (D, x^*_{D,\varepsilon}, \theta *, \gamma *)^2 .$$

The elements $I^{ij}(\theta)$ and $J^{kl}(\theta, \gamma)$ denote the components of the inverse matrices of $I(\theta)$ and $J(\theta,\gamma)$ respectively.

APPLICATION

A Poisson model as described above has been fitted to a partial duration series over 86 years. A detailed descripition of the hydrological conditions, the estimation procedures and the results of goodness-of-fit tests is given in Nachtnebel and Konecny (1987). The 234 selected flood peaks exhibited a pronounced seasonality in the occurence times and in their magnitudes. The events are generated by snow melt in spring and by rainfall in the early fall. A few events are caused by coinciding snowmelt and rainfall. An intensity function $\lambda(t;\theta)$ of the form with q=3 was fitted and the function $\beta(t;\gamma)$ was composed of four frequencies (r = 4). The ML-estimates for the system parameters are given in the table 1.

TABLE 1. Maximum Likelihood-Estimates of the System Parameters

$\theta_0 =$	0.0046	$\gamma_0 =$	0.0119
$\theta_1 =$	0.9670	$\gamma_1 =$	0.4230
$\theta_2 =$	1.2886	$\gamma_2 =$	0.3252
$\theta_3 =$	0.0268	$\gamma_3 =$	0.0
$\theta_4 =$	0.8661	$\gamma_4 =$	0.1999
$\theta_5 =$	0.3832	$\gamma_5 =$	-0.5289
$\theta_6 =$	2.0310	$\gamma_6 =$	0.0470
		$\gamma_7 =$	0.0
		$\gamma_8 =$	-0.1323

Based on these parameters the design exceedances are compared in table 2 for the seasonal and non-seasonal model and are estimated for various return periods and design life times in table 3. The seasonal model is equivalent to a weighted sum of exponential distributions and results therefore in a heavier tailed distribution. Due to the increased number of parameters the associated estimation uncertainty is greater compared to the i.i.d. model.

TABLE 2.Comparison of Design Discharges Obtained from the Seasonal and the I.I.D. Model

	Seasonal Model		I.I.D. Model	
T (years)	x_T (m^3/s)	st.dev. (m^3/s)	x_T (m^3/s)	st.dev. (m^3/s)
10	370	42.3	346	25.2
30	523	56.7	462	33.5
50	595	63.3	516	36.9
100	694	72.5	588	41.9

TABLE 3. Design Lifetime, Design Discharge and Flood Risk

T (years)	x_T (m^3/s)	r_T
100	677	.367
50	577	.366
30	505	.365
10	355	.358

The reliability is smaller than e-1 = 0.36788 for any design period T = 10, 30, 50, 100 and tends to this value for increasing T.

The design exceedances $x^{*}_{D,\varepsilon,p}$ were computed by these ML-estimates for various design periods T and risk values ε. The results are given in Table 4.

TABLE 4. Design Exceedance in m^3/s for a Design Period D (years) and a Design Probability 1-ε

D/ε	.20	.10	.05	.01
100	897	1011	1121	1374
50	794	906	1015	1266
30	719	829	937	1126
10	561	668	773	1018

The corresponding standard deviation of the asymptotic distribution is given in Table 5.

TABLE 5. Standard Deviation in (m^3/s) of the Design Exceedance

D/ε	.20	.10	.05	.01
100	84.90	105.30	119.45	152.50
50	79.39	922.58	105.80	137.58
30	70.73	83.53	96.37	127.35
10	58.70	66.34	77.01	106.14

Note that asymptotic variance of $X_{D,\varepsilon,p}$ corresponding to p years observations is given by

$$V(x_{D,\varepsilon,p}) = \frac{\sigma^2_{D,\varepsilon}}{p}.$$

DISCUSSION AND CONCLUSION

A PDS from an Austrian alpine river basin was investigated. The series obtained from 86 years of observations consisted of 234 flood peaks which exhibited a pronounced seasonality with an increased frequency of floods during spring and early fall. The flood generating mechanisms are completely different in the various seasons. To describe the flood series in more detail an inhomogeneous marked Poisson process was applied. Due to the continuously varying empirical intensity function a classification of the floods into several seasons with stationary intensity functions could not be justified. On the other hand, the continously time dependent model with exponentially distributed exceedances increases the number of model parameters.

Various design quantities were estimated from the extreme value process rather than from the distribution of the maximum annual exceedances. The advantage of this approach is seen in the fact that

- the pdf of the extreme value process is time dependent and provides the possibility to include easily cyclic and longterm trends in the intensity funktion;
- different flood generating mechanisms can be incorporated into the model;
- an economically which for instance is useful in regions with intensive agriculture endangered by floods.

It was shown, that design values based on the expected occurrence time of large exceedances lack reliability and a considerable flood risk must be taken during the project lifetime. Therefore, when a high reliability is undispensable, a design procedure with specified flood risk e in a given time horizon T is required.

Both quantities x_T and $x_{D,\varepsilon}$ were derived from the whole extreme value process instead of the series of maximal annual exceedances. Thus, the annual cycles in flood frequency and magnitude werde taken into account It turned out that the computation of the design values $x_{D,\varepsilon}$ from the extreme value process is easier compared with the computation of the design value x_T associated with the return period T.

The uncertainty in the design values was investigated in the context of uncertainties in parameter estimation due to a limited observation series of p years. The maximum likelihood method was used for the estimation of the parameters and the normal approximation of the design exceedance $X_{D,\varepsilon,p}$ was expressed in terms of the asymtotic variances and covariances of the maximum likelihood estimators.

Alternatively, Bayes estimators could be used. According.to a result of asymptotic parametric statistics, maximum likelihood estimators and Bayes estimators are, under mild assumptions, which are encountered in most applications, asymptotically equivalent and have the same limit distributuion. The standard deviation of both estimators tends to with $1/\sqrt{p}$.to zero for $p \to \infty$.

REFERENCES

Ashkar, F. and Rousselle, J. (1981) "Design Discharge as a Random Variable: A Risk Study", Water Resources Res., 17 (3), 577-591.

Karr, A. (1976) "Two extreme value processes arising in hydrology"", J. Appl. Probability, 13, 190-194.

Konecny, F. (1987) "The asymptotic properties of maximum likelihood estimators for marked Poisson processes with a cyclic intensity measure, Metrika 34, 1243-155.

Nachtnebel, H.P. and Konecny, F. (1987) "Risk Analysis and time-dependent flood models", J. Hydrology, 91, 295-318.

North, M. (1980) " Time-dependent Stochastic Model of Floods", J. Hydraul. Div.Am.Soc.Civ.Eng., HY5, 649-666.

Rasmussen, P.F. and Rosbjerg, D. (1989) "Risk estimation in partial duration series". Water Resour.Res. 25, (11), 2319-2330.

Rasmussen, P.F. and Rosbjerg, D. (1991 a) "Prediction uncertainty in seasonal partial duration series". Water Resour.Res. 27, (11), 2875-2883.

Rasmussen, P.F. and Rosbjerg, D. (1991 b) "Evaluation of risk concepts in partial duration series". Stochastic Hydrol. Hydraul. 5, 1-16.

Todorovic, P. (1978) "Stochastic Models of Floods", Water Resources Res., 14 (2), 345-356.

Todorovic, P. and Zelenhasic, E. (1970) "A stochastic model for flood analysis, Water Resources Res., 6 (6), 1641-1648.

ESTIMATION OF DESIGN QUANTITIES UNDER MISSPECIFIED MODELS

F. KONECNY
Institute of Mathematics and Applied Statistics
Universität für Bodenkultur, Wien
A-1180 Gregor Mendel-Strasse 33
Austria

Partial duration series (PDS) have been used to estimate design discharge of a given return period. A stochastic model for a PDS is a marked Poisson process which can be described by an intensity function (occurrence rate of peaks) and the distribution of the exceedances. In practice the true intensity and exceedances distribution is unknown and inference is based on some parametric model, which may be misspecified. Under general conditions given here, the maximum likelihood estimates converge strongly to parameters of an approximation of the true model. With additional regularity conditions the estimate of the design exceedance is asymptotically normal. With misspecification, however, the standard error differs from that obtained when the model is correctly specified.

INTRODUCTION

A partial duration series consists of the local maxima in a streamflow hydrograph above a certain theshold Q_0. Let (τ_n) denote the times of maxima (peaks) of the hydrograph and ξ_n the amount by which the n-th peak exceeds the base level. The most widely used model assumes that the occurrences of the peaks over the suitably high threshold are the points of an (inhomogeneous) Poisson process and the exceedances are independent or conditionally independent given their occurrence time.

The subject of this paper is the maximum likelihood estimation of the design exceedance under departures of the true intensity function of the Poisson process or the true exceedance distribution from those specified by the parametric model.

The need for such analysis results from the desirability of using estimators which are robust under small departures from the model assumptions. This analysis is familiar in other contexts, e.g. regression analysis (White, 1981) and diffusion processes (McKeague, 1984)

K. W. Hipel (ed.),
Stochastic and Statistical Methods in Hydrology and Environmental Engineering, Vol. 1, 85–94.

MAXIMUM LIKELIHOOD ESTIMATION UNDER MISSPECIFIED MODELS

We assume the τ_n are the points of inhomogeneous Poisson process with a cyclic trend. That is the intensity function $\lambda(t)$ is periodic with one-year period. Moreover we assume each exceedance ξ_n is conditionally independent of the whole past $(\tau_k, \xi_k; k = 1, 2, ..., n-1)$ given its occurrence time τ_n. The dependency of ξ_n on its occurrence time is expressed by the conditional density function

$$\chi(x|t) = \frac{d}{dx} P(\xi_n \leq x | \tau_n = t).$$

Assuming $\chi(x|.)$ has one-year period, we see that (τ_n, ξ_n) is a two-dimensional Poisson process with a cyclic trend.

Suppose a parametric model is used to estimate the intensity function and the time-dependent exceedances distribution. A family of functions $(f(t;\theta); \theta \in \Theta)$ with period one, parametrized by a vector-parameter θ, is provided for intensity estimation. Another family of time-dependent densities $(h(x|t;\gamma); \gamma \in \Gamma)$, dependent on a vector- parameter γ, is fitted to the exceedances. Inference is based on this parametric Poisson model of a PDS. The corresponding log-likelihood function is given by

$$l_p(\theta, \gamma) = \sum_{\tau_n < p} log[f(\tau_n;\theta) h(\xi_n | \tau_n; \gamma)] - \int_0^p f(t;\theta) dt, \tag{1}$$

where p is the size of the sample in years. Maximum likelihood estimators (MLEs) calculated from $l_p(\theta,\gamma)$ are denoted by $\widehat{\theta}_p$ and $\widehat{\gamma}_p$ respectively.

Assume that

$$\int_0^1 \{\lambda(t) log[f(t;\theta)] - f(t;\theta)\} dt$$

as a function of θ and

$$\int_0^1 \int_0^\infty log[h(x|t;\gamma)] \chi(x|t) \lambda(t) dx dt$$

as a function of γ, have a unique maximum $(\theta^*, \gamma^*) \in \Theta \times \Gamma$ and the partial derivatives w.r.t. the parameters are zero at (θ^*, γ^*). The following theorems describes the asymptotic properties of the MLEs, when the observed process have the intensity function $\lambda(t)$ and the time-dependent exceedances distribution is $\chi(x|t)$. The technical conditions and the proofs are given in section 5.

Theorem 1. Under Conditions (C.1) and (C.2), $(\widehat{\theta}_p, \widehat{\gamma}_p) \to (\theta^*, \gamma^*)$ with probability one (w.p.1) as $p \to \infty$

We introduce the following matrices for notational convenience (a prime $'$ denotes transposition of a matrix):

$$J_f(\theta) = -E[\frac{\partial^2 l_1}{\partial\theta\partial\theta'}]$$
$$I_f(\theta) = E[\frac{\partial l_1}{\partial\theta}(\frac{\partial l_1}{\partial\theta})']$$
$$J_h(\gamma) = -E[\frac{\partial^2 l_1}{\partial\gamma\partial\gamma'}]$$
$$I_h(\gamma) = E[\frac{\partial l_1}{\partial\gamma}(\frac{\partial l_1}{\partial\gamma})']$$

Theorem 2. Under conditions (C.1) -(C.5) the asymptotic distribution of $\sqrt{p}(\widehat{\theta}_p - \theta^*, \widehat{\gamma}_p - \gamma^*)$ is a multivariate normal distribution with mean vector 0 and the covariance matrix

$$V = \begin{pmatrix} J_f(\theta)^{-1} I_f(\theta) J_f(\theta)^{-1} & 0 \\ 0 & J_h(\gamma)^{-1} I_h(\gamma) J_h(\gamma)^{-1} \end{pmatrix}$$

Theorem 3. In the case that the intensity function and the distribution of the exceedances has been correctly specified, i.e. $\lambda(t) = f(t;\theta_0)$ and $\chi(x|t) = h(x|t;\gamma_0)$ for some $\theta_0 \in \Theta$ and $\gamma_0 \in \Gamma$, then

(i) $(\widehat{\theta}_p, \widehat{\gamma}_p) \to (\theta_0, \gamma_0)$ w.p.1 as $p \to \infty$;

(ii) The asymptotic distribution of $\sqrt{p}(\widehat{\theta}_p - \theta_0, \widehat{\gamma}_p - \gamma_0)$ is a multivariate normal distribution with mean vector 0 and the covariance matrix

$$V = \begin{pmatrix} I_f(\theta_0)^{-1} & 0 \\ 0 & I_h(\gamma_0)^{-1} \end{pmatrix}$$

BEHAVIOUR OF DESIGN QUANTITIES

In the context of PDS *design exceedance* can be defined (loosely) as the level $Q_0 + x$ which is exceeded on the average once in a return period of T years. The mean number of exceedances per year which are greater than $x, x > 0$, is

$$\int_0^1 R(x|t)\lambda(t)\,dt,$$

where

$$R(x|t) = P(\xi_n > x|\tau_n = t).$$

A more precise definition is as follows: the design exceedance x_T is implicitly given by

$$\int_0^1 R(x|t)\lambda(t)\,dt = \frac{1}{T}. \tag{2}$$

Using the parametric model of section 2 we get the corresponding value x_T^* by the equation

$$\int_0^1 (1 - H(x|t;\gamma))f(t;\theta)dt = \frac{1}{T}. \tag{3}$$

The last equation defines x_T^* implicitly as a function of the model parameters, i.e. $x_T^* = \phi_T(\theta, \gamma)$. Inserting the MLEs $\widehat{\theta}_p$ and $\widehat{\gamma}_p$ we obtain the MLE $\widehat{x}_{T,p}$ of the design exceedance. The following proposition is an obvious consequence of Theorem 2.

Proposition. Suppose the conditions (C.1)-(C.4) hold true. Then the MLE $\widehat{x}_{T,p}$ of the design exceedance is asymptotically normal with the mean $x_T^* = \phi_T(\theta^*, \gamma^*)$ and the variance σ^2/p, where

$$\sigma^2 = (\frac{\partial\phi_T}{\partial\theta})' J_f^{-1} I_f J_f^{-1} \frac{\partial\phi_T}{\partial\theta} + (\frac{\partial\phi_T}{\partial\gamma})' J_h^{-1} I_h J_h^{-1} \frac{\partial\phi_T}{\partial\theta}. \tag{4}$$

Consequently, $\widehat{x}_{T,p}$ is asymptotically biased estimate of x_T unless x_T coincides with x_T^*.

The partial derivatives in (4) are easily obtained by implicit differentiation of equation (3).

SPECIAL CASES

Suppose that a PDS follows a Poisson model with the intensity function $\lambda(t)$ and the time-dependent d.f. $\chi(x|t)$ for the exceedances. We discuss the properties of the MLEs taking some (misspecified) model.

(i) *Homogeneous Poisson process with exponentially distributed exceedances.* This model is specified by

$$f(t;\theta) = \theta,\ \theta > 0, \text{ and } \quad h(x|t;\gamma) = \gamma e^{-\gamma x}, \gamma > 0.$$

In this case only two scalar parameters θ and γ are to be estimated. The following expressions characterize the model with the least distance to the true distribution

$$\theta^* = \int_0^1 \lambda(t)dt$$

and

$$\gamma^* = \left(\int_0^1 \lambda(t)dt \right) \left(\int_0^1 \int_0^\infty x\chi(x|t)\lambda(t)dx\,dt \right)^{-1}.$$

The asymptotic variances of the MLEs are given in terms of the following functions

$$I_f(\theta) = J_f(\theta) = \frac{1}{\theta^2} \int_0^1 \lambda(t)dt$$

$$I_h(\gamma) = \int_0^1 \int_0^\infty (x - \frac{1}{\gamma})^2 \chi(x|t)\lambda(t)dx\,dt$$

$$J_h(\gamma) = \frac{1}{\gamma^2} \int_0^1 \lambda(t)dt$$

(ii) *Subdivision of the year in s seasons.* We choose s points $0 < t_1 < t_2 < ... < t_s < 1$ and set

$$f(t;\theta) = \theta_i \text{ for } \quad t_{i-1} < t \le t_i,$$
$$h(x|t;\gamma) = \gamma_i e^{-\gamma_i x} \text{ for } \quad t_{i-1} < t \le t_i.$$

In this model intensity and exceedance distributions are constant within each season. We obtain

$$\theta_i^* = \frac{1}{t_i - t_{i-1}} \int_{t_{i-1}}^{t_i} \lambda(t)dt,$$

$$\gamma^* = \left(\int_{t_{i-1}}^{t_i} \lambda(t)dt \right) \left(\int_{t_{i-1}}^{t_i} \int_0^\infty x\chi(x|t)\lambda(t)dx\,dt \right)^{-1},$$

$$I_f(\theta)_{ii} = J_f(\theta)_{ii} = \frac{1}{\theta_i^2 (t_i - t_{i-1})} \int_{t_{i-1}}^{t_i} \lambda(t)dt,$$

$$I_h(\gamma)_{ii} = \frac{1}{t_i - t_{i-1}} \int_{t_{i-1}}^{t_i} \int_0^\infty (x - \frac{1}{\gamma})^2 \chi(x|t)\lambda(t)dx\,dt,$$

$$J_h(\gamma)_{ii} = \frac{1}{\gamma^2 (t_i - t_{i-1})} \int_{t_{i-1}}^{t_i} \lambda(t)dt.$$

For $i \neq j$ $I_f(\theta)_{ij} = J_f(\theta)_{ij} = I_h(\gamma)_{ij} = J_h(\gamma)_{ij} = 0.$

(iii) *Cyclic model of Lewis (1970).* The intensity of the Poisson process is a continuous, unimodal and periodic functions w.r.t. t given by

$$f(t;\theta) = \frac{\theta_1}{I_0(\theta_2)} exp\, [\theta_2 sin(2\pi t + \theta_3)],$$

where I_0 is the modified Bessel function of the first kind of zero order. The distribution of the exceedances is exponential with time-dependent parameter, i.e.

$$\begin{aligned} h(x|t;\gamma) =& b(t;\gamma) exp[-b(t;\gamma)x], \\ b(t;\gamma) =& \gamma_1 exp[\gamma_2 sin(2\pi t + \gamma_3)]. \end{aligned}$$

Thus the mean of the exceedances given by $1/b(t;\gamma)$ is a continuous, unimodal and periodic function w.r.t. t. Setting

$$A = \int_0^1 cos(2\pi t)\lambda(t)\, dt \text{ and } \quad B = \int_0^1 sin(2\pi t)\lambda(t)\, dt$$

we obtain

$$\theta_1^* = \int_0^1 \lambda(t) dt.$$

θ_2^* is the unique solution of the equation

$$\frac{1}{\theta_1^*}\sqrt{A^2 + B^2} = \frac{I_1(\theta_2)}{I_0(\theta_2)},$$

where I_1 is the modified Bessel function of the first kind of first order, and

$$\theta_3^* = arctan \frac{B}{A}.$$

The values γ_1^*, γ_2^* and γ_3^* can be found by solving the equations

$$\int_0^1 \int_0^\infty \sin(2\pi t + \gamma_3)\{1 - x\gamma_1 exp\, [\gamma_2 sin(2\pi t + \gamma_3)]\}\chi(x|t)\lambda(t)\, dx dt = 0,$$

$$\int_0^1 \int_0^\infty \cos(2\pi t + \gamma_3)\{1 - x\gamma_1 exp\, [\gamma_2 sin(2\pi t + \gamma_3)]\}\chi(x|t)\lambda(t)\, dx dt = 0,$$

where

$$\gamma_1 = \frac{\int_0^1 \lambda(t) dt}{\int_0^1 \int_0^\infty x exp\, [\gamma_2 sin(2\pi t + \gamma_3)]\chi(x|t)\lambda(t)\, dx dt}.$$

PROOFS

The log -likelihood function can be written

$$l_p(\theta,\gamma) = l_p^{(f)}(\theta) + l_p^{(h)}(\gamma)$$

where

$$l_p^{(f)}(\theta) = \sum_{\tau_n<p} logf(\tau_n;\theta) - p\int_0^1 f(t;\theta)dt$$

and

$$l_p^{(h)}(\gamma) = \sum_{\tau_n<p} log\ h(\xi_n|\tau_n;\gamma).$$

It is convenient to use the counting integral notation (cf. Konecny,1987)

$$\int_0^p logf(t;\theta)dN_t = \sum_{\tau_n<p} logf(\tau_n;\theta),$$

and

$$\int_0^p\int_0^\infty log\ h(x|t;\gamma)\nu(dt,dx) = \sum_{\tau_n<p} log\ h(\xi_n|\tau_n;\gamma).$$

Technical Conditions

(C.1) $f(t;\theta)$ is continuous in (t,θ) and differentiable with respect to θ. There exists a bounded function $K(t)$ such that $|f(t;\theta)| + |\partial f/\partial\theta| \leq K(t)$, $0 \leq t \leq 1$.

(C.2) $h(x|t;\gamma)$ is continuous in (x,t,γ) and differentiable with respect to γ. There exists a bounded function $H(x,t)$ such that $|h(x|t;\gamma)| + |\partial h/\partial\gamma| \leq H(x,t)$, $0 \leq x < \infty$, $0 \leq t \leq 1$.

(C.3) $f(t;\theta)$ is three times continuously differentiable with respect to θ. There exists a bounded function $K(t)$ such that

$$|\frac{\partial^3}{\partial\theta_i\partial\theta_j\partial\theta_k} logf(t;\theta)| \leq K(t)$$

for all $\theta \in \Theta$ and all subscripts i,j and k.

(C.4) $h(x|t;\gamma)$ is three times continuously differentiable with respect to γ. There exists a bounded function $H(x,t)$ such that

$$|\frac{\partial^3}{\partial\gamma_i\partial\gamma_j\partial\gamma_k} logh(x|t;\theta)| \leq H(x,t)$$

for all $\gamma \in \Gamma$ and all subscripts i, j and k and

$$\int_0^1 \int_0^\infty H(x,t)\chi(x|t)\lambda(t)\, dxdt < \infty.$$

(C.5) The MLEs of θ and γ satisfy the equations

$$\frac{\partial}{\partial \theta} l_p(\theta) = 0 \quad \text{and} \quad \frac{\partial}{\partial \gamma} l_p(\gamma) = 0.$$

Proof of Theorem 1. By the periodicity of $\lambda(t)$ and $f(t;\theta)$ and the strong law of large numbers we get

$$\begin{aligned}\lim_{p\to\infty} \frac{1}{p} \int_0^p log f(t;\theta) dN_t &= E(\int_0^1 log f(t;\theta) dN_t) \\ &= \int_0^1 log[f(t;\theta)]\lambda(t) dt\end{aligned}$$

with probability 1. We obtain

$$\lim_{p\to\infty} \frac{1}{p} l_p^{(f)}(\theta) = \int_0^1 \{log[f(t;\theta)]\lambda(t) - f(t;\theta)\} dt \tag{6}$$

in the same way we get

$$\lim_{p\to\infty} \frac{1}{p} l_p^{(h)}(\theta) = \int_0^1 \int_0^\infty log\ h(x|t;\theta)]\chi(x|t)\lambda(t)\, dx dt \tag{7}$$

According to assumptions the right hand sides of (6) and (7) have unique maxima $\theta^* \in \Theta$ and $\gamma^* \in \Gamma$ respectively, and since $(\widehat{\theta}_p, \widehat{\gamma}_p)$ maximizes $(1/p) l_p(\theta, \gamma)$, it is easily seen that $(\widehat{\theta}_p, \widehat{\gamma}_p) \to (\theta^*, \gamma^*)$ w.p.1 as $p \to \infty$.

For the proof of theorem 2 we need the following

Lemma 1. Under the conditions of Theorem 2 we have

$$\frac{1}{\sqrt{p}} \frac{\partial l_p(\theta^*)}{\partial \theta} \to N(0, I_f(\theta^*))$$

and

$$\frac{1}{\sqrt{p}} \frac{\partial l_p(\gamma^*)}{\partial \gamma} \to N(0, I_h(\gamma^*))$$

The proof of this lemma is virtually the same as the proof of Lemma 3.2 in Konecny (1987).

Proof of Theorem 2. The usual Taylor expansion of $\partial l_p/\partial\theta$ about θ^* gives

$$0 = \frac{\partial l_p(\widehat{\theta}_p)}{\partial\theta} = \frac{\partial l_p(\theta^*)}{\partial\theta} + \frac{\partial^2 l_p(\theta^*)}{\partial\theta\partial\theta'}(\widehat{\theta}_p - \theta^*)$$
$$+ \text{ terms of higher order in } (\widehat{\theta}_p - \theta^*),$$

where $\partial^2 l_p/\partial\theta\partial\theta'$ is the matrix of second order derivatives with respect to the components of θ. Thus, ignoring the higher-order terms, we have the well-known result

$$\sqrt{p}(\widehat{\theta}_p - \theta^*) = -\left(\frac{1}{p}\frac{\partial^2 l_p(\theta^*)}{\partial\theta\partial\theta'}\right)^{-1}\left(\frac{1}{\sqrt{p}}\frac{\partial l_p(\theta^*)}{\partial\theta}\right). \tag{8}$$

From Lemma 1

$$\frac{1}{\sqrt{p}}\frac{\partial l_p(\theta^*)}{\partial\theta} \to N(0, I_f(\theta^*)). \tag{9}$$

By the strong law of large numbers

$$\frac{1}{p}\frac{\partial^2 l_p(\theta^*)}{\partial\theta\partial\theta'} \to -J_f(\theta^*). \tag{10}$$

Both (9) and (10) yield

$$\sqrt{p}(\widehat{\theta}_p - \theta^*) \to N(0, J_f^{-1} I_f J_f^{-1}).$$

The proof concerning $\sqrt{p}(\widehat{\gamma}_p - \gamma^*)$ is the same.

Theorem 3 follows from Theorems 1 and 2.

CONCLUDING REMARKS

The results presented in this paper are designed to provide answers concerning the effect of misspecification on the resulting estimates of design quantities. Any departure from the true probability law leads to a systematic error. At least in the case of large samples good approximations of the systematic error and the standard error of design exceedances can be obtained. For the case of small samples the analysis could be refined to provide higher order approximations for bias and variance. Throughout the paper we assumed that only a cyclic trend is present. In principle our results could be generalized to the case of cyclic plus monotonic trend at the price of more complicated expressions.

REFERENCES

Ashkar, F. and Rousselle,J.(1981) "Design Discharge as a random variable: A risk study", Water Resour.Res. 17, 577-591.

Konecny, F. (1987) "The Asymptotic Properties of Maximum Likelihood Estimators for Marked Poisson Processes with a Cyclic Intensity Measure", Metrika 34, 143-155.

Konecny, F. and Nachtnebel H. P.(1985) "Extreme value processes and the evaluation of risk in flood analysis", Appl.Math.Modelling 9, 11-15.

Lewis, P.A.W.(1970) "Remarks on the theory, computation and application of the spectral analysis of series of events", J.Sound Vib. 12, 353-375.

McKeague, I.W.(1984) "Estimation for diffusion processes under misspecified models", J.Appl.Prob. 21, 511-520.

Rosbjerg, D.(1993) Partial Duration Series in Water Resources. Institute of Hydrodynamics and Hydraulic Engineering,Technical University of Denmark,Lyngby.

Todorovic, P. and Zelenhasic, E.(1970) "A stochastic model for flood analysis", Water Resour.Res. 6, 1641-1648.

White, H.(1981) "Consequences and Detection of Misspecified Nonlinear Regression Models", J. Am. Stat. Assoc. 76, 419-433.

FIT OF PROBABILITY DISTRIBUTIONS TO UPPER SUB-SAMPLES OF PARTIAL DURATION SERIES

ARIE BEN-ZVI
Israel Hydrological Service
P.O.Box 6381, Jerusalem 91063
Israel

Partial duration series provide highly valuable information about floods which might disturb the normal activities around rivers. Recurrence intervals of floods important for hydrologic works are, in many cases, longer than the lengths of records to which probability distributions are fitted. Selection of a distribution which follows the trend in the upper tails of available records would improve the credibility of predictions of rare floods. Tail's trend is defined here by its thickness, which is determined through a display of cummulative mean exceedances against threshold discharges. The selected distribution is fitted to all the sub-samples which maintain the upper tail's trend. Goodness of fit is determined from the Anderson-Darling test, including the modification with respect to sample size, and for the case where the parameters are estimated from the sample. The best fit is recommended for application.

The above methodology has been applied to complete duration series of peak discharges for eight rivers located in the arid and the semi-arid regions of Israel. Thin tails are found for seven of the eight rivers. Yet, the displays indicate that thick tails may also be found in the study area. For reasons of uniformity, the generalized Pareto distribution is selected for description of magnitudes in the present case study because it can be fitted to both the trends of thin and thick tails. The distribution is well fitted to sub-samples associated with a range of threshold discharges for each one of the rivers. The best fits are attained for samples associated with middle-valued thresholds. For most of the rivers the significance level of rejection of the best fit is higher than 0.25. The sizes of sub-samples, for which best fits are attained, vary from 14 to 70 events, and their mean occurrence frequencies vary from 0.36 to 1.92 year^{-1}.

INTRODUCTION

Establishment of reliable relationships between peak discharges and their occurrence probabilities is a common task in hydrologic works. A considerable minority of these works fulfill the task by partial duration series modeling which makes use of two probability

K. W. Hipel (ed.),
Stochastic and Statistical Methods in Hydrology and Environmental Engineering, Vol. 1, 95–107.

distributions; one describing the occurrences of events which exceed a pre-selected threshold value, and the other describing the magnitudes of these events. The magnitudes and times of occurrence of the events are assumed as being independent of each other.

The occurrence distribution is generally considered as being Poissonian (e.g. Shane and Lynn, 1964; Todorovic and Zelenhasic, 1970; Ashkar and Rousselle, 1987; Cruise and Arora, 1990; Rasmussen and Rosbjerg, 1991; Rosbjerg et al., 1992). Some refined versions of this distribution are also considered (e.g. Todorovic and Rousselle, 1971; Taesombut and Yevjevich, 1978; Kavvas, 1982; Nachtnebel and Konecny, 1987). Ben-Zvi (1991) has shown an advantage for the negative binomial over the Poisson distribution for partial duration series of runoff in Israel.

The negative exponential distribution is selected, in most of the works, for description of the magnitudes of peak discharges (e.g. Todorovic and Zelenhasic, 1970; Cunnane, 1973; North, 1980; Ashkar and Rousselle, 1983a; Nachnebel and Konecny, 1987; Choulakian et al., 1990; Rasmussen and Rosbjerg, 1991). Yet, the lognormal distribution is considered in some other works (e.g. Hiemstra et al., 1976; Hiemstra and Francis, 1981; Ben-Zvi, 1982; Rosbjerg, 1987). Rosbjerg et al. (1992) recommend a use of the negative exponential distribution for long (thick) tailed samples, and of a better representing distribution for short (thin) tailed samples.

Noting that the recurrence intervals, which are of interest in many hydrologic predictions, are longer than the lengths of records to which the selected distributions are fitted, a number of researchers prefer distributions which adhere to trends in the upper tails of recorded data (e.g. Bryson, 1974; Picands, 1975; Smith, 1987). Bryson (1974) defines the trend as being heavy when the observed density distribution approaches zero more slowly than the negative exponential distribution does, and as being light when the distribution approaches zero more rapidly. He tests the rapidity of approach through a variable called conditional mean exceedance, CME, which is defined by:

$$CME_x = E(X - x \mid X \geq x) \qquad (1)$$

where X is a considered variable and x is a pre-selected value of X.

Bryson (1974) shows that CME_x obtains an increasing trend for heavy tails and a decreasing trend for light tails. He proposes a graphical display as an aid for the determination of trend. In this display $\log[\underline{F}(q)]$ is plotted against q, where q values coincide with recorded peak discharges and $\underline{F}(q)$ is the empirical exceedance frequency of peak discharges with respect to q. In this display, heavy tailed series would bend upwards while light tailed series would bend downwards. Smith (1987) proposes another graphical display in which CME_q is plotted against q. In this display an unbounded thick (i.e. heavy) tail would render an upwards trend, while a bounded thin (i.e. light) tail would render a downwards trend.

The examination of trend of upper tails has been applied, so far, for a few annual maxima series. The present work examines trends of partial duration series. We believe that partial duration series provide a better background for prediction of recurrence intervals of extreme events, and that the observed trend of upper tails is a reasonable aid for the extrapolation of magnitudes. Yet, extrapolation of a bounded thin tail is limited to a value which is determined through statistical properties of the sample and might not be physically justifiable. This situation is not desired, at least in cases where that limit is close to the maximal observed discharge.

THRESHOLD SELECTION

Partial duration series modeling is used in hydrology for more than sixty years (Jarvis et al., 1936). These series contain all and only events of which the peak discharge exceeds a pre-determined threshold value (called also base level). The determination of threshold values is usually done through subjective considerations. Obviously, the higher the threshold the fewer are the events in the selected series. This results in a lower data handling and computational load, a higher effect of sampling variations and, in many cases, a lower dependence between the selected events (Bhuiya and Yevjevich, 1968).

The original USGS guidelines for selection of the base level is that it should be equal to the lowest annual flood on record, but in a long record the level is raised to include, on the average, three or four events per year (Langbein, 1949). Chow (1953) has proposed that the sample size be equal to the number of years on record. Cunnane (1973) finds advantage when the sample size is at least 1.65 times the number of years on record. Taesombut and Yevjevich (1978) find the advantage when the sample size is at least 1.95 times the number of years on record. Ashkar and Rousselle (1987) prefer samples for which the variance of the occurrence frequency is close enough to the mean, which, in their example, results in sizes two to three times larger than the number of years on record. Kavvas (1982) elaborates on the physical and engineering considerations for the selection of thresholds, whereas Ashkar and Rousselle (1983a) prefer statistical considerations for the selection. We prefer statistical considerations for regional works (e.g. Ben-Zvi, 1991) and physical considerations for particular predictions (e.g. Ben-Zvi and Cohen, 1983).

Ben-Zvi (1991) has examined the effect of threshold value on the fit of the Poisson and the negative binomial distributions to recorded occurrences of events. He concluded that the fit of either distribution to samples with frequent events does not guarantee the fit to sub-samples composed of events of lower frequency only. Owing to shortage in data for events exceeding high thresholds, he could not test the fit of those distributions to strictly flood samples (i.e. exceedance frequency ≤ 0.1 year^{-1}). Yet, indirect considerations indicate a good fit of either distribution, and particularly of the negative binomial, to the occurrence of such flood samples.

The cited statistical selections of threshold values consider the occurrence frequency of events in the samples. We propose here another aspect, in which the selected distribution is fitted to nested sub-samples, and the threshold associated with the best fit is selected. The first sub-sample is composed of the uppermost five peaks. The following sub-samples are composed of the uppermost six, seven, eight... peaks; such that each new sub-sample is one value larger than its preceding sub-sample. The process of composing new sub-samples is terminated when the parameters of the fitted distribution do not suit the trend determined for the upper tail. It is believed, though not shown, that this selection would contribute to the credibility of the predictions.

DATA

The applicabilty of the proposed methodology has been examined on series of actual data. The series are abstracted from records of eight hydrometric stations located on ephemeral rivers in Israel. These locations are shown on the map in Fig. 1. The selected watersheds are

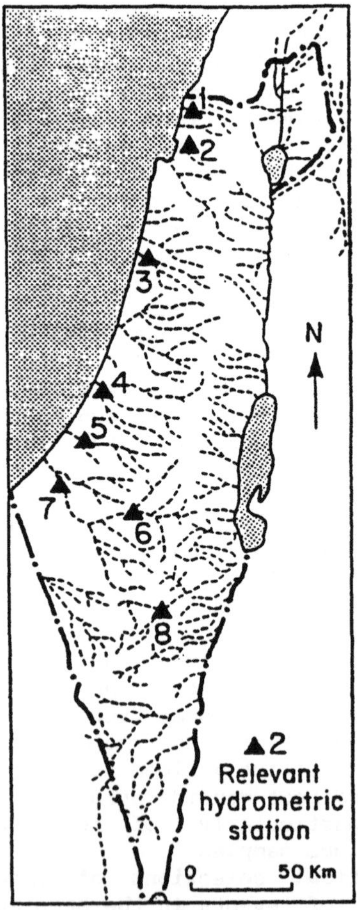

Figure 1. Hydrographic map of Israel.

exposed to winter precipitation and to a hot dry summer season. Each one of the stations is equipped with a continuous water level recorder, its rating curve provides for a fair to a good accuracy of discharges, and the processing follows standard techniques adopted by the Hydrological Service. Records of sufficient accuracy and completeness for at least 20 years are available for each one of the stations. The lengths of records, catchment areas and mean annual depths of precipitation are presented in Table 1.

The analysis of peak magnitudes begins with series composed of the peaks of all the runoff events (called complete duration series). These series have been abstracted from files of momentary values of the entire runoff records. A determination criterion, of no flow for at least 24 hours, has been chosen to distinguish between a multi-peaked event and a number of distinct events. An application of such a criterion, which requires a minimal time interval between occurrences of distinct events, might interefere with the fit of the Poisson distribution to the occurrence frequency (Ashkar and Rousselle, 1983b). Yet, it saves subjective decisions about multi-peaked records whether they are out-of-phase contributions of different tributaries or results of distinct rainfall events. In addition, it saves an application of a model for runoff separation which might add another source of error into the results.

TABLE 1. Watersheds and records

#	River	Station	Area km^2	Rain mm	N	n	CME_q m^3/s	Q_m m^3/s
1	Kziv	Gesher Haziv	131	830	39	130	15	13
2	Hillazon	Yavor	158	677	41	203	12	10
3	Alexander	Ma'aba* & Elya*	505	606	39	207	45	27
4	Lahish	Yavne Rd.	992	451	36	271	40	13
5	Shikma	Bror Hayil	378	348	35	195	25	13
6	Sheva	Nevatim	405	175	24	91		
7	Bsor	Re'im	2630	192	20	66	95	44
8	Zin	Zin Valley	660	80	22	70	45	26

Legend: # is station number, Area is catchment area for the station, Rain is mean annual depth of precipitation over the catchment, N is number of years of record, n is number of events in these years, CME_q is mean exceedance discharge for medium thresholds, Q_m is the lower edge of medium discharges.

ANALYSIS

For each one of the eight series, two graphical displays have been prepared. One display is for log[$\underline{F}$(q)] vs q (Bryson, 1974) and the other is for CME_q vs q (Smith, 1987). In view of these displays, it seems that the decision about bending directions (first display) is less certain than that about trend directions of the upper tails (second display). Therefore, our decisions about tail thicknesses are based upon the displays of CME_q vs q. These displays, for the entire case study, are presented in Fig. 2.

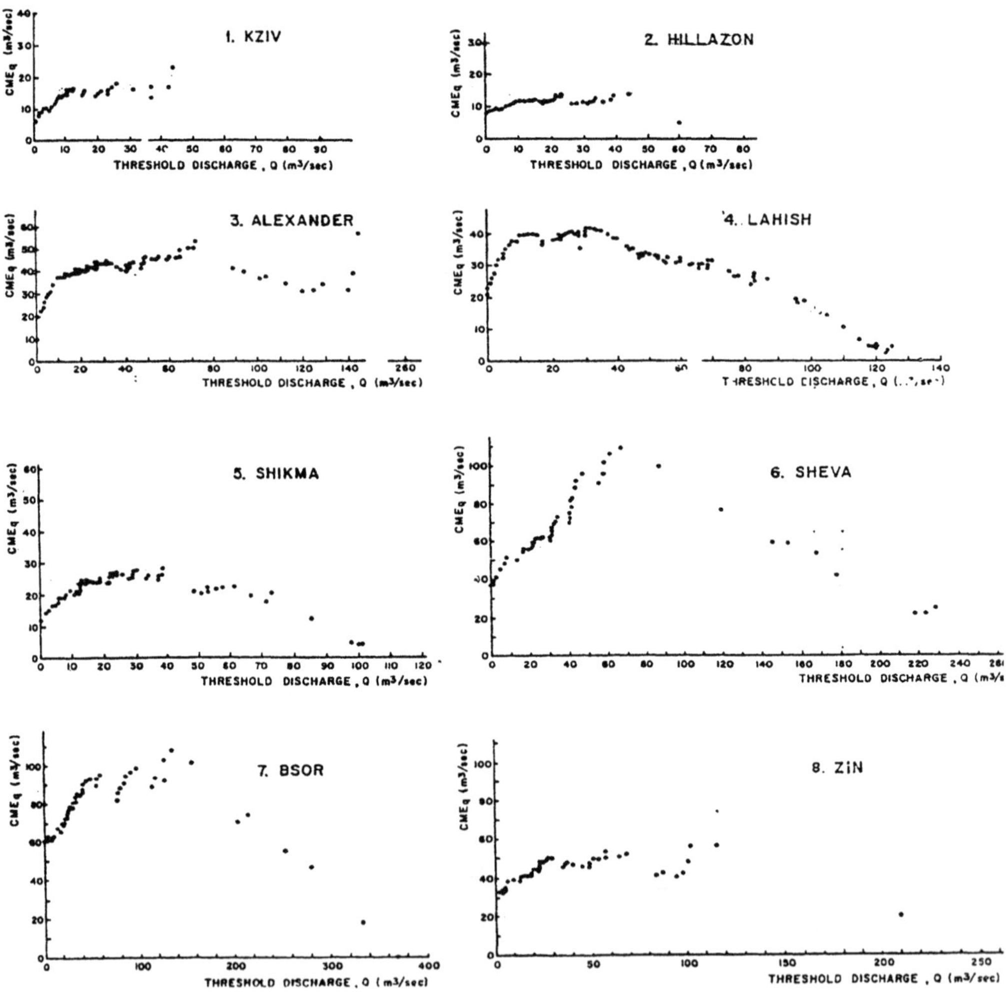

Figure 2. Cummulative mean exceedance of peak discharges.

In general, CME_q is found rising with q for low values of q, attaining a stable value for medium values, and declining with q for high values of q. For one of the rivers (Sheva), the stable segment is missing in the display, and for another river (Kziv) the declining segment is missing. A bump appears on the displays of two rivers (Alexander and Zin), where outlier discharges have been recorded. Conclusions about tail thickness can easily be reached for six of the eight rivers. This is not the case for the Hillazon and Zin rivers where the declining segments on the displays are composed of a few data points only.

Viewing the displays of CME_q vs q we conclude that the upper tails of peak discharges of runoff events in the present case study are generally thin. Yet, the possibility of a thick tail, or of a stable one, should not be ruled out. Thus, if we wish to fit the same distribution to the entire case study, we ought to select one which is flexible with respect to tail thickness. The generalized Pareto distribution (symbolized by GP) does have this flexibilty (e.g. Hosking and Wallis, 1987), and therefore it is selected for the description of magnitudes in the present case study. This selection does not mean to acknowledge, however, that the GP is the only distribution which possesses this flexibility.

The selected distribution has been fitted to all the upper sub-samples which are composed of at least five members and maintain the trend which had been determined for their upper tail. For reasons of computational convenience, the paramters have been estimated by the method of moments. The goodness of fit of the distribution has been determined through the Anderson-Darling test, modified with respect to the sample size, and concluded for the case where the parameter values are estimated from the sample (Stephens, 1986). The mathematical expressions of these computations are presented below.

The GP distribution is formulated by (e.g. Hosking and Wallis, 1987):

$$F(Q) = 1 - (1 - k(Q - q)/\alpha)^{1/k} \quad \text{for } k < 0 \text{ or } k > 0 \text{ and } Q \geq q \tag{2}$$
$$F(Q) = 1 - \exp(-(Q - q)/\alpha) \quad \text{for } k = 0 \text{ and } Q \geq q$$

in which F(.) is the cummulative distribution function, Q is peak discharge of a runoff event, q is a pre-selected threshold value, k is the shape parameter and α is the scale parameter. A thick tail is characterized by $k < 0$, a thin tail by $k > 0$ and a stable tail by $k = 0$ (the negative exponential distribution).

The domain of variation of $Q^* = Q - q$ is:

$$0 \leq Q^* < \infty \quad \text{for } k \leq 0 \tag{3}$$
$$0 \leq Q^* \leq \alpha/k \quad \text{for } k > 0$$

The parameter values are estimated by the method of moments (Hosking and Wallis, 1987):

$$\alpha\hat{} = .5\ \underline{Q}^*\ (\underline{Q}^{*2}/S^2 + 1) \qquad (4)$$
$$k\hat{} = .5\ (\underline{Q}^{*2}/S^2 - 1)$$

in which $\alpha\hat{}$ and $k\hat{}$ are estimated values of the parameters, $\underline{Q}^*$ is sample mean and S^2 is sample variance of Q^*.

The quantile estimation is (Hosking and Wallis, 1987):

$$Q^*(F) = \alpha\hat{}\ (1 - (1 - F)^{k\hat{}})/k\hat{} \qquad \text{for } k < 0 \text{ or } k > 0 \qquad (5)$$

$$Q^*(F) = -\alpha\hat{}\ \log\ (1 - F) \qquad \text{for } k = 0$$

in which F is a pre-selected non-exceedance probability of Q^*.

The Anderson-Darling test is preferred here because according Stephens (1986, p. 167) "it tends to lead the others because it is effective in detecting departures in the tails". Its formula is (Stephens, 1986, p. 101):

$$A^2 = -n -(1/n)\Sigma_i[(2i-1)\ln(Z(i))+(2n+1-2i)\ln(1-Z(i))] \quad \text{for } 1 \le i \le n \qquad (6)$$

in which A^2 is the test statistic, n is the number of members in the series, i is the rank in an ordered series, Z(i) is the estimated non-exceedance probability of Q_i^*, and Q_i^* is the i'th value, in the ordered series, of Q^*.

The value of A^2 is modified with respect to the sample size. In case that the parameters are estimated from the sample, the modified value, A^{2*}, for testing the extreme value and the Weibull distributions is (Stephens, 1986, p. 146):

$$A^{2*} = A^2\ (1 + 0.2/\sqrt{n}) \qquad (7)$$

This modification is considered here as fitting to the GP distribution as well. The values of A^{2*} for the 0.25 and 0.01 significance levels of rejecting the fit of those distributions are 0.474 and 1.038, respectively (Stephens, 1986). These values are applied also for the GP distribution.

The results indicate that for each one of the eight rivers, there have been found a number of sub-samples for which $A^{2*} \le 0.474$. This means that for each station there exist a number of partial duration series to which the GP distribution is well fitted. The multitude of good fits allows us to prepare a number of good, and probably different, predictions on the recurrence interval of rare events at a station. This situation may have some scientific merits, but it might cause an ambiguous application. An objective selection of one of these sub-samples would clear up the application.

Our selection procedure is composed of two steps. First we delete from consideration those sub-samples for which $0 < \alpha\hat{}/k\hat{} < 1.1\ Q^*_{max}$, where Q^*_{max} is the maximal observed value of Q^*. This limitation is made in order to avoid a prediction of an infinite recurrence interval for any observed discharge. In the second step, we select the sub-sample for which A^{2*} attains its minimal value, i.e. the partial duration series to which the GP distribution is "best fitted". The threshold discharge associated with this sub-sample and the parameter values obtained from the best fit are recommended for prediction applications. These values, for each one of the rivers, are presented in Table 2. The best fits for the eight rivers are displayed in Fig. 3.

The plotting positions used for the recorded data are as in Hosking and Wallis (1987):

$$F(i) = (i - 0.35)/n \tag{8}$$

in which F(.) is non-exceedance frequency and i is rank of peak discharge in an increasing order.

TABLE 2. Properties of best fitted samples

#	q m^3/s	n	$\hat{k}$	$\hat{\alpha}$ m^3/s	Q_a m^3/s	Q_{max} m^3/s	A^2	f $year^{-1}$
1	15.7	15	-.118	13.79		90	.193	.38
2	10.7	44	.010	12.60	1296	70	.182	1.07
3	48.0	24	.028	48.94	1825	260	.283	.62
4	24.7	70	.363	56.10	179	134	.423	1.94
5	22.4	30	.142	31.07	228	110	.200	.86
6	46.5	16	.405	141.93	397	271	.632	.67
7	43.3	24	.054	99.81	1888	370	.224	1.20
8	27.1	25	.003	52.15	19254	250	.375	1.14

Legend: q is threshold discharge, n is sample size, $\hat{k}$ and $\hat{\alpha}$ are parameter values of the fitted GP, $Q_a = \hat{\alpha}/\hat{k} + q$ is asymptotic value of the fitted distribution, Q_{max} is maximal observed discharge, A^2 is Anderson - Darling test statistic, and f is mean number of events per year.

DISCUSSION OF RESULTS

The length of records for the different stations vary from 20 to 41 years, and on the average it is 32 years per station. If we define flood events of hydrologic interest as those occurring at frequencies lower than 0.1 $year^{-1}$, and presume that the recorded occurrence frequency fairly represents the distribution of probable events, we might not obtain, in the present case, any sample composed of at least five high flood events. In this situation, conclusions about the fit of any distribution to strictly high flood samples can be derived only through its application to larger samples.

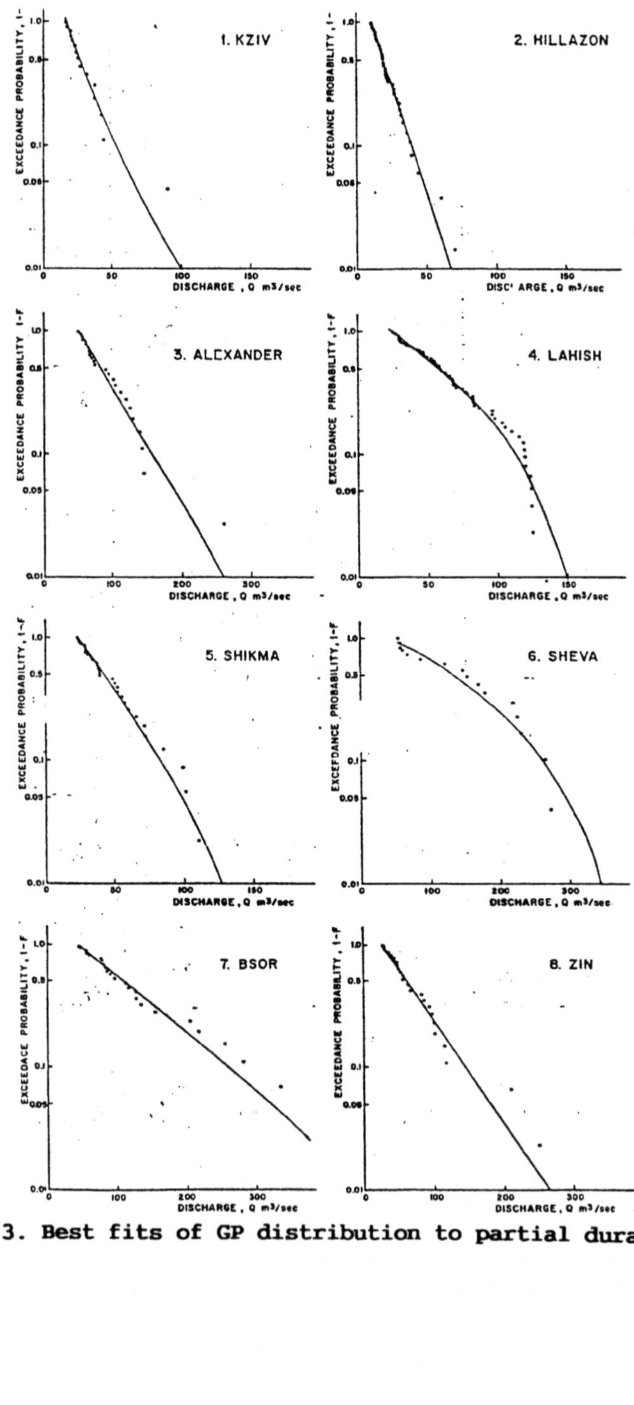

figure 3. Best fits of GP distribution to partial duration series.

Noting our interest in extrapolating the fit in order to predict the recurrence intervals of rare discharges, we prefer a distribution which follows the trends observed in upper tail of the series. The general trend in the present case is of thin tails, with an occasional appearance of a thick tail. Deciding to select the same distribution for the entire case study, we have selected the GP distribution because it is flexible enough to account for both the thicknesses. This selection does not rule out the suitability of any other distribution which is as flexible.

The fit of GP distribution to samples is limited here to the following criteria: threshold values coincide with observed discharges, each sample is composed of at least 5 events, it contains all the peak discharges which exceed the associated threshold, the sign of shape parameter $\hat{k}$ corresponds to the upper tail thickness concluded from Fig. 1, the value of shape parameter $\hat{k}$ is smaller than 0.5 (Hosking and Wallis, 1987) and the asymptotic exceedance discharge is higher than 1.1 times the maximal observed exceedance discharge.

The best fits to the different rivers are found for sub-samples composed of 15 to 70 events, and on the average 31 events per station. The mean occurrence frequency for the best fitted series, varies from 0.38 to 1.94 year^{-1}, and on the average it is 0.99 year^{-1}. These frequencies are considerably lower than those recommended by previous researchers (Langbein, 1949; Cunnane, 1973; Taesombut and Yevjevich, 1978; Ashkar and Rousselle, 1987). It should be recalled, however, that the previous conclusions were reached with respect to considerations which are quite different from the present ones.

The values of threshold discharges, which are associated with the sub-samples of best fit for each river, are found in the middle range of discharges, where CME_q attains a stable value. The lower bound of this range, determined through subjective considerations, seems to increase with the drainage area and to decrease with the mean annual depth of precipitation. These observations may be used for construction of a regional model of floods, a task which lies beyond the scope of the present work.

A hydrologic prediction of recurrence intervals of high floods, by use of partial duration series, should consider both the frequency and magnitude distributions of events. The distributions selected in most of the works are the Poisson and the negative exponential ones (e.g. Todorovic, 1970; Todorovic and Zelenhasic, 1970; Ashkar and Rousselle, 1983a; Rasmussen and Rosbjerg, 1991; Rosbjerg et al., 1992). These distributions are, respectively, particular cases of the negative binomial distribution, which has been found preferable in our previous work (Ben-Zvi, 1991), and of the GP distribution which is selected here. Consequently, our selections do not contradict the commonly accepted practices, but rather make them more flexible and better fitting to observed data.

SUMMARY AND CONCLUSIONS

Thickness of upper tails of partial duration series is considered here as an important criterion in the selection of a distribution through which rare flood discharges are predicted. The thickness is determined from a display of cummulative mean exceedances against threshold discharges. Series extracted from records of eight rivers, located in the semi-arid and the arid regions of Israel, serve as a case study for examining the prediction methodology. Their displays reveal that the CME_α attains a rising trend for low threshold values, a stable trend for medium values, and for seven of the eight rivers a declining trend for high thresholds. This is taken to indicate that the upper tails of most of the series in the case are thin. Yet, the possible occurrence of a thick tail should not be ruled out.

The combination of the GP distribution for description of magnitudes, and of the negative binomial distribution for description of frequencies (Ben-Zvi, 1991), provides a model which is more flexible than the commonly applied combination of the Poisson and the negative exponential distributions. This model fits better to the observed series, but its predictive power has not yet been assessed.

REFERENCES

Ashkar, F. and Rousselle, J. (1983a) "Some remarks on the truncation used in partial duration series models", Water Resour. Res. 19, 477-480.
Ashkar, F. and Rousselle, J. (1983b) "The effect of certain restrictions imposed on the interarrival times of flood events on the Poisson distribution used for modeling flood counts", Water Resour. Res. 19, 481-485.
Ashkar, F. and Rousselle, J. (1987) "Partial duration series modeling under the assumption of Poissonian flood count", J. Hydrol. 90, 135-144.
Ben-Zvi, A. (1982) "Flow events in the Negev - a regional quantitative model", Water International 7, 127-133.
Ben-Zvi, A. (1991) "Observed advantage for negative binomial over Poisson distribution in partial duration series", Stoch. Hydrol. Hydraul. 5, 135-146.
Ben-Zvi, A. and Cohen, O. (1983) "Maximal discharges in the lower Yarkon River", Rep. 4/83, Israel Hydrol. Service, Jerusalem, Israel (in Hebrew).
Bhuiya, R.K. and Yevjevich, V. (1968) "Effects of truncation on dependence in hydrological time series", Hydrol. Paper 31, Colorado State Univ., Fort Collins CO, USA.
Bryson, M.C. (1974) "Heavy-tailed distributions: properties and tests", Technometrics 16, 61-68.
Choulakian, V., El-Jabi, N. and Moussi, J. (1990) "On the distribution of flood volume in partial duration series analysis of flood phenomena", Stoch. Hydrol. Hydraul. 4, 217-226.
Chow, V.T. (1953) "Frequency analysis of hydrologic data with special application to rainfall intensities", Bull. 414, Engrg. Exp. Station, Univ. of Illinois, Urbana IL, USA.
Cruise, J.F. and Arora, K. (1990) "A hydroloclimatic application

strategy for the Poisson partial duration model", Water Resour. Bull. 26, 431-442.
Cunnane, C. (1973) "A particular comparison of annual maxima and partial duration series methods of flood frequency prediction", J. Hydrol. 18, 257-271.
Hiemstra, L.A.V. and Francis, D.M. (1981) "Run hydrographs for prediction of flood hydrographs", J. Hydr. Div. ASCE 107, 759-775.
Hiemstra, L.A.V., Zucchini, W.S. and Pegram, G.G.S. (1976) "A method of finding the family of run hydrographs for given return periods", J. Hydrol. 30, 95-103.
Hosking, J.R.M and Wallis, J.R. (1987) "Parameters and quantile estimation for the generalized Pareto distribution", Technometrics 29, 339-349.
Jarvis, C.S. and "others" (1936) "Floods in the United States magnitude and frequency", Water Supply Paper 771, USGS, Washington DC, USA.
Kavvas, M.L. (1982) "Stochastic trigger model for flood peaks, I. Development of the model", Water Resour. Res. 18, 383-398.
Langbein, W.B. (1949) "Annual floods and the partial duration flood series", Trans. AGU 30, 879-881.
Nachtnebel, H.P. and Konecny, F. (1987) "Risk analysis and time-dependent flood models", J. Hydrol. 91, 295-318.
North, M. (1980) "Time dependent stochastic model of floods", J. Hydr. Div. ASCE 106, 649-665.
Pickand, J. (1975) "Statistical inferences using extreme order statistics", Annals Statis. 3, 119-131.
Rasmussen, P.F. and Rosbjerg, D. (1991) "Evaluation of risk concept in partial duration series", Stoch. Hydrol. Hydraul. 5, 1-16.
Rosbjerg, D. (1987) "Partial duration series with log-normal distributed peak values", in V.P. Singh (ed.), Hydrologic Frequency Modeling, Riedel, 117-129.
Rosbjerg, D., Madsen, H. and Rasmussen, P.F. (1992) "Prediction in partial duration series with generalized Pareto-distributed exceedances", Water Resour. Res. 28, 3001-3010.
Shane, R.M. and Lynn, W.R. (1964) "Mathematical model for flood risk evaluation", Proc. Hydr. Div. ASCE 90, 1-20.
Smith, J.A. (1987) "Estimating of the upper tail of flood frequency distributions", Water Resour. Res. 23, 1657-1666.
Stephens, M.A. (1986) "Tests based on EDF statistics", in R.B. D'Agostino and M.A. Stephens (eds.) Goodness-of-fit Techniques, Dekker, New-York NY, 97-193.
Taesombut, V. and Yevjevich, V. (1978) "Use of partial flood series for estimating distribution of annual flood peaks", Hydrol. Paper 97, Colorado State Univ., Fort Collins CO, USA.
Todorovic, P. (1970) "On some problems involving random number of random variables", Annals Math. Statis. 41, 1059-1063.
Todorovic, P. and Rousselle, J. (1971) "Some problems of flood analysis", Water Resour. Res. 7, 1144-1150.
Todorovic, P. and Zelenhasic, E. (1970) "A stochastic model for flood analysis", Water Resour. Res. 6, 1641-1648.

REGIONAL ANALYSIS OF PARTIAL DURATION FLOOD SERIES IN QUEBEC.

S. BIRIKUNDAVYI, J.R. CORRÉA and J. ROUSSELLE,
Civil Engineering Department, École Polytechnique de Montréal
C.P. 6079 Succ. Centre-Ville, Montréal (Québec), Canada H3C 3A7

This case study deals with the use of partial duration series (PDS) for regional flood frequency analysis (RFFA). Data used are abstracted from continuous records at 45 hydrometric stations in Quebec. Some graphical tools for exploratory analysis of these data are presented. The index flood method is the approach used for RFFA. Results from this model are presented and some conclusions are drawn.

INTRODUCTION

Among the approaches commonly used for statistical inference on extreme flood events, two distinct ones, based respectively on annual maximum series (AMS) and partial duration series (PDS) may be identified. In annual maximum series, the highest flood peak observed in each year is recorded into a series that has as many elements as there are years in the data record. In contrast, the PDS consists of all those events that exceed a specified base or threshold level. This level is chosen so that a distribution suitable for extremes (e.g. exponential or generalized Pareto distribution) applies to the exceedance values.

The magnitude of events in AMS are denoted X_1, X_2, ..., X_N where the subscripts indicate years. It is assumed that the X_i's are independent and identically distributed. This assumption implies that the processes leading to the flood phenomena are stationary over time. In PDS, elements are denoted $\xi(\tau_1)$, $\xi(\tau_2)$, ..., $\xi(\tau_v)$ where τ_i indicates the time when flood $\xi(\tau_i)$ occur in the time interval (0,t]. It is also assumed that PDS values are independent and drawn from a common cumulative distribution function (CDF).

This case study deals only with the PDS approach to flood frequency analysis, with a special emphasis on using partial duration flood data for regional flood frequency analysis. The mathematical model used here was developed by Todorovic and his co-workers in early 70's (Todorovic and Zelenhasic, 1970; Todorovic and Rousselle, 1971) and is based on the well known theory of Poisson processes. The analysis conducted in

K. W. Hipel (ed.),
Stochastic and Statistical Methods in Hydrology and Environmental Engineering, Vol. 1, 109–121.

this study acknowledges only variations between years in the number of flood peaks but ignores variations between seasons. A model which takes into account the seasonality for regional flood frequency analysis with PDS is still investigated.

THE POISSON - EXPONENTIAL MODEL FOR PDS ANALYSIS

The intensity and magnitude of floods in PDS are usually analyzed using the theory of Poisson process. A large experience of applying this approach to flood analysis is observed throughout the world (Todorovic, 1978; North, 1980; Versace and al., 1981; Miquel, 1984; Goel and al., 1987; Reimann, 1989) and also for Canadian conditions (Waylen and Woo, 1984; Irvine and Waylen, 1986; El-Jabi et al., 1986; Caissie and El-Jabi, 1991). The model is described as follows.

If the series satisfies the criteria for a Poisson process, the number $\eta(t)$ of events within the time interval (0,t] is a stochastic variable described by a Poisson distribution:

$$P[\eta(t)=\nu]=p_\nu=\frac{[\lambda t]^\nu}{\nu!}\exp(-\lambda t)\,; \qquad \nu=0,1,2,\ldots, \tag{1}$$

with mean and variance given by $E[\eta(t)]=var[\eta(t)]=\lambda t$ and where p_ν is the probability of observing exactly ν exceedances in the time interval (0,t]. The parameter λ is the average number of flood events per time unit.

We also assume that the flood exceedance $\xi(\tau_\nu)$ is a stochastic variable exponentially distributed. That is:

$$H_\xi(x)=1-\exp(-x/\alpha) \tag{2}$$

where $\alpha=E[\xi]$; $\xi>0$, and $H_\xi(x)$ is the CDF of exceedance values.

Defining the largest exceedance over the base level X_0 in the time interval (0,t] by:

$$\chi(t) = \max_\nu \xi_\nu \tag{3}$$

and combining the assumption of independent flood exceedances to the Poisson process for flood occurrence as defined by equation (1), we get the following expression for the CDF of $\chi(t)$:

$$\begin{aligned} F_{\chi(t)}(x) &= P(\chi(t)\le x) \\ &= \exp\{-\lambda t\,[1-H_\xi(x)]\} \end{aligned} \tag{4}$$

Setting t=1 year in this expression and using (2), we get:

$$\begin{aligned} F_\chi(x) &= \exp\{-\lambda \exp[-x/\alpha]\} \\ &= \exp\{-\exp[-(x-\alpha \ln(\lambda))/\alpha]\} \end{aligned} \quad (5)$$

which is the well known double exponential or Gumbel distribution. The flood quantile function is then given by:

$$\chi(T) = X_0 + \alpha^{-1}[\ln\lambda + C_T] \quad (6)$$

where $C_T = -\ln[-\ln(1-1/T)]$.

ANALYSIS OF FLOOD COUNT DATA

Given a truncation level X_0, the question is whether to accept or reject the assumption of flood occurrence according to a Poisson process. This is usually checked using the chi-square test of fit. The problem is that this test is not very sensitive to small deviations from the Poisson distribution. In this section, we examine the potential use of the probability generating function (pgf) as a graphical tool for selecting the distribution of flood count data for a given truncation level. The main advantage of this approach is that the pgf enables us not only to fit a Poisson distribution to observed data, but also suggests alternatives to this assumption for flood count modeling.

The probability generating function

Let $\eta_1, \eta_2, ..., \eta_N$ be a random sample from a discrete distribution $G(.;\lambda)$ over 0, 1, 2,... The probability generating function (pgf) associated to $G(.;\lambda)$ is denoted by $\phi_\lambda(s)$ and is given by:

$$\phi_\lambda(s) = E(s^\eta) = \sum_{\nu=0}^{\infty} p_\nu s^\nu \quad (7)$$

in which the coefficient of s^ν is the probability p_ν, for ν=0, 1, 2, ... The possible values of s are of course limited to those for which (7) converges; at worst this is the interval $-1 \le s \le 1$, since the set $\{p_0, p_1, p_2, ...\}$ is a probability distribution. The empirical counterpart of $\phi_\lambda(s)$ is called *empirical probability generating function (epgf)* and is given by:

$$\Phi_N(s) = \frac{1}{N}\sum_{i=1}^{N} s^{\eta_i} \qquad |s| \le 1 \tag{8}$$

For the Poisson distribution with parameter λ, the expression of the pgf corresponding to (7) is:

$$\phi_\lambda(s) = e^{-\lambda(1-s)} \tag{9}$$

This expression (9) indicates that a plot of s against ln $\Phi_\lambda(s)$ is a straight line vanishing at s=1 and having intercept -λ. The advantage of using the empirical probability generating function as a graphical tool for flood count analysis is that the plot obtained can be used to capture deviations from the Poisson distribution as well as to suggest some alternative model such as binomial, negative binomial, truncated Poisson or other available discrete distributions. Figure 1 shows as an example how a Poisson distribution whith parameter λ=4 differs from a zero truncated Poisson distribution with the same parameter. Let's recall that the pgf of the zero truncated Poisson distribution with parameter λ is given by:

$$\phi_\lambda(s) = \frac{e^{-\lambda}}{1-e^{-\lambda}}(e^{-\lambda s}-1) \tag{10}$$

since in this case, the probability p_ν of observing exactly ν events in the time interval (0,t] is:

$$p_\nu = (1-e^{-\lambda})^{-1}\exp(-\lambda)\frac{\lambda^\nu}{\nu!} \tag{11}$$

for ν=1, 2, ...

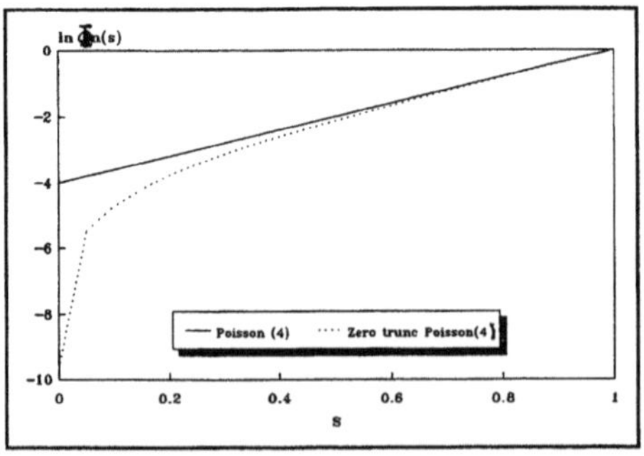

Figure 1. Probability generating functions of Poisson (4) and zero truncated Poisson (4) distributions

Beside the use of the epgf for testing the adequacy of the Poisson model, the method can be used for base level selection so as to be consistent with the assumption of a Poisson process for flood occurrence. Figures 2a and 2b exemplifies this potential use at two hydrometric stations in Quebec, respectively Station 02OE027: Eaton River at Saint-François and Station 02PJ007: Beaurivage River at Saint-Étienne. The plots of the empirical probability generating functions were obtained for different values of the base level and for these two stations. They indicate that the Poisson assumption for flood occurrence seems to be accurate when the value of λ is less than 2.5, since plots are roughly the straight lines in this range of values.

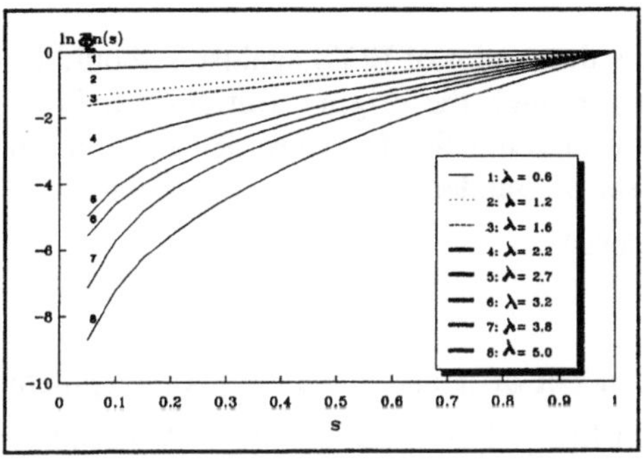

Figure 2a. Flood count data epgf as a fonction of the base level (02OE027)

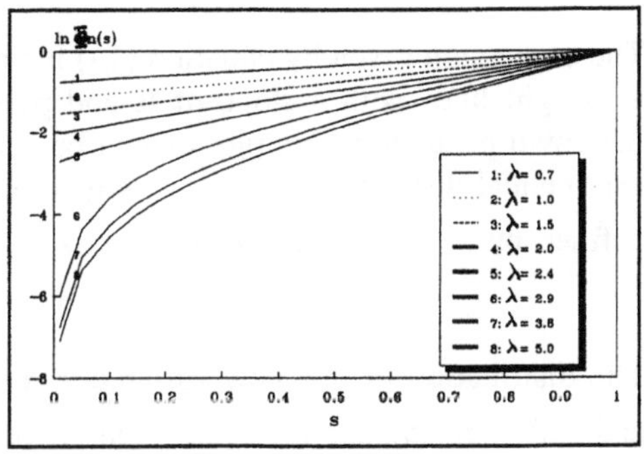

Figure 2b. Flood count data epgf as a fonction of the base level (02PJ007)

DISTRIBUTION OF FLOOD EXCEEDANCES

At this stage, we are interested in knowing if the underlying distribution of flood exceedances is in the attraction domain of extreme values. More precisely, we want to know if exceedance values are drawn from an exponential distribution or from some of it's possible generalizations (e.g. generalized Pareto or Weibull distributions). In this study, the case of exponentially distributed exceedances is discussed and a method based on the standardized distances of sample order statistics is presented as a graphical tool for testing if flood exceedances are drawn from this distribution.

Let $X_1, X_2, \ldots, X_M$ be a sample of size M=λN drawn from an exponential distribution with parameter α. The observations are assumed to be independent and their joint probability density function is given by:

$$f(x_1, \ldots, x_M; \alpha) = \alpha^{-M} \exp\{-\alpha^{-1} \sum_{i=1}^{M} x_i\} \tag{12}$$

Denoting by Y_i the i-th sample order statistic, it is well known that:

$$E[Y_i] = \alpha\delta_i; \qquad i=1,2,...,M \tag{13}$$

where δ_i is the expectation of Y_i if it is drawn from an exponential distribution with mean 1:

$$\delta_i = \sum_{j=1}^{i} \frac{1}{M-j+1} \tag{14}$$

The expression given by equation (13) indicates that a plot of $E[Y_i]$ against δ_i should be a straight line with a slope of α and passing through the origin. But since α is an unknown parameter, this plot cannot be used to measure the departure of data from an exponential distribution. This is why a new set of standardized random variables Z_i's defined by $Z_i = Y_i / \bar{Y}$ where $\bar{Y} = \sum_{i=1}^{M} Y_i / M$ is considered.

The distribution of this new variable is independent of α and since $\bar{Y}$ is a complete and sufficient statistic for α, one can prove that $E[Z_i]=\delta_i$ (see Basu, 1955). A plot of Z_i against δ_i should then be a straight line with slope 1 and passing through the origin if observed data are drawn from an exponential distribution. Figures 3a and 3b exemplify such plots for the two hydrometric stations we selected for illustration. It can be seen from these plots that the exponential distribution seems to be an acceptable assumption in these cases. A more quantitative measure of this acceptance should however be developped.

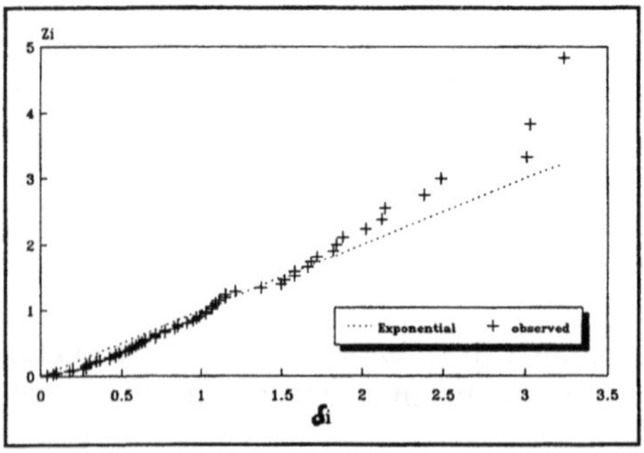

Figure 3a. Exponentiality test (02OE027: X_0=120 m^3/s)

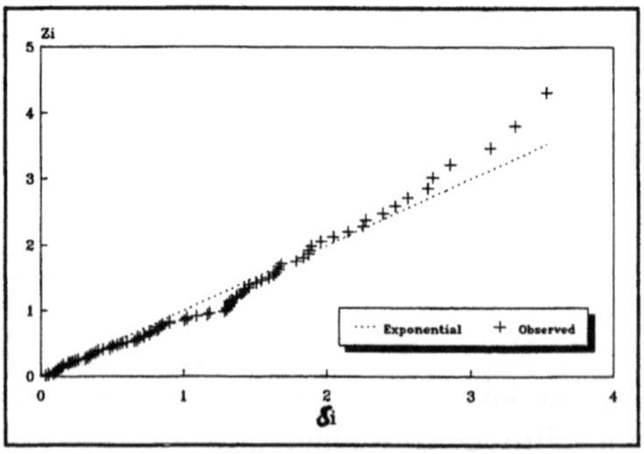

Figure 3b. Exponentiality test (02PJ007: X_0=113 m^3/s)

REGIONAL ANALYSIS

Regional flood frequency methods have been shown to offer the ability to reduce significantly uncertainties in quantile estimation of extreme events relatively to that inherent to conventional at-site frequency analysis (Wallis, 1980; Stedinger, 1983; Hosking and Wallis, 1988). In this section, we examine the use of the index flood method (Dalrymple, 1960) for regional flood frequency analysis of PDS and we present results from an application of the method to data from 45 hydrometric stations in Quebec (Table 1).

Preparation of PDS

Regional analysis of floods at different sites within a region implies comparison of flood behavior at these sites. For annual maximum series, this comparison is done in terms of statistical characteristics of observed flood magnitudes only. In the case of PDS, flood count data have also to be considered. This implies that the method for data abstracting from the continuous discharge records has to be the same at all the hydrometric stations within the region. For the purposes of this study, we assume that flood events at a site consist of all the flood peaks that exceed a level X_0 defined by:

$$X_0 = m + k.s \tag{15}$$

where m and s are respectively the mean and the standard deviation of annual maximum series and where the frequency factor k is set to a selected value (k=-1.2 in this study). This approach is quite similar to the one suggested by Dalrymple (1960), Waylen and Woo (1983) and Irvine and Waylen (1986) by which the base level for PDS is the flood peak value corresponding to a given return period from annual maximum series analysis. We assume in this study that the value of X_0 defined by equation (15) is only an indication or an approximation to the value of the threshold which is supposed to be the parameter of location of the two parameter exponential distribution used to model the complete flood peaks included in the series. In PDS preparation, two neighboring peaks were included in the series if (a) the flow between them dropped to less than 75% of the smallest of the two and if (b) the time between them exceeds 5+ln(A) where A is the drainage area expressed in square miles. These conditions are supposed to ensure independence of flood peaks included in the series (U.S. Water Resources Council, 1976).

TABLE 1. Selected hydrometric stations

No	Fed	Prov	LAT	LONG	AREA km2	START	YEARS
1	01BH007	10101	482840	643143	534	1972	16
2	01BG002	10801	480600	652742	2150	1966	17
3	01BG001	10901	481121	654843	1390	1962	19
4	01BF001	11201	480926	662058	1140	1965	23
5	01BD002	11507	480512	670602	2760	1970	18
6	01BH001	20601	485559	643808	748	1946	24
7	01BH005	20602	485847	644155	645	1970	18
8	02QC001	20802	491210	651744	1200	1968	20
9	02QC009	21407	490332	662912	783	1973	15
10	02QB011	21502	490323	664020	721	1967	21
11	02QA017	21915	483103	680944	492	1974	14
12	02QA002	22003	482448	683322	1610	1963	25
13	02PG004	22504	473457	694011	500	1957	22
14	02PJ007	23401	463933	711719	709	1926	62
15	02PJ030	23422	460951	703823	691	1965	22
16	02PL001	24001	462208	713721	1410	1923	46
17	02PL005	24003	461822	712705	919	1970	17
18	02PL007	24007	461141	721702	2330	1970	17
19	02OD003	30103	460323	721823	1540	1967	21
20	02OE027	30234	452802	713919	642	1954	34
21	02OE033	30242	452026	713404	197	1967	15
22	02OE032	30282	453450	712312	746	1975	13
23	02OG026	30316	455713	725134	342	1970	18
24	02OE018	31401	450241	712954	220	1949	40
25	02LG005	40830	470501	754513	6840	1975	13
26	02JB003	43009	482721	782140	1680	1949	17
27	02LC043	40212	462229	743012	39.9	1969	19
28	02OA025	46707	453907	740532	28	1972	16
29	02OA030	46403	454130	735617	18.9	1973	16
30	02NE007	50117	473503	724435	1850	1938	28
31	02NF003	50119	464109	735451	1390	1944	28
32	02NE011	50135	474602	724412	1570	1966	22
33	02PB006	50409	465818	715116	642	1966	22
34	02PC009	50701	464233	715229	355	1967	21
35	02PD002	51001	465339	710905	1100	1965	23
36	02PD012	51002	471618	710941	3.94	1966	22
37	02PD004	51005	471533	710813	269	1967	15
38	02PD006	51006	471936	710610	58.8	1967	15
39	02RH035	61020	481056	713843	1110	1972	16
40	02RC011	61801	484851	720252	1090	1976	12
41	02RF006	61906	491524	734230	4330	1962	26
42	02SC002	70401	483942	691001	2010	1974	14
43	02UA003	71401	491950	673921	1570	1974	13
44	02WA001	74601	501511	621244	2060	1963	25
45	02XB001	76101	513645	584209	5750	1968	14

The Regional Model

We assume that the 45 selected hydrometric stations belong to the same homogeneous region. This means that flood exceedances observed in the regions are drawn from the same distribution with same parameters and flood count data are also drawn from a same Poisson distribution. Under these assumptions, a perfect homogeneity is achieved if every site j flood quantile is defined by the product of the site j index flood and the regional growth curve. Thus

$$X_j(F) = IF_j \; x(F) \tag{16}$$

where $X_j(F)$ is the site j flood quantile function; IF_j is the site j index flood and x(F) is the regional flood growth curve.

Using the Poisson-exponential model and expressing the frequency F as a function of the return period T, the expression of the regional model defined by equation (16) becomes:

$$\begin{aligned} X_j(T) &= X_{0j} + \alpha_j^{-1}[\ln(\lambda_j) + C_T] \\ &= IF_j[X_{0R} + \alpha_R^{-1}(\ln(\lambda_R) + C_T)] \end{aligned} \tag{17}$$

where X_{0R}, α_R and λ_R are the regional parameters to be estimated. In this study, the index flood IF_j is assumed to be the mean value β_{0j} of the complete flood peaks included in the series.

Regional homogeneity under this model is then achieved if the following conditions are satisfied:

$$(i) \quad \lambda_j = \lambda_R \qquad \forall j \tag{18.a}$$

$$(ii) \quad \frac{\alpha_j}{IF_j} = \alpha_R \qquad \forall j \tag{18.b}$$

since $\alpha_j + X_{0j} = \beta_{0j} \equiv IF_j$.

Condition (i) implies that flood count data are drawn from the same Poisson distribution; differences observed being simply result of sampling variability. Condition (ii) refers to the possibility of eliminating the influence of the scale factor α_j by a simple standardization of observed data. This condition is equivalent to associating a regional hydrological signification to the threshold level relatively to flood exceedances since the mean value of flood peaks is $\beta_{0j}=X_{0j}+\alpha_j$. Let's also recall that we accepted as a basic assumption the fact that flood exceedances are drawn from an exponential distribution at all sites. This is why we don't care about the problem of regional flood distribution identification.

Estimation of the regional parameters

As outlined in the previous section, a perfect homogeneity is achieved if (i) flood rate is the same accross the region and (ii) flood exceedances are drawn from a same family distribution (e.g. exponential). Condition (i) implies that observed flood count data are drawn from a same Poisson distribution. Figures 4a and 4b present respectively the enveloppe curve for the regional variability of flood count data and that of a replicated regional sample when a Poisson distribution is assumed. The parameter of this distribution is equal to the weighted average of flood count parameters estimated at the 45 stations from the observed data. In this case study, the weight used is N, the number of years and the resulting value for this parameter λ_R of the regional Poisson distribution is 1.9.

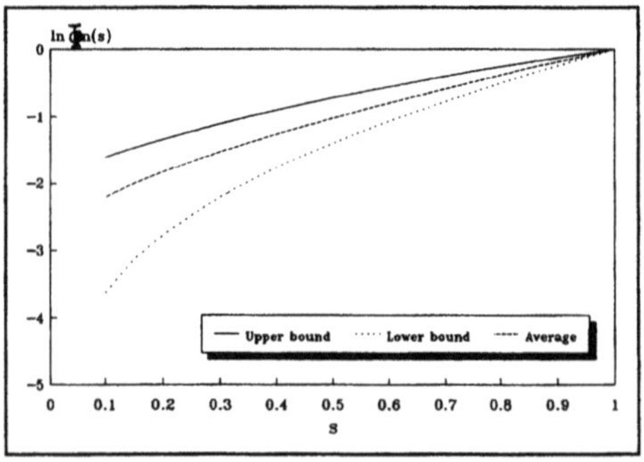

Figure 4a. Sample variability of observed flood count data within the region using the epgf.

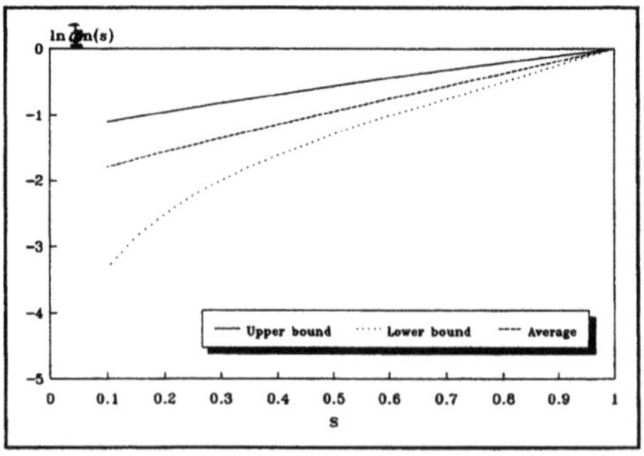

Figure 4b. Replicated sample from a Poisson (1.9) distribution

The other parameters X_{0R} and α_R of the regional model are obtained using the weighted average moments of the observed flood peaks following the methodology outlined by Wallis (1980). Hence, using for example the probability weighted moments method, the parameters of the regional two parameter exponential distribution used in this study are given by:

$$\alpha_R = 4\beta_{1R} - 2\beta_{0R}$$
$$X_{0R} = \beta_{0R} - \alpha_R \qquad (19)$$

where $\beta_{iR}=M_{1i0}/M_{100}$, i=0,1 are the weighted average of the standardized probability weighted moments estimated from observed data.

Results

The method was applied to the selected set of PDS and results obtained are presented on Figures 5 and 6. Figure 5 compares at-site to regional estimation of the 100-year flood. This plot indicates that results from these two models are not very different. Figure 6 compares the efficiencies of the two models (at-site and regional) in terms of sampling variance of the T-year flood. Results are presented for station 02PJ007 (62 years) and station 02OE027 (34 years). Sampling variances for X(T) were obtained from a simulation experiment using the bootstrap method.

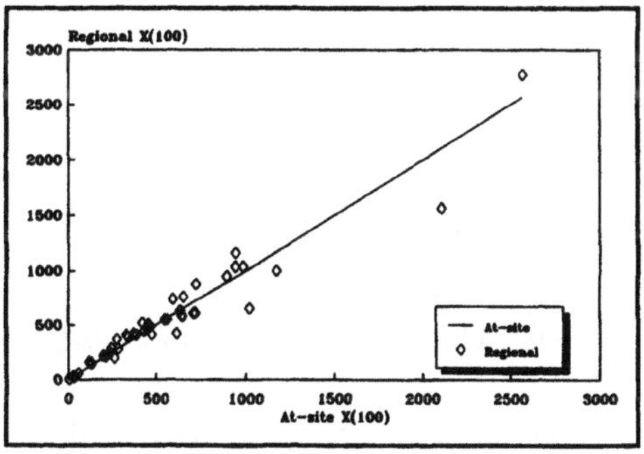

Figure 5. Regional vs at-site flood quantile estimates.

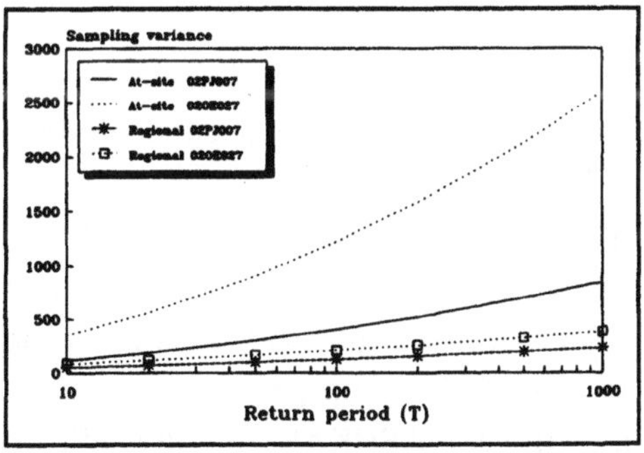

Figure 6. Efficiencies of the models: a comparison.

In order to estimate flood quantiles at ungauged sites, the index flood IF_j was related to drainage area A_j. The expression of this relation is given by:

$$IF_j = e^{-0.718} A_j^{0.880} \tag{20}$$

with a coefficient of regression R^2=0.981 and a standard error on estimate of σ=0.243.

This equation (20) have to be used jointly with the regional flood growth curve (equation (17)) in this given region for flood quantile estimation at ungauged sites. Parameters α_R and X_{0R} for the regional model are respectively 0.322 and 0.668.

SUMMARY AND CONCLUSIONS

In this study, partial duration series abstracted from 45 hydrometric stations in Quebec were used for regional flood frequency analysis using the index flood method. Before doing this, we presented some useful graphical tools for exploratory analysis of PDS data, such as the empirical probability generating function for flood count data analysis and the use of order statistics for testing the exponentiality of flood exceedances. Results from regional frequency analysis were then compared to those from at-site analysis and an equation for estimating flood quantiles at ungauged sites was provided.

In order to complete this study, we will have to evaluate the performances of the graphical tools presented in this study for model identification. Performances of the Poisson-exponential model for RFFA will also be compared to those of other available approaches.

REFERENCES

BASU, D. (1955) "On Statistic Independent of a Complete Sufficient Statistic", Sankhya 15, 377-380.

CAISSIE, D. and EL-JABI, N. (1991) "A Stochastic Study of Floods in Canada: Frequency Analysis and Regionalization", Can. J. of Civ. Eng. 18, 225-236.

DALRYMPLE, T. (1960) "Flood Frequency Analysis", U.S. Geol. Surv. Water Supply Pap. No. 1534A, 60 pages.

EL-JABI, N., ASHKAR, F., ROUSSELLE, J. and HOANG, V.D. (1986) "Étude stochastique des crues au Québec", Ministère de l'Environnement, Québec, Canada, H.P.-59.

GOEL, N.K., SETH, S.M. and CHANDRA, S. (1987) "Design Flood Estimation for Narmada Sagar Project Using Partial Duration Series: A Case Study", in V.P. Singh (ed.), Application of Frequency and Risk in Water Resources, D. Reidel Publishing Company, pp. 113-122.

HOSKING, J.R.M. and WALLIS, J.R. (1988) "The Effect of Intersite Dependance on Regional Flood Frequency Analysis", Water Resour. Res. 24(4), 588-600.

IRVINE, K.N. and WAYLEN, P.R. (1986) "Partial Series Analysis of High Flows in Canadian Rivers", Revue Canadienne des Ressources en Eau 11(2), 83-91.

MIQUEL, J. (1984) Guide Pratique d'estimation des probabilités de crues. Collection de la direction des études et recherches d'Électricité de France. Eyrolles, Paris.

NORTH, M. (1980) "Time Dependent Stochastic Models of Floods", J. Hydraul. Div. ASCE 106(HY5), 649-655.

REIMANN, J. (1989) Mathematical Statistics with Application in Flood Hydrology. Akadémiai Kiado; Budapest; 330 pages.

STEDINGER, J.R. (1983) "Estimating a Regional Flood Frequency Distribution", Water Resour. Res. 19(2), 503-510.

TODOROVIC, P. (1978) "Stochastic Models of Floods", Water Resour. Res. 14(2), 345-356.

TODOROVIC, P. and ROUSSELLE, J. (1971) "Some Problems of Flood Analysis", Water Resour. Res. 7(5), 1144-1150.

TODOROVIC, P. and ZELENHASIC, E. (1970) "A Stochastic Model for Flood Analysis", Water Resour. Res. 6(6), 1641-1648.

U. S. Water Research Council (1976) "Guidelines for Determining Flood Flow Frequency", Hydrology Committee, Bulletin 17, Washington, D.C.

VERSACE, F.; FIORENTINO, M. and ROSSI, F. (1981) "Analysis of Flood Series by Stochastic Models", in A.H. El-Shaarawi et S.R. Esterby (eds.), Time Series Methods in Hydrosciences, Elsevier, Amsterdam, pp. 315-324.

WALLIS, J.R. (1980) "Risk and Uncertainties in the Evaluation of Flood Events for the Design of Hydraulic Structures", paper presented at Seminar on Extreme Hydrological Events: Floods and Droughts, Erice, Italy, 33 pages.

WAYLEN, P.R. and WOO, M.K. (1983) "Stochastic Analysis of High Flows in Some Central British Columbia Rivers", Can. J. of Civ. Eng. 10(2), 639-648.

WAYLEN, P.R. and WOO, M.K. (1984) "Regionalization and Prediction of Floods in the Fraser River Catchment, B.C.", Water Resour. Bull. 20(6), 941-949.

PART III

DROUGHTS

STOCHASTIC FEATURES OF DROUGHT IN KENYA, EAST AFRICA

T.C. Sharma
Department of Agricultural Engineering
University of Nairobi
Nairobi, P.O Box 30197
Kenya

Annual rainfall sequences from six raingauge stations in Kenya, namely, Thika, Naivasha, Narok, Kibwezi, Makindu and Voi were used to determine the stochastic features of critical drought duration and severity. The critical drought duration and severity mean respectively the maximum expected drought duration and deficit sums at the desired truncation or water demand level. The drought generating mechanism is abstracted from an observed time series and simple analytical formulation of the probabilistic behaviour of extreme droughts. The sequence was assumed to constitute a stationary random process on which the theory of extremes of random number of random variables combined with the theory of runs was applied.

The study revealed that spells of drought and wet years at the truncation levels ranging from -1 to 0 in terms of standardized annual rainfall follow a random process. Under the random structure of the spells, the occurrence of droughts can be regarded to follow a Poisson probability distribution. The drought severity defined as the deficit sum can be approximated through a normal probability distribution. The duration and severity of the critical (longest) drought can be predicted for a desired time span (10, 25, 50, 75, 100, or 200 years), truncation level (x_0 = -1, -0.8, -0.6, -0.4, -0.2, 0) representing the acceptable level of water demand for rainfed crops and the probability quantile of drought associated with the chosen truncation level. At the truncation level x_0 = -0.2 (representing a mild drought) the critical drought duration ranged from 2 to 6 years over a time span ranging from 10 to 200 years. The corresponding drought severity ranged from 1.5 to 4.5 units of the standardized annual rainfall. The critical values of the drought severity should be used as a basis for designing a water storage system for supplemental irrigation. The whole analysis can be carried out by simply knowing the three parameters, i.e. the mean and coefficient of variation of annual rainfall and the probability of any year being a drought year at a desired truncation level, which can be derived from the standard normal probability tables.

INTRODUCTION

Droughts are of great concern throughout the continent of Africa because of the devastating effects they have inflicted on the economies of some of the countries. For

K. W. Hipel (ed.),
Stochastic and Statistical Methods in Hydrology and Environmental Engineering, Vol. 1, 125–137.

example, one estimate (Oyabande, 1990) suggests that droughts in the last two decades have threatened the lives of more than 20 million people and uprooted millions of others in Africa. Kenya in East Africa is no exception. Droughts are natural phenomena and cannot be prevented. But there are possibilities of predicting them to provide the necessary information for early warning and preparedness. A drought forecast includes four vital components, duration, severity, regional spread and its time of occurrence. Currently there is no available method which produces a feasible time forecast for several months or a couple of years ahead. Forecasting limitations would suggest a shift of emphasis to investigations for predictability of droughts in frequency domain. It is against this background that the present investigation was initiated to study the duration and severity aspects of critical droughts based on annual rainfall totals in Kenya. Critical drought is defined here as one which extends continually for the longest duration (years) and experiences the maximum deficit sum over a desired span of years at a given demand level of water. The severity is equivalent to deficit sum at a given demand level and its unit of measure is millimeters of water.

For the purpose of the present study six rainfall stations representing the semi-arid regions of Kenya were chosen for analysis. These stations, Thika, Naivasha, Narok, Kibwezi, Makindu and Voi possess the rainfall data of nearly 60 years and lands around these gauging stations are cultivated for maize, coffee, vegetables and fodder production (Fig. 1). The mean annual rainfall ranges from 550 to 800 mm with the coefficient of variation of 0.27 to 0.38. It should be noted that the agricultural activities in the domain of the above raingauge stations are largely rain fed and plans are underway to develop an irrigation system for watering cash crops such as vegetables and coffee. It is, therefore, important to plan and design irrigation systems for the region in relation to the duration and severity of drought. Notions of run lengths, run sums, transitional probabilities, Poisson process and extremes of random number of random variables were used to derive the stochastic structure of drought duration and severity.

The severity of the drought has been expressed through indices such as decile range (Gibbs and Maher, 1965), aridity index (Thornthwaite, 1947) and standardized index (Gibbs, 1975). The standardized index is equivalent to the standard normal variate in terms of the annual rainfall. The droughts have been categorised as mild if the standardized index is less than or equal to -0.2, moderate if the index is less or equal to -0.5 and severe, if the index is less than or equal to -0.8. The standardized index > -0.2 means no drought conditions exist at least for agricultural purposes (Downing et al., 1988). In the present study droughts have been analysed using the notion of standardized index.

DROUGHT SPELL ANALYSIS

Yevjevich (1967) gave a definition of drought using the theory of runs and the stationary structure of time series. If a time series such as annual rainfall x_i (standardised form) is truncated at a truncation level x_o, then the events of surplus ($x_i > x_o$) and deficits ($x_i \leq x_o$) would emerge along the time axis (Fig. 2). The truncation level x_o could be a measure of the demand level and would vary for different activities. The truncation level can be designated by its probability quartile as q = P ($x_i \leq x_o$) where P() denotes probability quantile such as 0.2, 0.5, etc. Any uninterrupted sequence of deficits can be regarded as a drought of length equal to the number of deficits in the sequence. In the ensuing analysis this length in years will be referred to as the drought duration or negative run length, denoted by L_j. The sequence of surplus (positive, wet) or deficits (negative, drought) runs are denoted by 1 or 0. Every drought duration is associated with a deficit sum denoted by D_j (Fig. 2). A drought of the longest duration is generally associated with the maximum deficit sum or severity as the duration and severity are strongly correlated (Sen, 1977). The statistical and mathematical aspects of drought duration are well investigated by Yevjevich (1972), Gupta and Duckestein (1975), Sen (1977, 1980), Chander et al. (1981) and Lee at al. (1986). The severity aspect is far less developed although thoughts have been presented by Yevjevich (1972), Sen (1980) and Chander et al. (1981). The relationships for drought duration and severity documented in Sen (1980) appear to be simple yet sensible and are reproduced briefly as follows:

The probability density function (pdf) of the number of drought occurrences (N) displaying lag 1 Markov persistence at any truncation level (x_o) during the time interval of n years (0, n) can be regarded to follow Poisson probability law with an average rate of occurrence nq(1 - r), in which r = conditional probability of any year being a drought year given the past year was also a drought year and q = simple probability that any year is a drought year at a given truncation level.

If the drought pulses are independent or random then the conditional probability r = simple probability q.

Thus in mathematical terms

$$P(N = i) = \exp[-nq(1 - r)]. \frac{[nq(1 - r)]^i}{i!} \qquad \ldots(1)$$

The probability P ($L_n \leq j$), i.e. that the probability of the longest duration L_n is equal to or less than an integer value j, can be determined from the relationship

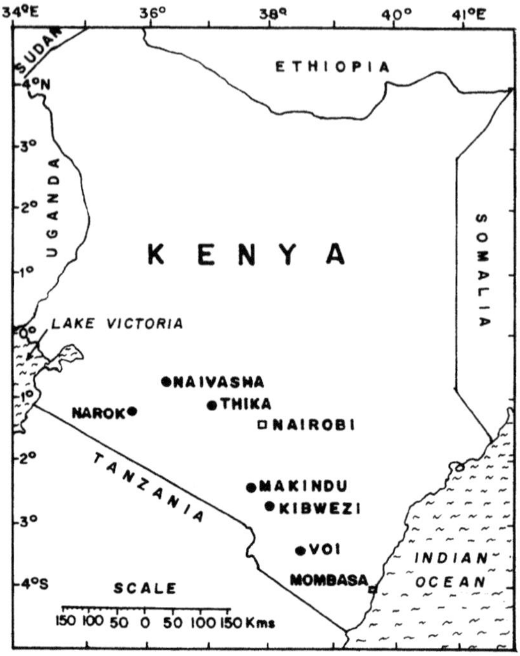

FIG. 1 — RAINGAUGE STATIONS USED FOR DROUGHT STUDY IN KENYA

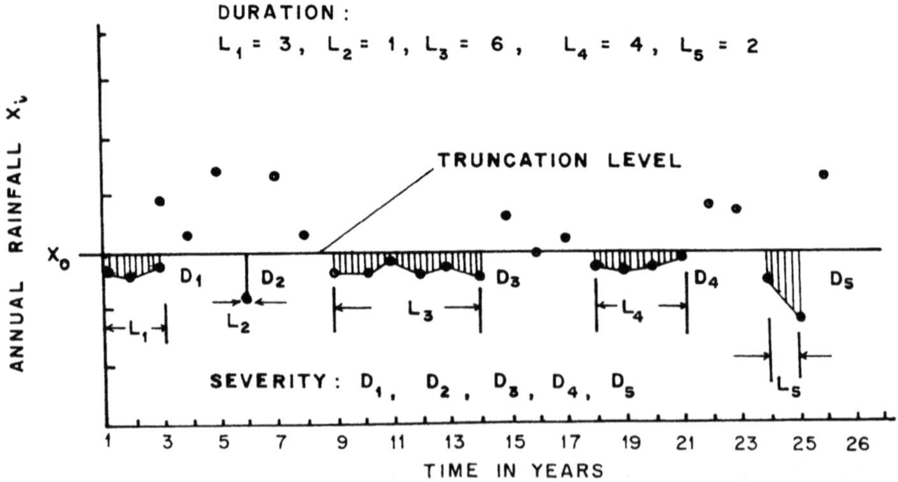

FIG. 2 — DROUGHT DURATION AND SEVERITY SEQUENCES AT TRUNCATION LEVEL X_0

$$P(L_n \leq j) = P(N = 0) + \sum_{i=1}^{\infty} [P(L \leq j)]^i \cdot P(N = i) \qquad \ldots(2)$$

Eq. (2) can be simplified by involving expressions for P(N=i) and P(L≤ j). For lag 1 Markov process, it has been shown by Sen (1977) that

$$P(L \leq j) = 1 - r^{j-1} \text{ or } P(L > j) = r^{j-1} \qquad \ldots(3)$$

The algebraic simplification of Eq. (2) involving Eqs. (1) and (3) yields the following

$$P(L_n \leq j) = \exp[-nq(1 - r) \cdot P(L > j)] \qquad \ldots(4)$$

$$P(L_n \leq j) = \exp[-nq(1 - r)\ r^{j-1}] \qquad \ldots(5)$$

$$P(L_n = j) = \exp[-nq(1 - r)\ r^{j-1}] \cdot [\exp\{nq(1 - r)^2\ r^{j-1}\} - 1] \qquad \ldots(6)$$

The expected value of the longest drought duration $E(L_n)$ can be obtained as

$$E(L_n) = \sum_{j=1}^{n} j \cdot P(L_n = j) \qquad \ldots(7)$$

Probability distribution function of maximum deficit sum $P(D_m \leq s)$ to be less than or equal to any s value can be obtained in the same way as Eq.(4) in which P(L>j) is replaced by P(D>s). Alternatively one can use Eq.(2) and replace $P(L_n \leq j)$ by $P(D_m \leq s)$ and P(L≤j) by P(D≤s). The mathematical simplification of Eq.(2) with the aforesaid substitution results into

$$P(D_m \leq s) = \exp[-nq(1 - r) \cdot P(D > s)] \qquad \ldots(8)$$

$$\text{for } (0 \leq s \leq +\infty)$$

in which P(D ≤ s) = pdf of deficit sum or drought severity to be equal or less than a given value s. The pdf of D can be assumed to be normal in view of the central limit theorem as it is the sum of several deficits. For a standardized normal sequence, the deficit sum can be approximated to obey truncated normal distribution whose parameters mean (μ_D) and variance (σ_D^2) can be obtained from the following (Sen,1980):

$$\mu_D = \frac{\exp\left(-\frac{x_o^2}{2}\right)}{(1 - q)\sqrt{2\pi}} + x_o \qquad \ldots (9)$$

$$\sigma_D^2 = 1 + \frac{x_o \exp\left(-\frac{x_o^2}{2}\right)}{(1 - q)\sqrt{2\pi}} - \frac{\exp(-x_o^2)}{(1 - q)^2 \; 2\pi} \qquad \ldots (10)$$

The mean (μ_s) and variance (σ_s^2) of the deficit sum of duration i for a lag 1 Markov drought spells can be estimated from the following relationships documented in Sen(1980).

$$\mu_s = i \; \mu_D \qquad \ldots (11)$$

$$\sigma_s^2 = \sigma_D^2\left\{i + 2\rho \; \frac{i(1-\rho) - (1-\rho^i)}{(1-\rho)^2}\right\} \qquad \ldots (12)$$

In which ρ = *serial correlation at lag* 1

$$= \sin\{\pi(r - 0.5)\} \qquad \ldots (13)$$

It should be noted that for a standardized normal sequence, the values of q corresponding to a desired truncation level x_0 can be obtained from the normal probability tables as shown in Table 1.

The value of i for the longest drought duration can be approximated as

$$i = \frac{1}{2}\left\{E(L_n) + \frac{1}{1-q}\right\} \qquad \ldots (14)$$

If the drought years evolve randomly and show the lack of persistence then r = q in all equations from 1 through 13. Note $x_o = 0$ in terms of standardized variable and is equivalent to 100% of the mean annual rainfall and likewise $x_o = -0.2$ corresponds to 100 x_m $(1+x_o.cv)$ percent of mean (x_m) annual rainfall. If the cv(coefficient of variation) is 0.27 then $x_o = -0.2$ means 94.5% of mean annual rainfall.

In order to apply the above relationships, it is important to investigate the stationarity of the annual sequences and their pdf. If the annual rainfall sequence x_i can be shown to be normally distributed and stationary then the assumption of normality of deficit sum is reasonable and the analysis for drought durations is valid.

The second point to be investigated is the dependence structure in the occurrence of drought years at varying truncation levels. This analysis can be done by evaluating the simple and transitional probabilities and testing them for equivalence. A chisquare statistic based on likelihood ratio test can be used to test the null hypothesis of equivalence (Medhi, 1982). The transitional matrix for lag 1 Markov persistence can be written as

$$\begin{bmatrix} pp & 1 - pp \\ 1 - r & r \end{bmatrix} \qquad \ldots(15)$$

in which pp = conditional probability that the present year is wet given the past year is also wet and r is defined earlier. Under the assumption of independence pp = 1 - q and r = q and the matrix takes the form

$$\begin{bmatrix} 1 - q & q \\ 1 - q & q \end{bmatrix} \qquad \ldots(16)$$

Note 1 - q can also be written as p and p + q = 1

Thirdly the validity of relationships for E (L_n) and $E(D_m)$ should be checked with the observed values. In the present exercise annual rainfall data for 60 years are available and therefore 6 sets of observed $E(L_n)$ and $E(D_m)$ are available at truncation levels x_o = 0, -0.2, -0.4, -0.6, -0.8 and -1.0. The predictions using Eqs. (1) through (14) should be compared with the observed ones. It is only after the above comparisons are found satisfactory that one should proceed to predict $E(L_n)$ and $E(D_m)$ for sample sizes n = 10, 25, 50, 75, 100 and 200 years.

RESULTS AND DISCUSSIONS

The annual rainfall sequences in semi-arid and semi-humid environments with mean annual rainfall exceeding 500 mm tend to follow normal distribution and can be regarded trend free (Sharma, 1990). Therefore the assumption of normality and stationarity of annual rainfall sequences is generally true for the rainfall stations chosen for the study. The transitional matrices were computed and tested against the simple probabilities, q using a chisquare statistic (Table 1). Invariably, in almost all cases the chisquare statistic indicated the lack of persistence in the occurrence of drought spells (critical value of chisquare is 3.84 at 5% level of significance and 1 degree of freedom for lag 1 Markov persistence). Therefore the drought spells can be regarded to follow

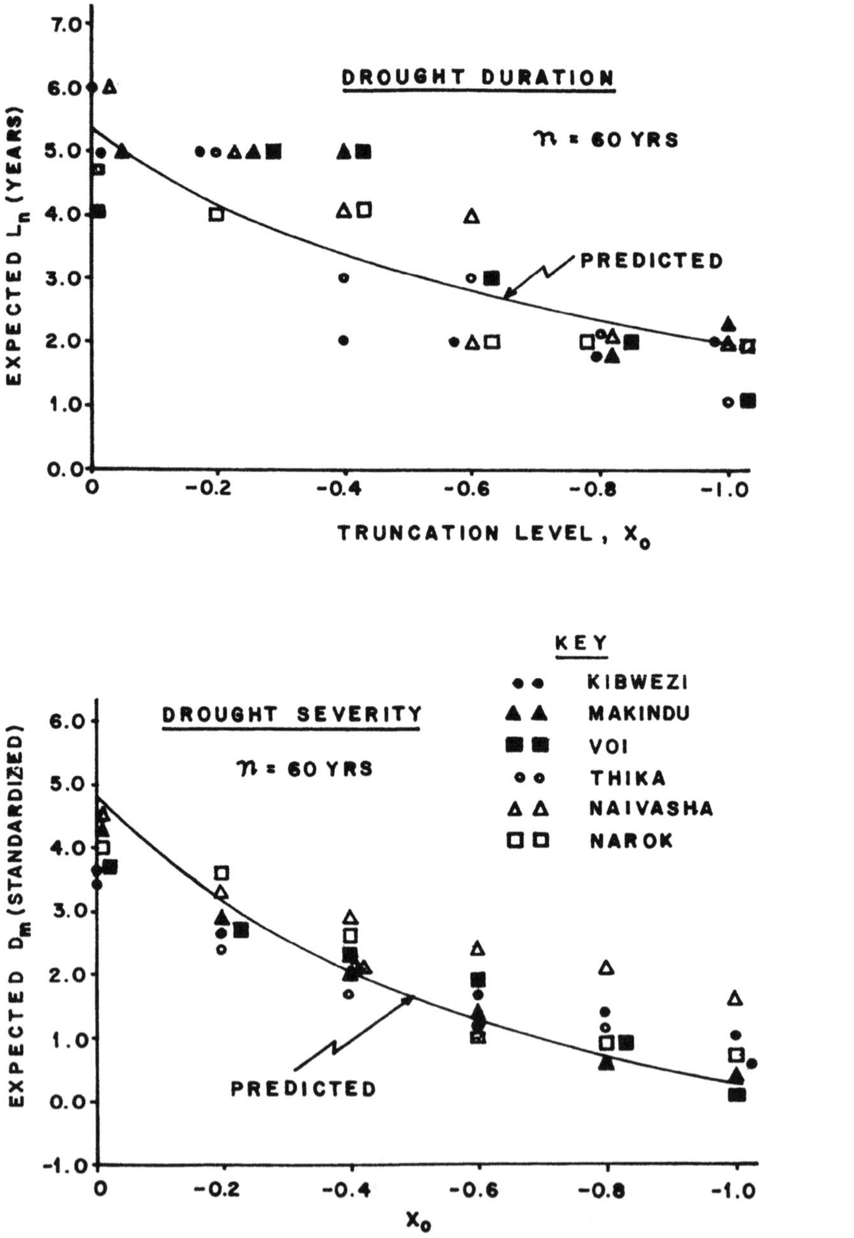

FIG. 3 — COMPARISON OF PREDICTED CRITICAL DROUGHT INDICES WITH OBSERVATIONS IN SEMI ARID KENYA

a random process. Finally $E(L_n)$ and $E(D_m)$ were computed and are plotted in Fig. 3 along with the observed points. The close correspondence between the predicted and observed points in Fig. 3 is indicative of the validity of relationships for predicting the expected longest drought duration $E(L_n)$ and the maximum deficit sum or the critical severity $E(D_m)$.

Note that the points in Fig.3 on the duration graph are discrete such as 5,4, 3, 2 or 1. They have been plotted around the discrete values in order to avoid the clustering of points. Ideally they should all plot on a discrete value such as 5, 4, 3, 2 or 1.

Table 1: Values of chisquare statistics indicating the lack of persistence in the drought spells

Station name and sample size	Statistics of annual rainfall		Truncation level (standardized) and probability quantile, q (Values of chisquare are shown on each column below)					
	x_m (mm)	cv	x_o=0.0 q=0.5	x_o=-0.2 q=0.42	x_o=- 0.4 q=0.34	x_o=-0.6 q=0.27	x_o=-0.8 q=0.21	x_o=-1.0 q=0.16
Thika, 58	811.64	0.27	0.53	0.28	0.01	0.03	1.41	1.63
Naivasha, 62	687.55	0.30	4.02	2.34	0.22	0.35	1.04	0.01
Narok, 61	760.84	0.27	0.02	0.32	0.09	0.99	0.57	0.23
Kibwezi, 62	643.21	0.38	0.47	0.26	0.73	0.59	0.77	1.35
Makindu, 62	552.44	0.38	0.38	1.15	0.17	0.12	1.06	0.17
Voi, 61	557.87	0.38	0.35	0.15	0.0	0.11	0.83	3.13

In view of the satisfactory performance of the model equations, predictions were made for the expected longest drought durations and maximum severity for the time span ranging from 10 to 200 years and are depicted in Fig. 4. The following points are of interest in relation to Fig. 4.

(1) At a truncation level x_o = -0.2, which corresponds to 90 to 95% of mean annual rainfall for the semi-arid region under question, the critical duration of the mild drought may extend from 2 to 6 years over a time span of 10 to 200 years. However, at the low level of truncation (x_o = -0.5) which corresponds to 80 to 86% of the mean annual rainfall the moderate drought may persist for 2 to 4 years over the aforesaid time span. The duration of the severe drought (x_o = -0.8) corresponding with 63 to 73% of the mean annual rainfall is expected to range from 1 to 3 years over the above time span.

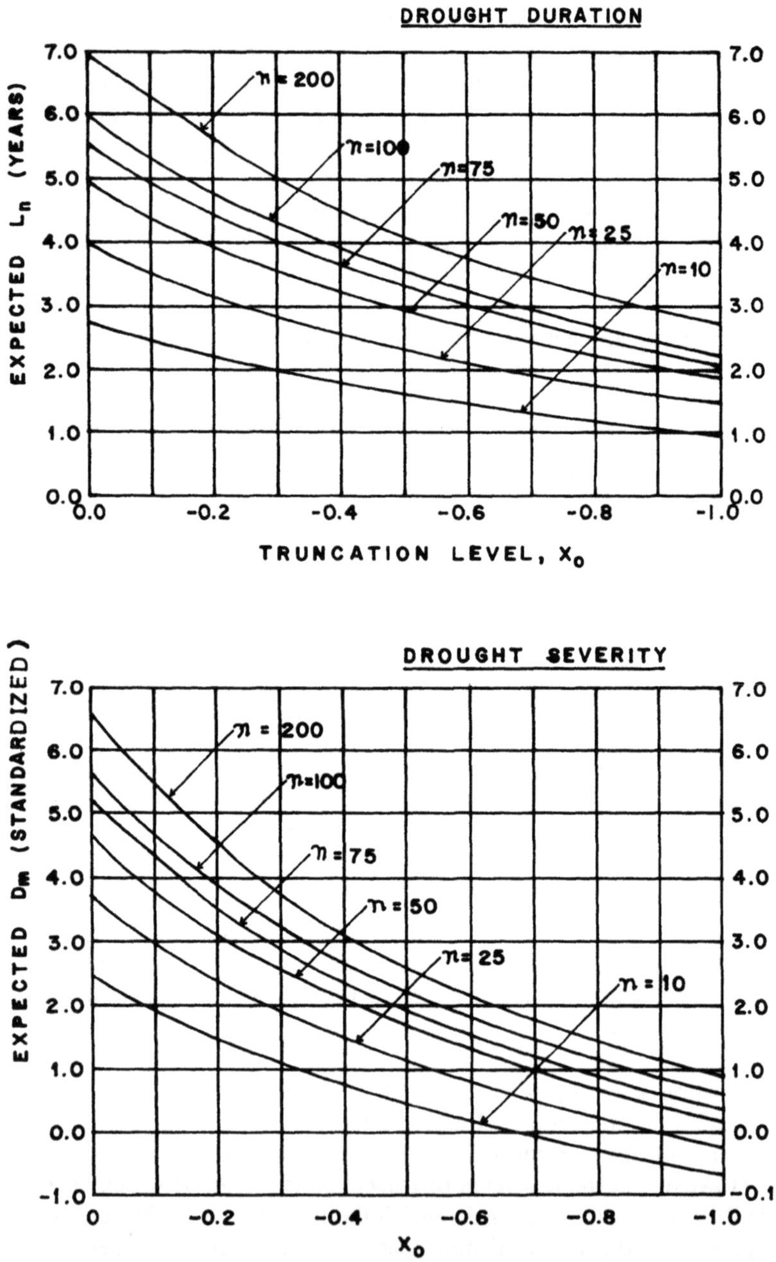

FIG. 4 — PREDICTIONS OF CRITICAL DROUGHT DURATIONS AND SEVERITY IN SEMI ARID KENYA

(2) A record of rainfall for 100 years (1884 - 1993) in semi-arid Kenya (Downing et al., 1984) indicates that twice droughts have occurred lasting for 4 years once from 1896 to 1900 and, then from 1907 to 1911. The return period of the droughts is 50 years and therefore confirms the findings of the model. Likewise, an extended drought of 6 years duration occurred from 1913 - 1919. The return period of this drought can be regarded as 200 years. Both these droughts had the truncation level x_o = -0.2, which incorporates all years when the standardized index is less than or equal to -0.2.

(3) Looking at the critical severity, one should expect severity ranging from 1.5 to 4.0 units of the standardized annual rainfall over the time span of a decade to a century. Taking x_o = -0.2 for a mild drought one should expect the critical shortage of 1.5 units of standardized rainfall persisting 2 years once every 10 years. That means for Thika region, this severity is equivalent to $x_m(1+1.5\ cv)$ = 811.64 (1+1.5 * 0.27) = 1140 mm. The critical drought duration is approximately 2 years at x_o = -0.2 over the time span of a decade. Therefore, if a water storage facility is to be developed, it must be able to supply 1140 mm of water over a critical duration of 2 years or 570 mm/yr. Such a system may comprise a reservoir built across a stream. On the other hand, if a big irrigation scheme made up of a high dam and a network of canals is to be built, then one might think of critical severity over a time span of 100 years. Therefore, the system should be able to supply the water equivalent to 811.64(1+4*0.27) = 1688 mm during an extended period of 5 years or 338 mm/yr.

(4) For a severe drought (say x_o = -1.0) occurring once in 10 years, the storage system should be designed to supply 811.64 (1-0.77 * 0.27) = 643 mm during the drought period of 1 year or it should be capable of supplying 643 mm/year. However, to stand a severe drought of 100 years return period, the storage system should be capable of supplying 811.64 (1+0.50 * 0.27) = 921.21mm over 2 years or nearly 460 mm/yr. Thus, the storage system should be designed for severe drought conditions at the desired level of probability (return period). A design based on such logic would be able to supply the water for mild to moderate drought conditions. In the above case, for a return period of 10 years, the storage system should be designed for 643 mm/yr (severe drought) as against 570 mm/yr for a mild drought. Likewise for a period of 100 years, it should be designed for 460mm/yr (severe drought) as opposed 338 mm/yr (mild drought).

It should be remembered that in the above analysis the total annual rainfall has been taken as an index of drought and its seasonal variation has not been considered. Despite this simplification, the analysis demonstrates that prolonged droughts are highly probable in the environments of Kenya. The values of severity are based on normal pdf of deficit sums which do not hold for low truncation levels. Yet, the values of severity provide some basis for designing storage systems for supplemental irrigation during drought years.

CONCLUSIONS

Spells of drought and wet years at truncation levels x_0 = -1 to 0 follow random processes in the climatic environments of semi-arid regions of Kenya. The duration of the longest drought is dependent upon the truncation level (demand), the time span chosen for the study and the probability of any year being drought year at the level of truncation x_0. A mild drought ($x_0 \leq -0.2$) may extend for a period of 5 years over a period of century whereas the return period of severe drought of one year ($x_0 \leq -0.8$) may be close to a decade. The corresponding critical drought severity aspect can be modelled using the normal probability distribution. The expected value of the critical severity shall be 1.5, 3, 4.0, and 4.5 units of the standardized annual rainfall over the time span of 10, 50, 100 and 200 years. A water storage system should be designed using the above values of critical severities for the desired return periods. The model structure can be formed using three parameters : the mean and coefficient of variation of annual rainfall and the probability value of any year being drought year at a desired truncation level, which can be obtained from the normal probability tables.

ACKNOWLEDGEMENTS

The resources and facilities for carrying out the study were extended by the Department of Agricultural Engineering, University of Nairobi. The funds for participation in the International Conference in Waterloo, Canada were provided by the SAREC project currently in operation in the Department of Agricultural Engineering, University of Nairobi, Kenya. The author gratefully acknowledges the financial and material support from these organizations.

REFERENCES

Chander, S., Kambo, N.S. Spolia, S.K. and Kumar, A. (1981) "Analysis of surplus and deficits using runs", Journal of Hydrology 49, 193 - 208.

Downing, T.E., Gitu, K.W., Kamau, C.M., and Barton J. (1984) "Drought in Kenya" in T.E. Downing, K.W. Gitu, and C.M. Kamau (eds.), Coping with Drought in Kenya: National and Local Strategies, Lynne Rienner Publishers, London, U.K.

Gibbs, W.J. and Maher, J.V. (1965) "Rainfall deciles as drought indicator", Bulletin No. 48. Bureau of Meteorology, Melbourne, Australia.

Gibbs W.J. (1975) "Drought, its definition, delineation and effects", Drought - Special Environmental Report No. 5, WMO, Geneva, 1 - 40.

Gupta, V.K. and Duckestein, L.(1975) "A stochastic analysis of extreme droughts", Water Resources Research 11(2), 221-228.

Lee, K.S., Sadeghipour, J. and Dracup, J.A. (1986) "An approach to frequency analysis of muiltiyear drought durations", Water Resources Research 22(5), 655 - 662.

Medhi, J (1982) Stochastic Processes, Wiley Eastern Limited, New Delhi, India.

Oyabande, L. (1990) "Drought policy and drought planning in Africa", International Journal of Water Resources Development 6(4), 260 - 269.

Sen, Z. (1977) "Run sums of annual flow series", Journal of Hydrology 35, 311 - 324.

Sen, Z (1980) "Statistical analysis of hydrologic critical droughts", Journal of Hydraulics Division, ASCE 106 (HY 1),99-115.

Sharma, T.C. (1990) "A mathematical rainfall simulator", Proceedings Kenya Society of Agricultural Engineers Annual Seminar, Nairobi, Kenya, pp. 138-148.

Thornthwaite, C.W. (1947) "Climate and moisture conservation". Annals Assoc. Amer. Geogr. 37(2), 87 - 100.

Yevjevich, V. (1967) "An objective approach to definitions and investigations of continental hydrologic droughts", Hydrology Paper 23, Colorado State University, Fort Collins, Colorado, USA.

Yevjevich, V. (1972) Stochastic Processes in Hydrology, Water Resources Publications, Fort Collins, Colorado, USA.

DERIVED FREQUENCY DISTRIBUTION FOR LOW FLOWS

J.G. PERZYNA and L. GOTTSCHALK
Institute of Geophysics
University of Oslo
P.O Box 1022, Blindern, 0315 Oslo
Norway

ABSTRACT. The study reveals the usefulness of the derived distribution approach for low flow frequency analysis. A physically based distribution for instantaneous summer minimum flows (Gottschalk & Perzyna, 1989) is generalized in order to consider stochastic variation in initial moisture conditions at the onset of a recession period. Two alternative forms of generalized distributions are presented. The flexibility of the generalized models increase with respect to increases of the coefficient of variation of the moisture variable. This increase is mainly manifested in the shapes of the upper tails of the distributions. Comparison of the models to empirical distribution of low flow data showed, however, that the improvements achieved due to generalization of the original derived distribution are rather small.

INTRODUCTION

To derive a physically based distribution for instantaneous summer low flows, Gottschalk & Perzyna (1989) used the basin recession model of the simple exponential type:

$$Q_T = Q_0 \exp(-T/k) \tag{1}$$

where Q_T is flow at time T from the beginning of recession, Q_0 is initial low flow at time T=0, and k is a constant characterizing the rate of flow decay. The response function (1) was chosen for its ability to describe low flow having low frequency of occurrence. For a Norwegian summer season, it can be assumed that the main portion of such flows comes from the depletion of groundwater storage.

An essential step in the derivation of the low flow distribution was the specification of climatic inputs that cause low flows. Minimum flow during a summer season can be expected to occur at the onset of the rainfall event that terminates the longest dry spell for that season. It is now possible to assume that the duration of recession T, that leads to minimum flow, can be approximated by the duration of the longest maximum dry spell. Perzyna (1992a) analysed the various aspects of spatial and temporal characteristics of summer maximum dry spells in southeast Norway, and the influence they exert on river flows. This analysis showed, that during 'dry years', long dry spells have uniform and stable distribution over large areas. It was also verified, that for Norwegian

K. W. Hipel (ed.),
Stochastic and Statistical Methods in Hydrology and Environmental Engineering, Vol. 1, 139–149.

conditions, the distribution function of maximum dry spells followed the Extreme Value Type 1 (EV1) distribution.

The derived low flow distribution conditioned on $Q_0=q_0$, has the form (Gottschalk and Perzyna, 1989):

$$F_{Q|Q_0}(q|q_0) = 1 - \exp\left\{-\left[q\,\frac{\exp(u/k)}{q_0}\right]^{k/\alpha}\right\} \tag{2}$$

where q_0 and k are parameters of the response model (1), and characterize the dynamic properties of drainage basin, and α ,u are climatic parameters of the EV1 distribution of summer maximum dry spell samples. The first two moments of the distribution are:

$$E(Q|Q_0=q_0) = \frac{q_0}{\exp(u/k)}\,\Gamma\left(1+\frac{\alpha}{k}\right) \tag{3}$$

$$Var(Q|Q_0=q_0) = \frac{q_0^2}{\exp(2u/k)}\left[\Gamma\left(1+\frac{2\alpha}{k}\right) - \Gamma^2\left(1+\frac{\alpha}{k}\right)\right] \tag{4}$$

where $\Gamma()$ is the Euler Gamma function. The parameters of the derived distribution (2) are uniquely specified by values of the depletion constant k, initial low flow Q_0 and the two parameters, α and u, of the EV1 distribution fitted to a maximum dry spell series. The local recession parameters describe dynamic properties of the catchment. The parameters α and u supply the model with regional information about climatic conditions of recession. The distribution function (2) was derived under the assumption that the recession parameters, Q_0 and k, are constant for a particular basin.

Repeated tests of the distribution on Norwegian summer annual low flows series (Gottschalk & Perzyna, 1989, Perzyna, 1990, 1992b) showed that the distribution systematically underestimated low flows with high frequency of occurrence. This feature of the derived distribution was traced to properties of the applied basin response model (1), which is not robust enough to describe summer low flows dominated by quick flow components. There was also question regarding the performance of the distribution in its upper tail; whether it was biased because the model does not account for temporal variations in basin moisture conditions.

To answer this question, a generalization of the derived distribution that considers stochastic variations in the initial moisture conditions at the onset of recession periods is made. Two numerical solutions of the problem will be presented. Finally, the behavior of both models will be compared to empirical data sets.

GENERALIZED DERIVED DISTRIBUTIONS FOR LOW FLOW

To consider stochastic variation in initial moisture conditions we will treat the parameter Q_0 in the recession model (1) as a stochastic variable. Given the distribution of the variable Q_0, the unconditional distribution of the low flow Q can be determined by means of the formula of total probability (Kartvelishvili, 1967):

$$F_Q(q) = \int_{-\infty}^{\infty} F_{Q|Q_0}(q|q_0)\, f_{Q_0}(q_0) dq_0 \tag{5}$$

Insertion of the distribution equation (2) gives the expression:

$$F_Q(q) = \int_{-\infty}^{\infty} \left(1 - \exp\left\{ - \left[q \frac{\exp(u/k)}{q_0} \right]^{k/\alpha} \right\} \right) f_{Q_0}(q_0) dq_0 \tag{6}$$

The moments of the distribution function $F_Q(q)$ can be derived directly from the conditional moments of the distribution $F_Q(q|q_0)$ if the moments of the distribution of Q_0 can be expressed in a closed form (Bain & Engelhardt, 1987). The expected value E(Q) of the distribution function (6) can, with help of equation (3), be calculated as:

$$E(Q) = E[E(Q|Q_0)] = E\left[\frac{Q_0}{\exp(u/k)} \Gamma\left(1 + \frac{\alpha}{k}\right) \right] = E(Q_0) \left[\frac{1}{\exp(u/k)} \Gamma\left(1 + \frac{\alpha}{k}\right) \right] \tag{7}$$

The variance Var(Q) of the unconditional distribution of Q can be found from:

$$\mathrm{Var}(Q) = \mathrm{Var}[E(Q|Q_0)] + E[\mathrm{var}(Q|Q_0)] \tag{8}$$

If we substitute equations (3) and (4) for $E(Q|Q_0)$ and $\mathrm{var}(Q|Q_0)$, the result is:

$$\mathrm{Var}(Q) = \left\{ \frac{1}{\exp(u/k)} \Gamma\left(1 + \frac{\alpha}{k}\right) \right\}^2 \mathrm{Var}(Q_0) + \tag{9}$$

$$+ \frac{1}{\exp(2u/k)} \left[\Gamma\left(1 + \frac{2\alpha}{k}\right) - \Gamma^2\left(1 + \frac{\alpha}{k}\right) \right] E[Q_0^2]$$

The unconditional distribution $F_Q(q)$ was derived for two different, postulated distributions of the initial flow Q_0. First, we assume that the variable Q_0 is lognormally distributed with a probability density function $f_{Q_0}(q_0)$ equal to:

$$f_{Q_0}(q_0) = \frac{1}{\sigma_n q_0 \sqrt{2\pi}} \exp\left\{-\frac{1}{2\sigma_n^2}(\ln q_0 - \mu_n)^2\right\} \tag{10}$$

The unconditional distribution $F_Q(q)$ is obtained by solving the integral (11) with respect to q_0:

$$F_Q(q) = \int_0^\infty \left(1 - \exp\left\{-\left[q\,\frac{\exp(u/k)}{q_0}\right]^{k/\alpha}\right\}\right) \frac{1}{\sigma_n q_0 \sqrt{2\pi}} \exp\left\{-\frac{1}{2\sigma_n^2}(\ln q_0 - \mu_n)^2\right\} dq_0 \tag{11}$$

For the lognormally distributed Q_0, the expected value E(Q) of the unconditional distribution (11) of Q is:

$$E[Q] = \exp(\mu_n + \sigma_n^2/2)\left[\frac{1}{\exp(u/k)}\,\Gamma\left(1 + \frac{\alpha}{k}\right)\right] = \exp(\mu_n + \sigma_n^2/2 - u/k)\,\Gamma\left(1 + \frac{\alpha}{k}\right) \tag{12}$$

and the variance Var[Q] is:

$$\mathrm{Var}[Q] = \left(\exp(\mu_n + \sigma_n^2/2)\right)^2 \left\{\left(\frac{Cv^2}{\exp(2u/k)}\,\Gamma^2\left(1 + \frac{\alpha}{k}\right)\right) + \right. \tag{13}$$

$$\left. + \exp(\sigma_n^2 - 2u/k)\left[\Gamma\left(1 + \frac{2\alpha}{k}\right) - \Gamma^2\left(1 + \frac{\alpha}{k}\right)\right]\right\}$$

Second, we assume that Q_0 follows the two parameter gamma distribution with the probability density function $f_{Q_0}(q_0)$ expressed by:

$$f_{Q_0}(q_0) = \frac{1}{b^a \Gamma(a)}\, q_0^{\,a-1} \exp(-q_0/b) \quad ; \quad q_0 \geq 0 \tag{14}$$

The unconditional distribution is, in this case, equal to:

$$F_Q(q) = \int_0^\infty \left(1 - \exp\left\{-\left[q \frac{\exp(u/k)}{q_0}\right]^{k/\alpha}\right\}\right) \frac{1}{b^a \Gamma(a)} q_0^{a-1} \exp(-q_0/b) dq_0 \quad (15)$$

When Q_0 is gamma distributed then the first two moments of $F_Q(q)$ are:

$$E[Q] = \frac{ab}{\exp(u/k)} \Gamma\left(1 + \frac{\alpha}{k}\right) \quad (16)$$

$$Var(Q) = \frac{ab^2}{\exp(2u/k)} \Gamma^2\left(1 + \frac{\alpha}{k}\right) + \frac{ab^2(a + 1)}{\exp(2u/k)} \left[\Gamma\left(1 + \frac{2\alpha}{k}\right) - \Gamma^2\left(1 + \frac{\alpha}{k}\right)\right] \quad (17)$$

APPLICATION EXAMPLE.

Parameter estimation

The parameters of the derived distributions have been determined in accordance with the methodology developed by Perzyna (1992b). Recession parameters, Q_0 and k, are estimated from available discharge data. A minimum of 10 years of discharge data should be used (Perzyna, 1990).

Table.1 Physical parameters of the derived distribution function

Basin No.	Area km^2	$\bar{Q}_0$ l/skm^2	k days	u days	α
423	865	9.3	15.2	15.5	6.0

The parameter Q_0 has been defined above as the initial low flow at time t=0 (equation (1)). The mean value of Q_0 is calculated through a standardized procedure as a constant equal to 2 x mean value of summer annual low flow series.

The recession constant k has been estimated by fitting a linearized form of equation (1) to the log-transformed discharge records, and using an ordinary least square regression. For the analysis, we have used recession limbs of minimum 5 days duration and an initial discharge below the mean value of Q_0.

Finally, the parameter k was calculated as a mean value of the sample estimates of k. Due to the fact that samples of k contained theoretically unjustifiable high values, before averaging, 20% of the highest estimates of k have been removed from original samples.

The parameters α and u of the EV1 distribution of maximum dry spell samples have been regionally mapped (Perzyna, 1992a) and their areal values are taken directly from the maps.

For the assumptions of lognormal and gamma distribution of Q_0 (see equations (11) and (15) respectively) the parameters μ_n, σ_n, a and b have been calculated by the method of moments.

Analysis of the results and discussion.

Summer annual low flow series from basins located in southeast Norway were selected to evaluate the performance of the generalized models. The theoretical models were compared graphically to empirical data. The results were similar for all tested cases. Due to the method of estimation of recession parameters, the coefficient of variation (CV) of moisture conditions could not be explicitly determined. To demonstrate the flexibility of the generalized models several values were assumed for the coefficient of variation of Q_0 , i.e.,: CV=(0.2, 0.4, 0.6).

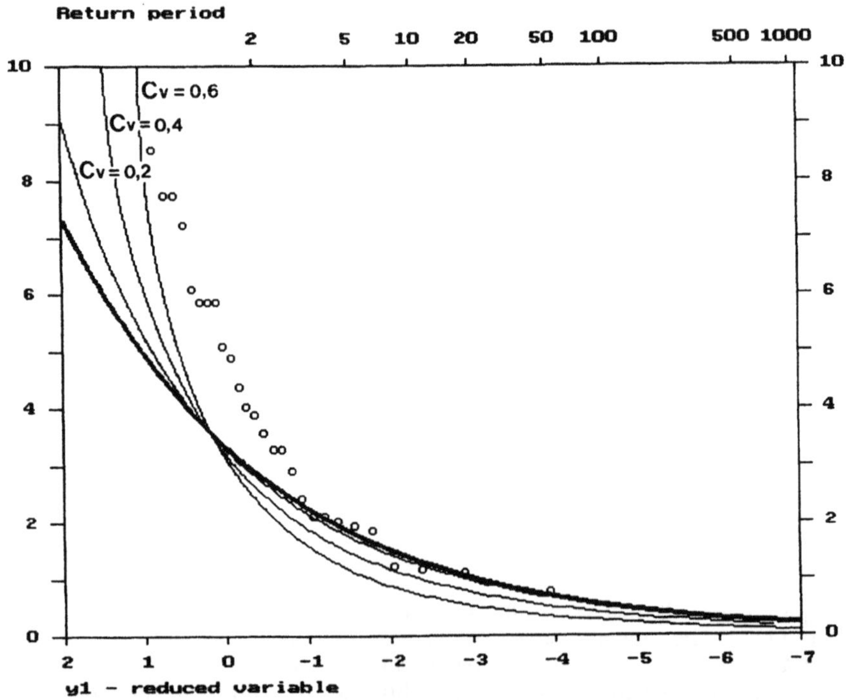

Figure 1. Distribution function curves for a generalized model derived on the assumption of a lognormally distributed Q_0 (equation (11)), and compared to empirical data. Basin 423.

Examples of plots for basin 423 are shown in Figs. 1 and 2. Information about the values of the estimated recession and climatic parameters for basin 423 are given in Table 1. The figures illustrate behavior of the distributions, derived under the following assumptions: Fig.1 of lognormally distributed Q_0, Fig.2 of gamma distributed Q_0. For comparison, the previously derived distribution function (2), is also plotted (thick line).

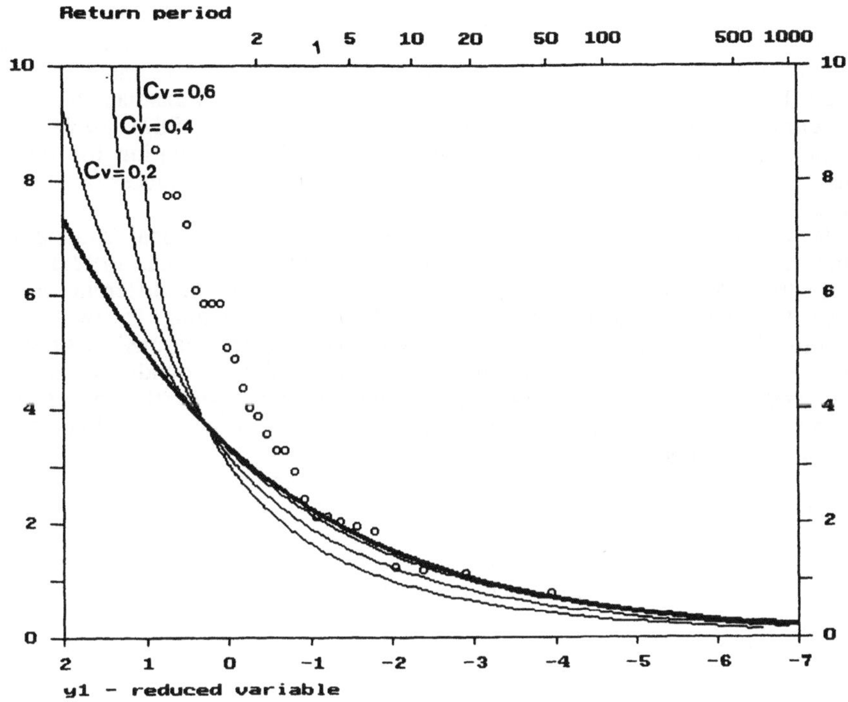

Figure 2. Distribution function curves for a generalized model derived on the assumption of a gamma distributed Q_0 (equation (15)), and compared to empirical data. Basin 423.

It was believed that the incorporation of stochastic variations in the basin initial moisture condition would improve the performance of the derived distribution in its upper tail. This expectation has been only partially fulfilled. Inspection of the plotted examples shows that the flexibility of the generalized models as expected increased. The fact that this increase is mainly manifested in the shapes of the upper tails of the distributions can be interpreted as a promising characteristic of the generalized models. However, it can also be observed that the shapes of the generalized distributions for low flows with a high probability of occurrence differ considerably from those represented by empirical data. The difference is also large for the highest value of the CV (0.6) of Q_0. This feature of the generalized distributions reinforces earlier opinions (Gottschalk & Perzyna, 1989; Perzyna, 1992b), that stated that poor fit of the former model (equation (2)) to low flows with high frequency of occurrence is due mainly to the nature of the chosen response model. The Norwegian low flow series contain typically, not only the events resulting from the depletion of groundwater storage, but also those dominated by direct flow components.

The nonhomogeneity patterns can be frequently identified in series of low flow events. Veltz & Gannon (1953) were some of the first to notice that annual low flow frequency graphs exhibit a brake in the frequency curve near the modal value. It is generally accepted that, in basins located in higher latitudes, where annual low may occur in the winter or in the summer, these events are generated by different processes (Gottschalk, 1985). For Canadian basins with similar climatic conditions to Norway, Waylen and Woo (1987) used a mixed distribution approach and applied a compounded Gumbel distribution for frequency analysis of annual low flows that are induced by winter coldness and summer evaporation.

It is a possible to divide annual low flows into the seasonal subpopulations for which frequency analysis can be performed separately. Perzyna (1992a) showed, however, that in southeast Norway the summer low flow in different years are governed by distinctly different weather patterns, and belong to two different populations. Extreme low flows occur during 'dry summer seasons' with spatially persistent dry spells of long duration, and include mainly groundwater discharge. During 'wet summer seasons' with short maximum dry spells that are spread in time, true recession conditions are lacking and low flows are dominated by quicker flow components.

All of the derived distribution functions (see equations (2), (11) and (15) are flexible enough to also describe low flow with a higher frequency of occurrence. However, the method used here for estimation of recession parameters, direct from the analysis of low flow recession records, then would have to be abandoned. This will be exemplified below applying the simplest of the models - equation (2).

The derived model (2) resembles the Weibull distribution function (Weibull, 1961):

$$F(x) = 1 - \exp[-\{(x - \varepsilon)/\phi\}^{1/\delta} \qquad (18)$$

where ε is the parameter defining the lower bound of the distribution, and ϕ and δ are scale and shape parameters. Equation (2) may be transformed into equation (18) by defining:

$$\varepsilon = 0, \quad \phi = Q_0/\exp(u/k) \quad \text{and} \quad \delta = k/\alpha. \qquad (19)$$

The procedure of fitting the derived distribution (2) to low flow data is therefore equivalent to the fitting of a two parameter Weibull distribution.

Fig. 3 shows the fit of the Weibull distribution to low flow series from the basin 423 compared to the fit of the physically based distribution (2) (thick line). The latter with the physically based parameters which are presented in Table 1. The moment estimates of the parameters ϕ and δ of the Weibull distribution are 5.11 and 1.38 respectively. From the moment estimates of ϕ and δ, the estimates of recession parameters Q_0 and k, can be obtained using the expressions (19) and the earlier estimated values of the climatic parameters α

and u (see Table 1). The calculated values of recession parameters corresponding, to $\alpha = 6.0$ and $u = 15.5$ are: $k = 8.28$ and $Q_0 = 33.2$.

Fig.3 illustrates, that the distribution function (2) with parameters estimated by method of moments, fits low flows with high frequency of occurrence well, but underestimates the events with low frequencies of occurrence. The parameters, k and Q_0, are not any more related to the dynamic properties of the basin, which are decisive for development of low flow with high return periods. The observed patterns suggest that the lower tail of the model is biased due to the influence of high low flows, generated during 'wet seasons' with suppressed recession conditions. In contradiction, due to the proprieties of the response model (1), the derived distribution with physical parameters performs well for the population of low flows governed by depletion of groundwater storage.

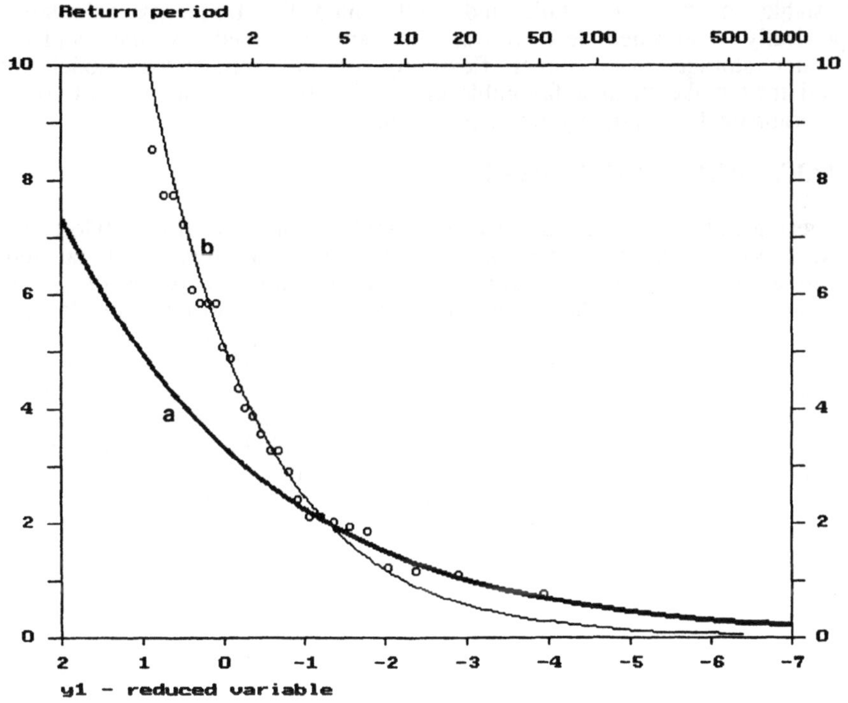

Figure 3. Derived distribution function (equation 2) fitted to empirical data from basin 423. (a) physically based parameters (Table 1) (b) recession parameters Q_0 and k estimated by method of moments.

Nathan and McMahon (1990), pointed out that low flows with higher frequency of occurrence can not be considered as drought flows, and therefore should be excluded from frequency analysis. Consequently, they used reduced data samples containing only those flows with exceedance probability of 80%. For the same purpose, Institute of Hydrology (1980) and Tollow (1987), suggested values of 65% and 70% respectively.

The derived distribution function with physically based parameters can be a viable alternative in frequency of low flows. The distribution can be used for extreme low flow, drawn from different populations. The fact that the distribution systematically underestimated low flow extremes with high probability of occurrence was not considered to be a drawback. The prediction of low flows with low return periods are of little importance for purposes of design and management. Referring to Fiering (1967), ''the statistical significance of one distribution or another, the degree of resolution of the tail of one distribution or another, represent operationally meaningless questions unless they are framed with the context of making decisions''. The derived distribution reflect causative factors of low flows. The developed methodology of the estimation of its parameters reduces the influence of the very high or very low summer minimum flows on the behavior of the distribution. This property makes the model stable in the lower tail, and well suited for frequency calculations of asymmetrically distributed low flows, that are governed by the depletion of groundwater storage. The higher flexibility of the generalized models in the upper tail may make them a favorable choice for frequency analysis of low flows if these originate from homogeneous populations.

SUMMARY AND CONCLUSIONS.

A generalization of the derived distribution for low flows (Gottschalk & Perzyna, 1989) is made to consider the stochastic variation in initial moisture conditions at the onset of recession period. Two alternative forms of generalized models are presented. The flexibility of the generalized models increase corresponding to increases of the CV of the variable Q_0, which represents initial flow at the beginning of recession period. This increase is mainly manifested in the shapes of upper tails of the distributions. The comparison of the generalized distributions to empirical distribution of low flow data showed that their shapes differ considerably for low flows with a high probability of occurrence. The improvements achieved due to generalization of the original model are rather small. The former model is characterized by stable behavior and, due to its properties, can be used to calculate frequency of low flow generated by the depletion of groundwater storage.

REFERENCES

Bain, L.J. & Engelhardt, M. (1987) Introduction to probability and mathematical statistics. 190-197. Duxbury Press, Boston.

Fiering, M.B. (1967) Streamflow synthesis, Macmillan, London.

Institute of Hydrology (1980) Low Flow Studies Report. Wallingford, U.K.

Gottschalk, L. (1985) Hydrological regionalization of Sweden. *Hydrol. Sci. J.*, **30**(1), 65-83.

Gottschalk, L & Perzyna, G. (1989) Physically based distribution for low flows. *Hydrol. Sci. J.* **35**(5), 559-573.

Gottschalk, L & Perzyna, G. (1993) Low flow along a river, to be published 1993.

Kartvelishvili, N.A.(1967) Teorija veroyatnostnykh protsessov v gidrologii i regulirovanii rechnogo stoka, Gidrometeoizdat, Leningrad.

Nathan, R.J. & McMahon, T.A. (1990) Practical aspects of low flow frequency analysis. *Wat. Resour. Res.* **26**(9), 2135-2141.

Perzyna, G. (1990) Parameter estimation from short observation series of low flows. Paper presented on XV EGS General Assembly in Copenhagen, April 1990.

Perzyna, G. (1992a) Spatial and temporal characteristics of maximum dry spells in southern Norway, to be published 1993.

Perzyna, G. (1992b) A dimensionless low flow frequency distribution, to be published 1993.

Tollow, A.J. (1987). Methods used in assessing low flow probabilities. Paper presented at National Hydrology Symposium, Br. Hydrol. Soc.,1987.

Veltz, C.J & Gannon, J.J (1953) Low flow characteristics of streams. *Ohio State Univ. Studies. Eng. Series.* **XXII**(4), 135-154.

Waylen, P.R & Woo, M. (1987) Annual low flows generated by mixed processes. *Hydrol. Sci. J.*, **32**(3), 371-383.

Weibull, W. (1961) Fatigue testing and analysis of results. Pergamon Press.

PART IV

STORMS

TIME DISTRIBUTION OF SHORT DURATION STORMS IN VENEZUELA

E. GUEVARA, and H. CARTAYA
Escuela de Ingenieria Civil, Universidad de Carabobo
Barbula, Valencia, Venezuela.

Based on the Intensity-Duration-Frequency curves of 81 raingauge stations in Venezuela, mean relations have been obtained between the 10-Year rainfall depths of 15, 30, and 45 minutes duration and the corresponding amount of one hour. The results agree with those obtained in the U.S.A. as well as with the world records enveloping curve. Therefore it can be stated that the extreme storm events of short duration (less than one hour duration) have a similar time distribution pattern all over the world, allowing the establishment of a unique relation to estimate short duration rainfall data which are usefull to estimate the design flood of small hydraulic structures as drainage and watershed conservation structures.

INTRODUCTION

This investigation deals with the deterministic analysis of the local geomorphological influence over the time distribution of short rainfall events. Short duration rainfall data, less than one hour, are very usefull to estimate the design flood of small hydraulic structures as drainage. However quite ofen this kind of information is not available because of the lack of adequate recording equipment in developing countries. Besides that, the Weather Services normaly supply the values of rainfall depths for duration equal to or greather than one hour, and the designer muss therfore estimate the short duration events on the basis of available data.

In this investigation, efforts have been made to establish the relation between the rainfall depth of less than one hour duration with the event of one hour duration. It is based on the Intensity-Duration-Frequency analysis of 81 raingauges located all over Venezuela. The results are

K. W. Hipel (ed.),
Stochastic and Statistical Methods in Hydrology and Environmental Engineering, Vol. 1, 153–162.

compared with those obtained for U.S.A. and the rest of the world. The extreme short rainfall events in Venezuela are only marginaly affected by the relief. The relation between the depth of rainfalls with duration less than one hour and the one hour duration rainfall remain almost constant. This conclusion agrees closely with those in the rest of the world. The relations resulting from this analysis are subject to errors of 11% to 20% in 95% of the analysed cases; the error decreases to half in 66% of the cases.

Hydrologists normaly try to establish the I-D-F-curves under homogeneus morphological conditions (analogic models). The results are given as a set of maps for each duration and frequency. As an example of this, we could mention the former Depth Duration-Frequency-Atlas of short events in Venezuela. Based on such Kind of information one could estimate Frequencys-Intensity funtions, except for mountainous areas where the interpolations are not reliable.

Many efforts have been made to relate the rainfall depths of duration shorter than one hour with the corresponding values of one hour. The U.S. Weather Bureau (Hersfield, 1961) recommends to use portions of 0.29, 0.45, 0.57 and 0.79 of the one hour value to estimate the corresconding values of 5, 10, 15, and 30 minutes duration respectively, for the same frequency. The estimation errors range from 5% to 8% (Hersfield, 1961). Avellán and Porras (1984) have also tried to find the same kind of relations, nevertheless, they limited the analysis only to one state of Venezuela (Lara state).

As mentioned above, the rainfall events of shorter duration (no more than two or three hours) appear to be less influenced by geographical location. They are originated as a result of convective local cells with similar characteristics all over the world. For this reazon a quasiuniversal relation could be used to estimate time distribution of short events.

METHODOLOGY

The records of 81 raingauge stations have been used in this study. The analysis is based on the 10-years recurrence interval values for duration of 15, 30, 45, and 60 minutes read from the corresponding Intensity-Duration-Frequency curves (INOS,1969). Table 1 gives the relative location of the raingauge network (see also Figure 1). The rainfall amounts for especific duration and frequencies are given in Table 2. The mean length of record is 14,7 years with a smalest being 7 years and a largest being 21 years; 85% of the records are equal or longer than 10 years. On this basis a 10 years record was adopted for the analysis. The values of the rainfall intensity were transformed to depth values and given as a fraction or reduction factor (R) of the rainfall depth of 60 minutes.

The probability density funtions (pdf) of R for rainfall durations of 15, 30, and 45 minutes are given in Graphs 1, 2 and 3 respectively.

Table 1. Relative location of the raingauge network.

State	Number of raingauges
Anzoátegui	3
Aragua	6
Bolívar	2
Carabobo	10
Cojedes	1
Dto. Federal	10
Falcón	4
Guárico	4
Lara	3
Miranda	13
Mérida	2
Monagas	2
Nva. Esparta	3
Portuguesa	2
Sucre	4
Táchira	3
Trujillo	3
T.F. Delta Amacuro	1
Yaracuy	1
Zulia	4
TOTAL	81

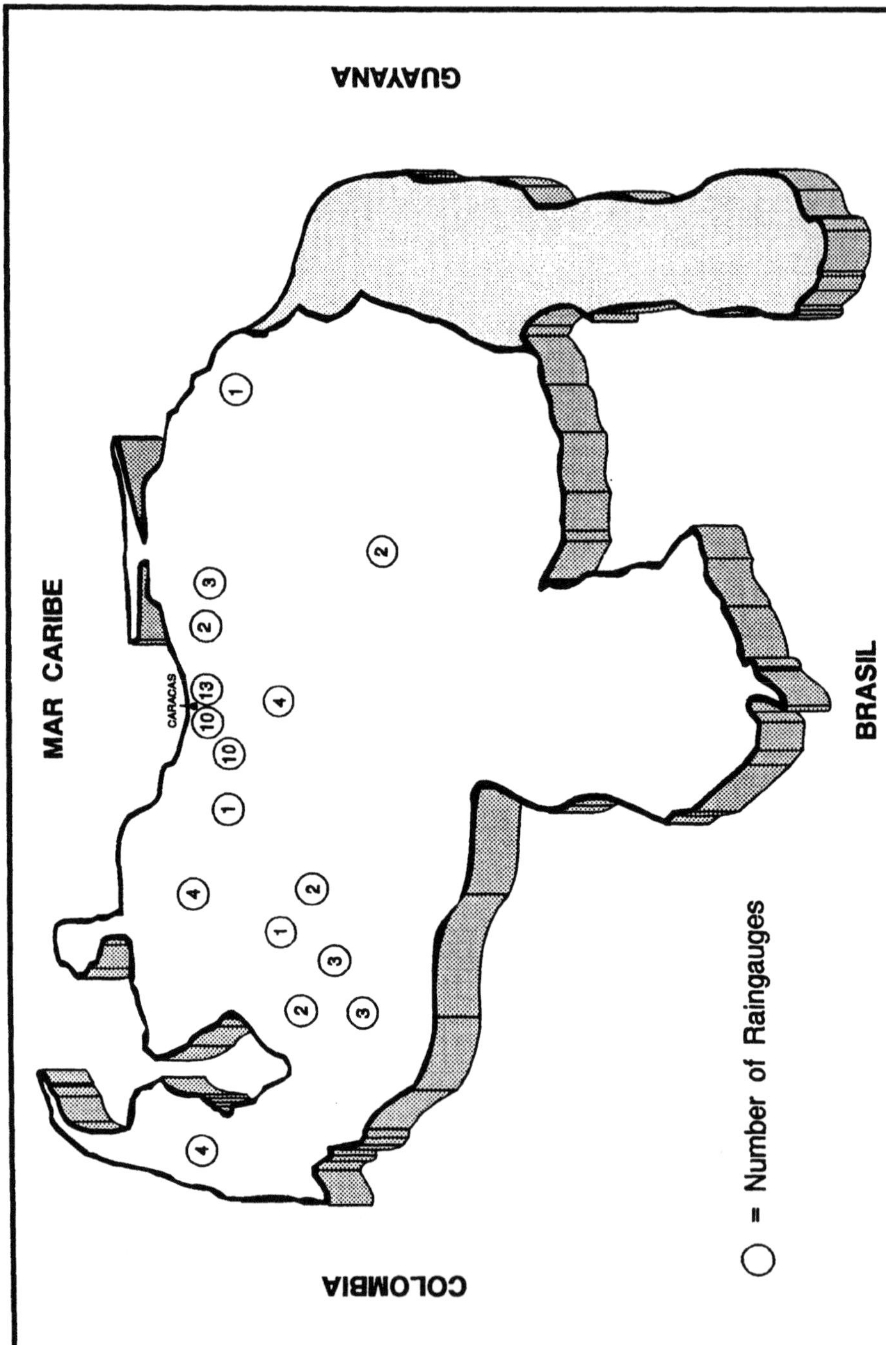

Figure 1. Relative location of raingauges

Table 2. Rainfall amounts for specific duration in min and frequence of 10 years.

State	Estation	period	Length	Duration 15	30	45	60
Az	El Tigre Las Mcedes	1956-66	11	132	95	78	66
Az	Pariguán	1951-67	17	140	98	79	69
Az	Pto. La Cruz	1949-66	18	101	71	53	45
Ar	Barbacoa	1961-67	7	100	71	58	48
Ar	La Cabrera	1952-67	16	100	81	69	58
Ar	Cagua	1959-67	9	98	78	60	50
Ar	Camatagua	1955-67	13	140	107	82	65
Ar	Guayabita	1955-67	13	140	98	76	60
Ar	La Victoria	1949-67	19	101	79	62	50
Bo	Puerto Ordaz	1953-67	15	101	80	60	46
Bo	Upata	1947-67	21	101	78	61	49
Ca	La California	1948-68	21	145	105	81	70
Ca	Campo Carabobo	1954-67	14	135	100	88	80
Ca	Guacara	1950-67	18	125	80	66	52
Ca	Guataparo Dique	1952-67	16	118	98	82	69
Ca	Montalbán	1954-67	14	100	79	62	49
Ca	Pto. Cabello	1948-67	20	97	82	70	54
Ca	San Diego	1960-67	8	107	70	51	42
Ca	El Torito	1947-67	21	130	93	82	71
Ca	Valencia	1947-67	21	135	105	74	61
Co	Vigirima	1951-67	17	117	81	69	56
Co	Cachinche	1962-69	8	117	88	64	55
DF	Caraballeda	1958-66	9	92	72	60	48
DF	Los Caracas	1954-67	14	125	91	79	62
DF	Caravaca	1956-64	9	142	115	82	68
DF	Caricuao	1953-66	14	89	68	55	45
DF	Catia	1954-67	14	90	71	52	41
DF	Cd. Universitaria	1948-67	20	101	70	52	44
DF	Macuto	1951-67	17	101	82	62	52
DF	Maiquetía	1948-67	20	88	69	60	49
DF	Las Mercedes	1949-67	19	135	96	80	62
DF	Montalbán	1959-67	9	89	67	54	44
Fa	Churuguara	1955-67	13	135	107	90	70
Fa	Pueblo Nuevo	1948-67	20	97	70	59	50
Fa	Puerto Cumarebo	1959-67	9	110	89	68	55
Fa	Punto Fijo	1953-67	15	130	105	76	67
Gu	Altagracia Oritoco	1956-67	12	140	99	68	53
Gu	Chaguaramas	1951-67	17	135	100	81	62
Gu	S. J. de Los Morros	1951-67	17	125	90	72	61
Gu	Zaraza	1951-67	17	142	105	88	64
La	Barquisimeto	1950-67	18	86	71	61	48
La	Carora	1950-66	17	108	90	69	52
La	Quibor	1958-67	10	128	83	67	54

Table 2.(Cont.)

Mi	Altos de Pique	1956-67	12	101	80	61	51
Mi	Baruta	1951-67	17	91	85	68	58
Mi	Caucagua	1949-65	17	127	107	75	64
Mi	Chacaito	1950-67	18	120	90	72	62
Mi	Charallave	1955-65	11	126	81	67	52
Mi	Cuarenas	1950-65	16	107	80	62	51
Mi	Guatire	1953-67	15	101	69	54	46
Mi	El Hatillo	1956-67	12	83	68	57	46
Mi	Lecherito	1958-67	10	114	80	61	51
Mi	Panaquire	1956-67	12	120	97	82	66
Mi	Petare	1948-67	20	115	80	62	50
Mi	Tacarigua	1961-67	7	105	88	71	56
Mi	Los Teques	1948-67	20	96	71	56	45
Me	Mérida	1950-67	18	91	72	58	44
Me	Timotes	1955-67	13	60	46	39	29
Mo	Maturín	1950-67	18	101	86	66	54
Mo	La Sabana	1860-67	8	100	78	51	40
NE	La Asunción	1952-66	15	82	69	54	45
NE	Juan Griego	1949-67	9	126	89	80	69
NE	Macanáo	1953-67	15	142	117	92	70
Po	Acarigua	1954-67	14	135	95	71	62
Po	Guanare	1952-67	16	127	95	82	70
Su	Carúpano	1953-67	15	94	78	62	55
Su	Cumaná	1950-67	18	125	100	70	57
Su	Cariaco	1958-69	12	150	110	88	71
Su	Qda. Seca	1959-67	9	150	115	102	85
Ta	Capacho	1956-67	12	84	62	52	45
Ta	La Grita	1950-69	20	80	51	47	34
Ta	San Cristobal	1951-67	17	87	75	55	45
Tr	Boconó	1956-67	12	87	64	49	38
Tr	Trujillo	1952-67	16	97	81	64	51
Tr	Valera	1950-67	18	135	90	71	59
TF	Delta Amacuro	1951-67	17	120	82	61	50
Ya	San Felipe	1950-67	18	145	120	98	73
Zu	Carretera N	1956-67	12	170	124	105	89
Zu	Maracaibo	1951-69	19	127	100	71	63
Zu	Plan Bonito	1956-67	12	137	112	88	76
Zu	El Venado	1956-67	12	155	115	90	70

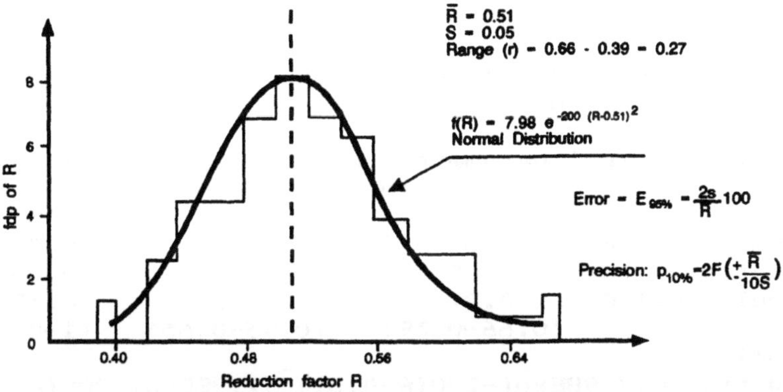

Graph 1. Probability density funtion of the reduction coefficient R from one hour rainfall depth to 15 minutes

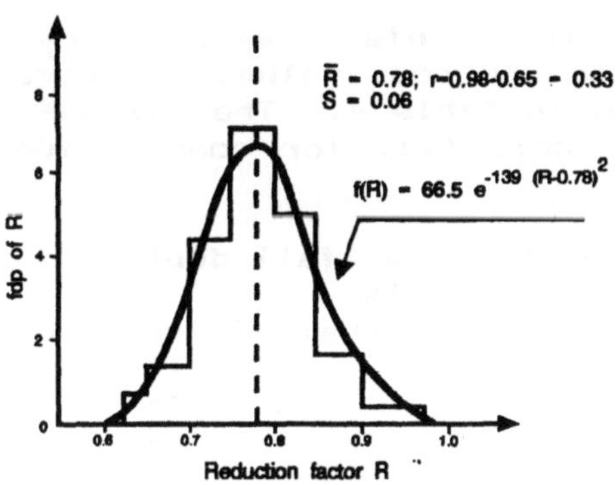

Graph 2. Probability density funtion of the reduction coefficient R from one hour rainfall depth to 30 minutes

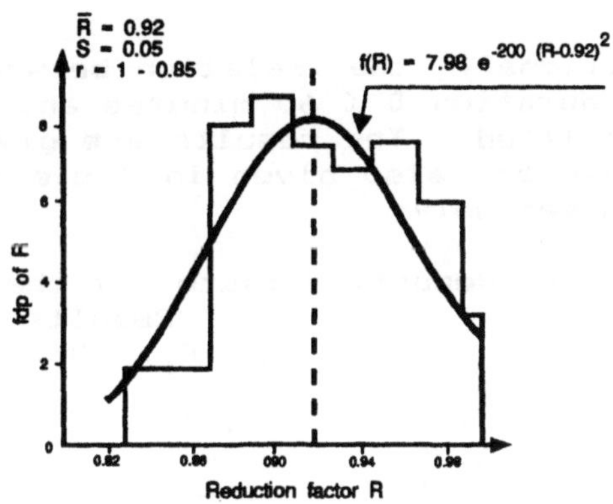

Graph 3. Probability density funtion of the reduction coefficient R from one hour rainfall depth to 45 minutes

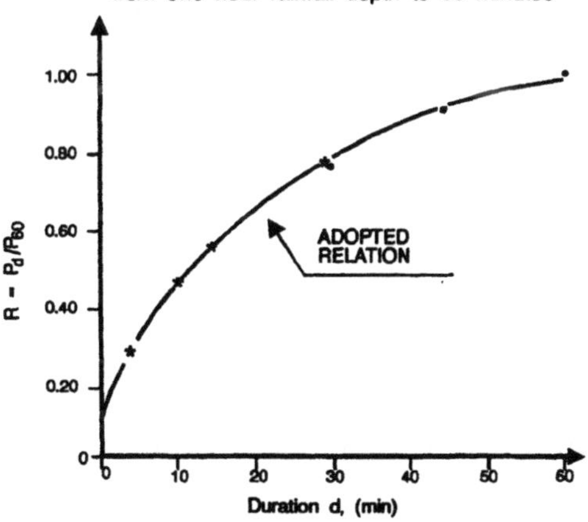

Graph 4. Graphical relation between duration and reduction coefficient R

d (min)	$R = P_d/P_{60}$ VENZ.	USWB
	●	*
5	–	0,29
10	–	0,45
15	0,51	0,57
30	0,76	0,79
45	0,92	–
60	1,00	1,00

RESULTS

From the pdf-curves of Graphs 1, 2, and 3, following information was extrated (Table 3).

Table 3. Statistical parameters of the reduction factor R

Paramater		Duration in minutes 15	30	45
Mean value	R	0.51	0.78	0.92
St. Desviation	S_R	0.05	0.06	0.05
Range (amplitude)	A	0.27 (0.66-0.39)	0.33 (0.93-0.65)	0.15 (1-0.85)
Math. model (Normal Dist. Assumed)	f(R)	$7.98Exp(-200(R-0.51)^2)$	$6.65Exp(-139(R-0.78)^2)$	$7.98Exp(-200(R-0.92)^2)$

Additionaly the relation between the rainfall amounts, P_D, for duration $D < 60$ minutes and the one hour value, P_{60} were calculated. The results are given in Table 4. The U.S.W.B. values are also given in Table 4, col. (2), for comparation purposes only.

Table 4. Reduction factor of the one hour rainfall depths

Duration (1)	USWB (2)	Results for Venezuela: $R = P_D / P_{60}$ (3)	95% of error (%) (4)	Level of Precision (%) (5)
5	0.29	-	-	-
10	0.45	-	-	-
15	0.57	0.51	20	69
30	0.79	0.78	15	81
45	-	0.92	11	94
60	1.0	1.0	-	-

Using the MSM analitical correlation following model was obtanied for the values of R in Table 4:

$$R = 0.14\ D^{0.49} \qquad (1)$$

$$r^2 = 0.98$$

Where $R = P_D / P_{60}$

P_D = Ranifall depth of duration D

P_{60} = Ranifall depth of one hour duration

D = Duration in minuts ($D < 60$)

The results are also presented graphically in Figure 4. The estimation errors range from 11% to 20% in the 95% of the analized cases. In the 68% of the cases, this errors are less than 10%. Col. (5) of Table 4 gives the probability that R of colum (3) varies in the range of - 10% and + 10%.

Additional effort was made to find a relationship including the USWB values together with those found for Venezuela. Following expresion was obtained:

$$R = 0.14 \quad D^{0.5} \quad \ldots\ldots\ldots\ldots\ldots\ldots \quad (2)$$

$$r^2 = 0.98$$

The similarity between equations (1) and (2) is obvious, confirming the initial assumption that the short rainfall events are independent of geographycal influences.

In order to compare equation (1) with the results obtanied for the world records following transformations were made (basic relations taken from (Hersfield, 1961):

Original equation (P in mm, D in hour)	Number of Equation	Transformed Equation (P in mm D in Munutes)	Number of Eq.
Paulhus $P = 421\ D^{0.475}$	(3)	$R = 0.14\ D^{0.475}$	(5)
Raudkivi $P = 389\ D^{0.486}$	(4)	$R = 0.14\ D^{0.486}$	(6)

Equations (5) y (6) strongly agree with equations (1) and (2) corroborating again the initial statment, allowing the use of universal empirical relationships as that developed for Venezuela.

CONCLUSIONS

The extreme rainfall event of short duration have a similiar time distribution pattern all over the world. Its values can be estimated using equation (1) with enough precision.

The found relationship is very usefull to estimate the design storm for small hydraulic works in Venezuela and other developing countries, spetially in regions where there is a lack of information.

REFERENCES

Avellan, H. y Porras, P., (1984) "Curvas de profundidad-frecuencia-duración". El Agua, No. 32. Revista Hidráulica. Marzo.
Carciente, J., (1980) "Carreteras, estudio y proyecto". pp. 338-345. Edc. Vega. Caracas.
Hersfield, D.M., (1961) "Rainfall frequency atlas of the united States", Weather Bureau.Techn.Paper No. 40.Washinton, D.C.
INOS. Div. Hidrología. (1969) "Análisis de intensidad-duración-frecuencia de lluvias en Venezuela". Caracas.
M.O.P., (1963) "Atlas de profundidad-duración-frecuencia de lluvias en Venezuela". Dirección de Obras Hidráulicas, Publicación Técnica No. 2. Caracas.
Raudkivi, A. J., (1979) "Hydrology", pp.82. Pergamon Press. London.

PROBABILITY OF MAXIMUM DISTRIBUTION RATE OF RAINFALL AND ITS APPLICATION TO DESIGN FLOOD AND RESERVOIR MANAGEMENT

K. TOMOSUGI and S. IKEBUCHI
Disaster Prevention Research Institute, Kyoto University
Gokasho, Uji, Kyoto, Japan 611

First, the results of analyses on some statistical properties of time distribution patterns of observed point rainfalls are summarized with a brief explanation on the adopted method which we call *the distribution-rate analysis*. Secondly, a method to estimate the joint and conditional probability distributions between a rainfall depth in a fixed period and a possible peak discharge is derived from the empirical probability distributions of *the maximum distribution-rate* of the moving sum type and the relation between a peak discharge and its concentration time based on the kinematic wave theory. Finally, a method to apply the estimated probability distributions to real-time flood control by a reservoir is shown. It is also shown that a peak discharge of the same return period as that of a design rainfall depth is obtainable in the estimation process of the probability distributions.

INTRODUCTION

The safety of the design of flood protection works such as embankment or spillways is in many cases in Japan evaluated indirectly by the return period of the total depth of *the design rainfalls* of the fixed duration that would be longer than the concentration times of flood runoffs in the catchment concerned, because the direct estimation of the return period of the peak discharge, which represents the actual safety, has some problems such as lack of data or statistical unstability of data caused by natural and artificial changes of flood runoff characteristics. In this design process, the designers should estimate the peak discharges applying a suitalbe runoff model to the design rainfalls that have different time distribution patterns of some observed heavy rainfalls adjusted to have the same total depth in the fixed duration by proportional distribution, and select one of them, e.g., the maximum or medium. In this situation it becomes a problem that the actual safety of the design, i.e., the return period of the selected peak discharge is not clear.

On the other hand, it is difficult to predict a time distribution of rainfall (hyetograph) that is necessary to predict the flood hydrograph or the peak discharge for a good management of a reservoir whose catchment area is relatively small(from 10 to 300km^2 or so), though it is relatively easy to predict roughly the rainfall depth in a time span(several hours to half a day or so) which would be longer than the consentration times of flood runoffs in such a catchment as stated above. In this case it seems regrettable that such information of the rainfall depth is not yet fully utilized because the lack of knowledge of the statistical properties of time distributions of

K. W. Hipel (ed.),
Stochastic and Statistical Methods in Hydrology and Environmental Engineering, Vol. 1, 163–176.

heavy rainfalls.

Considering these points, the authors have made the analyses of the statistical properties for the two types of rainfall distribution-rates, which represent some factors on the shape of a hyetograph normalized by its total depth, and have also studied on methods to apply the obtained results to such design and management problems as mentioned above in the 1970's(Ishihara et al., 1971, 1974; Tomosugi, 1977, 1978, 1980). The analyses were made mainly by comparing the observed statistical properties of *the maximum distribution-rates* of *the elementary type* and *the moving sum type* with the expected ones in a mathematical model which we call *the random distribution model.* These terms will be defined or explained later.

The concept of the distribution-rate of rainfall was first introduced and investigated by Ishihara et al.(1960) using a mathematical model which is the same one as ours in essence, however they failed to obtain the general form of the maximum distribution-rate of the elementary type in the mathematical model. After about ten years, the first author took over their work and succeeded in getting the general form through the different approach(Ishihara et al., 1971), i.e., through the analysis of *the occupancy problem*(see Feller, 1957) regarding rainfall units as undistinguishable balls and equally divided sections of rainfall time as distinguishable cells. Later, we found that the same general form had been found by Fisher(1929) under the different purpose from ours through the analysis on the volumes in the multi-dimensional Euclidean space which are the same approach as that of Ishihara et al.(1960) and that of our subsequent studies on the moving sum type.

The authors intend to present the essence of these somewhat old works in this opportunity, because they have never been published in English. The abstract is as follows:

First, after a brief explanation of the results of analyses on *the random distribution model* is given, the results of analyses on some statistical properties of time distribution pattern of observed point rainfalls in the Japan's Kinki District, especially on persistency or degrees of rainfall concentration in time, are summarized.

Secondly, a method to estimate the joint and conditional probability distributions between a rainfall depth in a fixed period and a possible peak discharge is shown using the empirical probability distributions of *the maximum distribution-rate* of the moving sum type obtained from the data of 12-hour-rainfalls in the Japan's Kanto District, and the relation between a peak discharge and its concentration time which was analyzed on the basis of the kinematic wave theory and obtained numerically in the upper basin of the Yura River by Ishihara et al.(1962). Some examples of the estimation are also shown.

Finally, a method to apply the probability distributions estimated above to a real-time flood control in an imaginary reservoir for the event of a predicted 12-hour-rainfall depth is used as an example. Moreover, for an application to design phase, it is shown that a peak discharge of the same return period as that of a design rainfall depth is obtainable in the same process as that of the estimation of the probabitity distributions mentioned above.

DISTRIBUTION-RATE ANALYSIS

Summary of the mathematical model analysis

Let us consider a rainfall of a fixed period T and its total depth R_T. If the period T is equally divided into n sections and $\{r_j\}(j = 1, 2, \ldots, n)$ denotes the time series of each rainfall depth distributed to the equally divided sections, the value $x_j = r_j/R_T$

is called the distribution-rate in n equally divided sections and clearly we have

$$x_1 + x_2 + \cdots + x_j + \cdots + x_n = 1 \qquad (x_j \geq 0) \tag{1}$$

All the points satisfying (1) exist on the $(n-1)$-dimensional Euclidean space of the volume $V_{n-1} = 1/(n-1)!$. When these points have an equal probability, we call this equation *the random distribution model.* Then, a probability density function(p.d.f.) of a variable is derived from dividing the volume V_{n-2} where the variable exists by the volume V_{n-1} mentioned above. In this way, the p.d.f.'s of some variables of our interest were obtaind as follows.

The elementary type:

When the variables $x_{(i)}$ $(i = 1, \ldots, n)$ denote the nonincreasing ordered one in $\{x_j\}$ $(j = 1, \ldots, n)$, that is, $x_{(1)}$ denotes the maximum one in all x_j for example, we have

$$f_n(x_{(i)}) = \textstyle\sum_{m=i}^{M}(-1)^{m-1}m(n-1)\binom{n}{m}\binom{m-1}{i-1}(1-mx_{(i)})^{n-2} \tag{2}$$

where M is the maximum integer not greater than $1/x_{(i)}$ and varies to $(n-1)$ for $i = 1$ and to n for $i \geq 2$. The region of their existence is $1 \geq x_{(1)} \geq 1/n$ and $1/i \geq x_{(i)} \geq 0$ for $i \geq 2$. We call $x_{(1)}$ the maximum distribution-rate of the elementary type and the parentheses in the subscripts is occasionally omitted for simplicity.

The 1st and 2nd moments of $x_{(i)}$ $(i = 1, 2, \ldots, n)$ consequently become as follows by using some binomial expansion formula and their derivatives, respectively

$$E_n(x_{(i)}) = \frac{1}{n}\sum_{m=i}^{n}\frac{1}{m} \quad (3), \qquad E_n(x_{(i)}^2) = \frac{2}{n(n+1)}\sum_{m=i}^{n}\frac{1}{m}\sum_{k=i}^{m}\frac{1}{k} \quad (4)$$

The moving sum type:

Let us consider the variable $y_{l,n}$ $(l = 2, \ldots, n-1)$ defined by

$$y_{l,n} = \max(x_1 + \cdots + x_l, x_2 + \cdots + x_{l+1}, \ldots, x_{n-l+1} + \cdots + x_n) \tag{5}$$

which we call briefly the maximum distributon-rate of the moving sum type. It is the point that the variable of this type may reflect serial properties of time series unlike those of the elementary type.

The general form of the p.d.f. of this variable was obtained only for $1 \geq y_{l,n} \geq 1/2$ as shown below, though complete solutions were obtained for the cases of $n \leq 6$,

$$g_{n,l}(y) = \textstyle\sum_{m=1}^{k} m(m+1)\binom{2k-m}{k}\binom{n-1}{2k-m}(1-y)^{2k-m-1}(2y-1)^{n-2k+m-1} \tag{6}$$

where $k = n - l$. The subscripts of y is hereafter occasionally omitted for simplicity. The expectation(the 1st moment) of y for $l \geq n/2$ is obtainable in spite of the incompleteness of (6), because the existence region of y is $1 \geq y \geq 1/2$ for the case. The result is

$$E_{n,l}(y) = \frac{l}{n} + \frac{1}{2n}\sum_{m=1}^{k}\frac{m^2(m+1)}{(2k-m)2^{2k-m}}\binom{2k-m}{k} \tag{7}$$

Incidentally, if $l = 1$, y is equvalent to $x_{(1)}$ from their definitions, so $x_{(1)}$ will be treated as one of the maximum distribution-rates of the moving sum type.

Summary of the data analysis

The abstract of the rainfall data:

The data used comes from information gathered from 6 points in Japan's Kinki District where hourly rainfalls have been observed for more than 20 years(the maximum is 64 years at Kyoto). The data sets of 12-hour-rainfalls were made by the following criteria; 1) a duration is longer than 11 hours, 2) a zero rainfall in a one hour does not continue more than 1 hour, and is not included more than once, 3) no overlap in time. The total numbers of the elements of the data set in each point were 140 to 881(at Kyoto).

The 12-hour-rainfalls were classified by their total depths R_{12} into three classes; Large($R_{12} > 40$ mm), Medium($40 \geq R_{12} > 20$ mm), and Small($20 \geq R_{12} > 0$ mm).

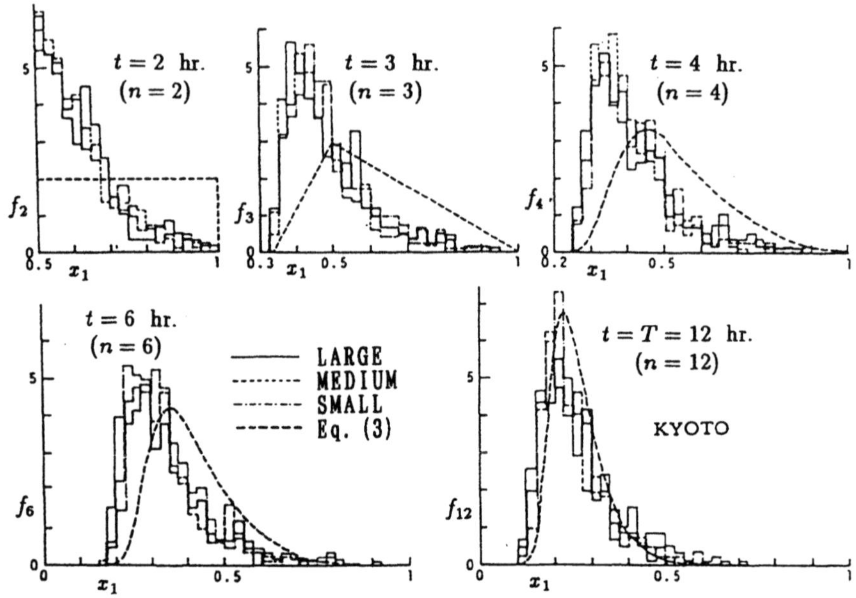

Figure 1. Comparisons between distributions of x_1 in each class of R_{12}, with each one given by letting $i = 1$ and $n = t$ in Eq.(2). In this case, t denotes a length of *the maximum rainfall sub-period.*

Matters for investigation:

1) Degrees of difference between the observation points on the probability distributions or the means of the maximum distribution-rates of the elementary type(x_1) and of the moving sum type(y).
2) Degrees of difference between the classes of R_{12} mentioned above on the same items as in 1).
3) Statistical properties of time distribution in the sub-periods to which the maximum rainfall of the moving sum type is distributed. We call such a period *the maximum rainfall sub-period.*
4) Degrees of difference between the means of the maximum distribution-rate of the moving sum type which can be estimated from point rainfall data of a recording unit of one hour and those which would be obtained from instantaneous rainfall records.

Point 3) was investigated by analyzing the maximum distribution-rates of the elementary type on the maximum rainfall sub-period. An example of the result of this analyis is shown in Figure 1.

Point 4) was carried out by analyzing the empirical relations between the time units $u(\geq 1)$ (hr) and the means of y obtained as time unit u , and by extrapolating the relations to a time unit shorter than one hour.

The main results corresponding to the matters above:

1) The maximum differnce of the means of x_1 and of y was 11% and 6%, respectively.
2) The same was 20% and 9%, respectively. Therefore, it is concluded that the effect of R_{12} is larger than that of a point, especially on the means of x_1.
3) Degrees of concentration into short time spans are cosiderably less than those in the case of the random distribution model(see Figure 1). In other words, persistency in the maximum rainfall sub-period becomes more remarkable with the shorter time length.
4) Ratio of the latter to the former will be 1.1 $\sim$ 1.2 for sub-period of 1 hour, but becomes nearly equal to 1.0 when the sub-period is longer than 3 $\sim$ 4 hours.

ESTIMATION OF PROBABILITY DISTRIBUTIONS BETWEEN RAINFALL DEPTH AND PEAK DISCHARGE

Abstract of the basic matters for composing the method

The relation between a peak discharge and its concentration time:

It was found by Ishihara et al.(1962) and has been verified by many hydrologists that the next relations (8) (known as *the rational formula*)and (9) approximately hold in such a relatively small basin(less than 300km^2 or so) that the runoff delay effect of a channel-network system is negligible in comparison with that of mountainside slopes and the spatial distribution of rainfall is nearly uniform.

$$Q_p = r_{mp} \cdot A/3.6 \tag{8}$$

$$t_{pc} = B \cdot r_{mp}^{p-1} \tag{9}$$

where A: basin area (km^2), Q_p: peak discharge of direct runoff (m^3/sec), r_{mp}(mm/hr): mean intensity of effective rainfall in t_{pc}(hr), concentration time of peak discharge, 3.6: coefficient of unit conversion. And $B = KL^p$ where K: parameter related to roughness and inclination of mountainside slope, L: representative length of basin slope, and p: parameter ranged in $0 < p \leq 1$ and $p = 0.6$ under the Manning's resistance law.

The cummulative distributions of y:

When R_{tm} denotes the rainfall depth in the maximum rainfall sub-period of length t (hr) derived from a T-hour-rainfall of total depth R_T, we have

$$y = R_{tm}/R_T \tag{10}$$

and the next relation generally holds for any time unit.

$$1 \geq y \geq 1/n \qquad (\text{ for } 1/(n-1) > t/T \geq 1/n) \tag{11}$$

where $n = 2, \ldots, T$ in this case, because recording time unit is 1 hour.

Although the cumulative distributions of this y obtained from the hourly rainfall data must be as smooth as possible for application purposes, such a large number of data could not be obtained only in a small basin. We obtained the smooth cumulative distributions of y of 12-hour-rainfalls as follows: the data from 38 rainstorms were collected from about 100 points in the relatively large basin, and were analyzed together as one set of samples. The number of the samples(12-hour-rainfalls) was 1748. The resut is shown in Figure 2, where $F_T(y \mid t)$ denotes the empirical cumulative distribution function(c.d.f.).

The conditional existence range of rainfall depth and peak discharge:

The existence range of a possible peak discharge to a given rainfall depth and of the reverse and the point of this method will be shown here.

Firstly, we assume that observed rainfalls are nearly equal to the effective rainfalls because we are interested in peak discharges caused by heavy rainfalls. Then we can use directly the relations (8) and (9) without any modification, and may discuss r_{mp} instead of Q_p.

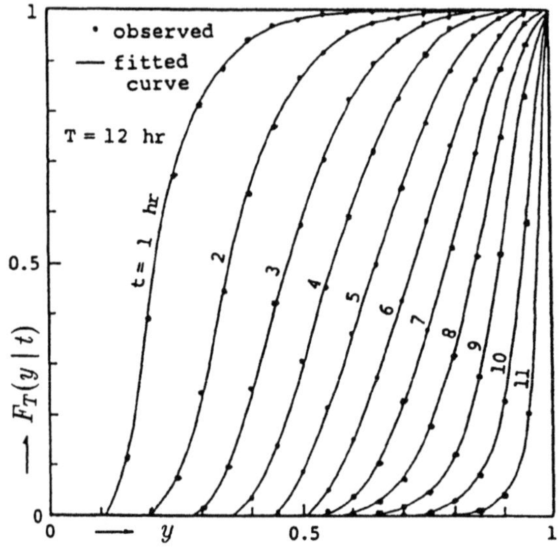

Figure 2. The empirical c.d.f. of y, on the 12-hour-rainfall data obtained at the upper Tone River basin.

Now let us consider the condition of occurrence of a peak discharge. We notice that the rianfall depth of $r_{mp} \times t_{pc}$ must be equal to R_{tm} for $t = t_{pc}$. Therefore, for $T > t_{pc}$ we have

$$y = t_{pc} \cdot r_{mp}/R_T \qquad (12)$$

Taking account of (9) we have consequently

$$r_{mp} = (yR_T/B)^{\frac{1}{p}} \qquad (13)$$

If we consider the existence range of y for $t = t_{pc}$ given by (11), we get the coaxial diagram illustrated in Figure 3, which was made for $T = 12$ (hr) by letting $p = 0.6$, $B = 17.7(\mathrm{mm}^{-0.4}\mathrm{hr}^{0.6})$ referring to Ishihara et al.(1962). In this figure, the segment $\overline{\mathrm{ab}}$ parallel to the r_{mp}-axis shows the existence range of r_{mp} at the R_T and the segment $\overline{\mathrm{cd}}$ shows the reverse. These segments correspond to the arc $\widehat{\mathrm{a'b'}}$ and to the segment $\overline{\mathrm{c'd'}}$ in the quadrant of $y - t$, respectively. Therefore, point e corresponds to point e'

The upper limit r_{mpu} and the lower limit r_{mpl} of the r_{mp} are given as follows.

$$r_{mpu} = (R_T/B)^{\frac{1}{p}} \tag{14}$$

$$r_{mpl} = (R_T/nB)^{\frac{1}{p}} \quad \left(\text{for } nB\{(n-1)B/T\}^{\frac{p}{1-p}} \le R_T < nB(nB/T)^{\frac{p}{1-p}} \right) \tag{15}$$

$$r_{mpl} = (nB/T)^{\frac{1}{p}} \quad \left(\text{for } nB(nB/T)^{\frac{p}{1-p}} \le R_T < (n+1)B(nB/T)^{\frac{p}{1-p}} \right) \tag{16}$$

where n is equivalent to that in (11).

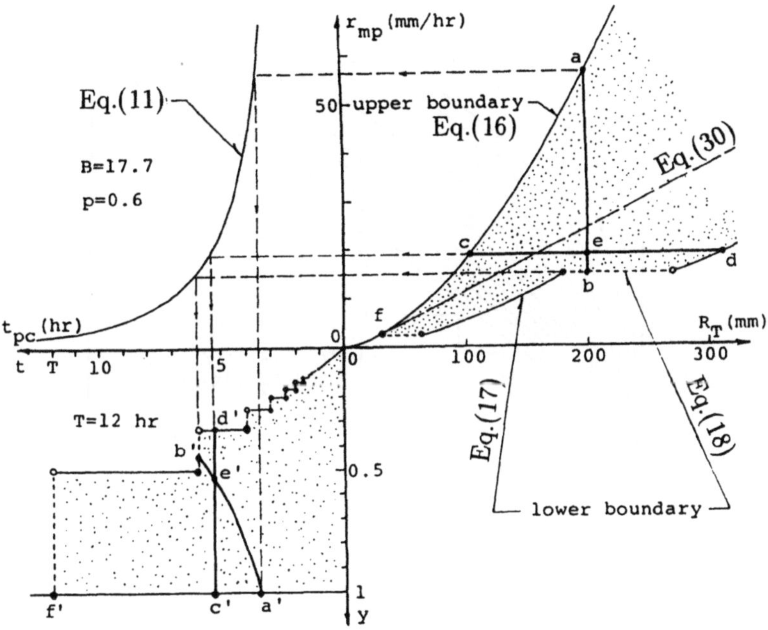

Figure 3. Illustration of the existence range of points (R_T, r_{mp}) and (y, t), and the relation between them.

Estimation method of probabilistic relations between R_T and r_{mp}

Let us consider the joint p.d.f $g(R_T, r_{mp})$, the conditional p.d.f.'s $h_1(R_T \mid r_{mp})$, and $h_2(r_{mp} \mid R_T)$. It is generally difficult to estimate these p.d.f.'s by the common statistical analysis because of the lack of data. On the other hand, with the marginal p.d.f.'s $g_1(R_T)$ and $g_2(r_{mp})$, it may be relatively easy in Japan. But even with the latter marginal p.d.f, it is difficult in such a basin that has as little data of the peak discharges as many of the small and minor basins.

Considering these conditions the estimation methods based on (13) are shown below for each unknown p.d.f. seperately.

The estimation of $h_1(R_T \mid r_{mp})$:

From Figure 3 it is clear that a point on $\overline{\text{cd}}$ corresponds to one point on $\overline{\text{c'd'}}$. Therefore we have

$$h_1(R_T \mid r_{mp}) = f_T(y \mid t) \cdot |\partial y/\partial R_T| = f_T(y \mid t) \cdot B \cdot r_{mp}^p / R_T^2 \tag{17}$$

where $f_T(y \mid t)$ denotes the conditional p.d.f of y at $t(= t_{pc})$. Because this $f_T(y \mid t)$ can be estimated from $F_T(y \mid t)$ shown in Figure 2, $h_1(R_T \mid r_{mp})$ can be estimated by this equation. However, values of $F_T(y \mid t)$ for any continuous value of t must be obtained by some interpolation method. Figure 4 was made from Figure 2 for the interpolation.

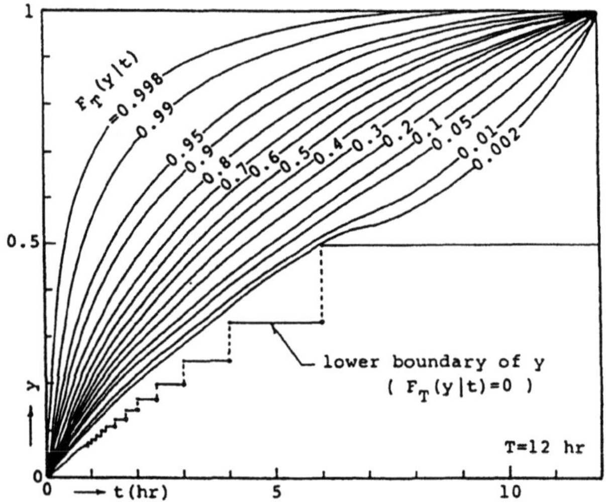

Figure 4. Iso-probability lines obtained from Fig.2 for the interpolation of $F_T(y \mid t)$.

The estimation of $h_2(r_{mp} \mid R_T)$ and $g(R_T, r_{mp})$ in the case that both $g_1(R_T)$ and $g_2(r_{mp})$ are known:

In this case, we must use the well-known relation (18) shown below, because the mapping of the segment $\overline{ab}$ to the $y-t$ plane is not parallel to y-axis unlike that of $\overline{cd}$ as shown in Figure 3.

$$g_1(R_T) \cdot h_2(r_{mp} \mid R_T) = g_2(r_{mp}) \cdot h_1(R_T \mid r_{mp}) = g(R_T, r_{mp}) \tag{18}$$

From this we have

$$h_2(r_{mp} \mid R_T) = g_2(r_{mp}) \cdot h_1(R_T \mid r_{mp}) / g_1(R_T) \tag{19}$$

Then h_2 can be estimated by using h_1 obtained by (17) and consequently g can be estimated by (18).

The estimation of $h_2(r_{mp} \mid R_T)$ in the case that only $g_1(R_T)$ is known:

This problem will be the most general and difficult one and is the focus of our interest. In this case, adding to (18) we must use the next relation (20) which must hold for p.d.f.

$$g_1(R_T) = \int_{rmpl}^{rmpu} g(R_T, r_{mp}) \cdot dr_{mp} \tag{20}$$

where r_{mpu} and r_{mpl} are the upper and lower limits previously defined by (14), (15) and (16). Then using (18) we have

$$g_1(R_T) = \int_{rmpl}^{rmpu} h_1(R_T \mid r_{mp}) \cdot g_2(r_{mp}) \cdot dr_{mp} \tag{21}$$

This is an integral equation belonging to the Fredholm type of the 1st class. An integral equation is commonly solved by the numerical method. That is, in this case, the method of letting the discrete values of g_1, g_2 and h_1 denote Δg_{1i}, Δg_{2j} and Δh_{1ij} ($i = 1, \ldots, m; j = 1, \ldots, m$), respectively, and solving the next simultaneous linear equation

$$\Delta g_{1i} = \textstyle\sum_{j=1}^{m} \Delta h_{1ij} \cdot \Delta g_{2j} \qquad (i = 1, 2, \ldots, m) \tag{22}$$

This method failed to obtain such a rational solution that satisfies the condition of probability(nonnegative values) because of the type of the integral equation. Another method was tried to succeed to some sufficient extent. The method is a kind of trial end error method as follows:
First, after the same conditions as those in Figure 3 about p, B and T were selected, the marginal c.d.f. of the annual maximum of R_{12} was estimated as following the Gumbel distribution

$$G_1(R_{12}) = \exp\{-\exp(-0.0245R_{12} + 1.6852)\} \tag{23}$$

Secondly, the marginal c.d.f. of the annual maximum of r_{mp} is assumed to be the Gumbel type as follows

$$G_2(r_{mp}) = \exp\{-\exp(\alpha \cdot r_{mp} + \beta)\} \tag{24}$$

Although there are two unknown parameters α and β in this formula, the degrees of freedom to determine them reduce to one, because $G_1(R_{12})$ must be equal to $G_2(r_{mp})$ at the point f in Figure 3. Lastly, to find the value of the parameter by the trial and error method, the following criterion was adopted. That is to minimize

$$\sigma_a^2 = \textstyle\sum_{i=1}^{N}(1 - \sum_{j=li}^{ui} \Delta\tilde{h}_{2ji})^2/N \tag{25}$$

where $\Delta\tilde{h}_{2ji}$ denotes the tentative value of Δh_{2ij} which is obtaned by substituting an assumed value of Δg_{2i} into the discrete expression of (19)

$$\Delta h_{2ij} = \Delta g_{2j} \cdot \Delta h_{1ij}/\Delta g_{1i} \tag{26}$$

The results of the estimations

The marginal c.d.f. of r_{mp} :

The following c.d.f. was obtained in the trial and error method mentioned above.

$$G_2(r_{mp}) = \exp\{-\exp(-0.1916r_{mp} + 1.4138)\} \tag{27}$$

This estimation was judged good enough because the σ_a^2 was 0.0021.

Incidentally, from this formula and (23), the locus of a point (R_{12}, r_{mp}) which satisfies $G_1(R_{12}) = G_2(r_{mp})$ becomes the following straight line

$$r_{mp} = 0.128R_{12} - 1.418 \tag{28}$$

This locus will take an important role in the desin phase.

The conditional p.d.f. of r_{mp} to a given r_{mp} :

Some examples of $h_2(r_{mp} \mid R_{12})\Delta r_{mp}$ obtained by the trial and error method are shown in Figure 5. These were redrawn by setting $\Delta r_{mp} = 0.5$ mm/hr after fitting smooth curves to the c.d.f of h_2 obtained by setting $\Delta R_{12} = 5$ mm, $\Delta R_{tm} = 2.5$ mm. These p.d.f. or c.d.f will take very impotant roles both at the design phase and at the management phase of flood control.

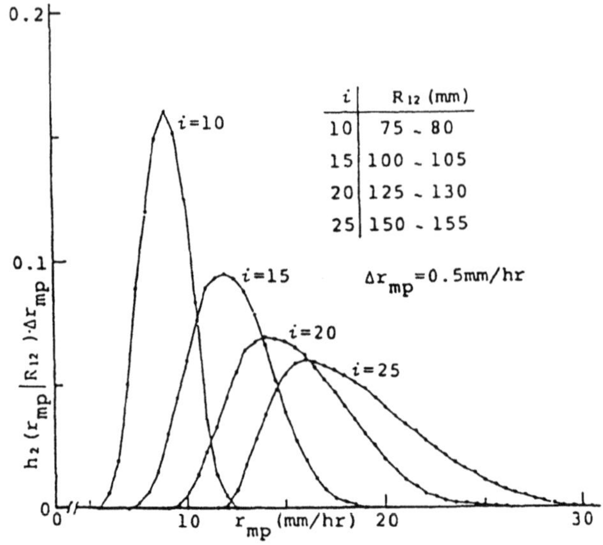

Figure 5. Examples of discrete expression of the conditional p.d.f. of r_{mp} to R_{12}.

APPLICATION TO A REAL-TIME FLOOD CONTROL BY A RESERVOIR

Assumption of shape of flood hydrograph

The assumed shape of flood hydrograph given the condition that R_T (mm) in T (hr) and Q_p(m^3/sec) are given is shown in Figure 6. Then from (8) and (9), t_{pc} (hr) becomes

$$t_{pc} = B(3.6Q_p/A)^{p-1} = Br_{mp}^{p-1} \tag{29}$$

Further, by assuming that the volume of this triangular hydrograph is equal to $R_T A$, we have

$$t_b = 2R_T A/(3.6Q_p) = 2R_T/r_{mp} \tag{30}$$

These relations show that such a triangular hydrograph is determined only by R_T and Q_p (or r_{mp}).

Criterion for making a change in the constant-rate release rule

Let us consider the case wherein the constant-rate release rule, which is usually adopted as a flood control rule by Japanese resevoirs, is applied to the assumed hydrograph as shown in Figure 6. In this figure, Q_0 means the discharge to begin the control and Q'_p represents the peak discarge after the control. In this case the control-rate c, is represented by

$$c = (Q'_p - Q_0)/(Q_p - Q_0) \tag{31}$$

and taking account of (29) and (30) the flood water volume stored in the reservoir $V_s(\mathrm{m}^3)$ is estimated by

$$V_s = 1800(1-c)(1-Q_0/Q_p)^2\{2(1-c)R_TA/3.6 + cB(A/3.6)^{1-p}Q_p^p\} \tag{32}$$

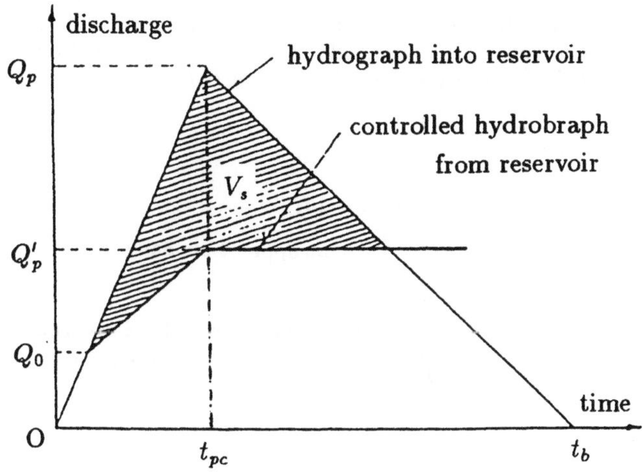

Figure 6. Definition sketch of the assumed flood hydrograph and the control rule for it.

Now let us consider the case in which a rainstorm with R_T (mm) is forecasted just before or at the beginning of a flood in a catchment of a reservoir. At the time, if there exists some vacant capacity(or available volume) $V_a(\mathrm{m}^3)$, the intended control as shown in Figure 6 is possible only if $V_a \geq V_s$. And the critical boundary is represented by

$$V_a = V_s(c, Q_0, Q_p, R_T) \tag{33}$$

If c and Q_0 are both fixed to some constants as is the present common rule in Japan, this critical line can be drawn on the $R_T - Q_p$ plane for any V_a. On the other hand, when Q_{pcr} denotes the critical discharge allowed at the downstream, $Q'_p \leq Q_{pcr}$ must hold for safety. Considering this critical line $Q'_p = Q_{pcr}$ with (31) and (33), we have

$$V_a = V_s(Q_0, Q_p, R_T) \tag{34}$$

From this relation the critical line can be also drawn on the $R_T - Q_p$ plane for any V_a if Q_0 is fixed. This line means that V_a are most effectively used in storing the flood by the control rule as shown in Figure 6.

Now it becomes possible to estimate such a probability that a point (R_T, Q_p) will lie over the critical line (33) or (34), using the conditional and joint p.d.f. and c.d.f. of Q_p and R_T estimated by the method shown above. Such probability will become a criterion for making a change of the constant-rate release rule or for warning.

Numerical examples

According to the above consideration, some computations on an imaginary reservoir were carried out by adding new conditions: the catchment area $A = 360(\mathrm{km}^2)$, the effective storage volume of the reservoir $V_e = 20(10^6\mathrm{m}^3)$, $Q_{pcr} = 1200(\mathrm{m}^3/\mathrm{sec})$, and $Q_0 = 0$ for simplicity. On the other hand, it is assumed that the forecasted rainfall R_{12} lies in the range of $100 \sim 150$ (mm) with an equal probability.

Some examples of the critical lines on the $R_T - Q_p$ plane are drawn in Figure 7. In this figure, the light broken curves show the variation with c for $V_a = 15$, the light dots-dash-lines show the variation with V_a for $c = 0.5$ both on (33). The three heavy solid curves show the variation with V_a on (34), and the heavy dot-dash-curve shows the critical line for $V_a = 20$ for the ideal peak cut($c = 0, Q_0 = Q'_p$) to compare with others.

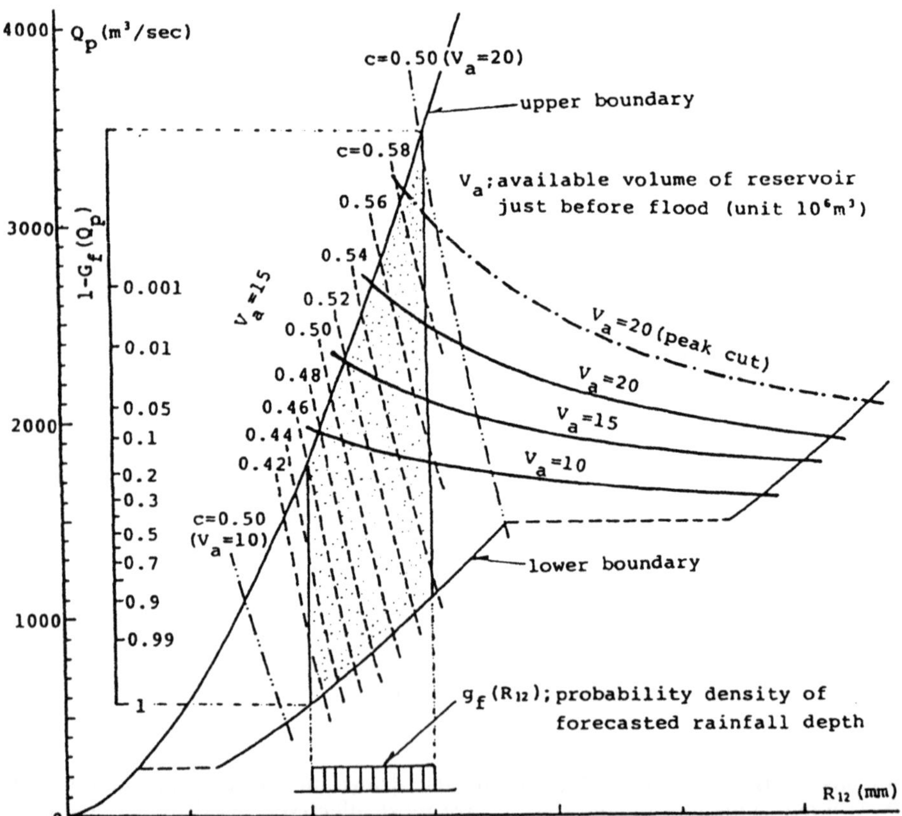

Figure 7. Some examples of the critical lines for the estimation of probability of flood risk in reservoir operation based on a forecasted rainfall depth with some range.

Estimations of probability of the risk that point (Q_p, R_T) lie over those critical lines were done through the estimation of the joint c.d.f. by means as shown above in the shaded region(the existence region of point (Q_p, R_T)). The results were as follows. In the case of (33) and $V_a = 15$, the pair (c, p) was as (0.5, 0.55), (0.54, 0.18), and (0.56, 0.03). This result suggests that the effect of change of control-rate c on probability of risk is remarkable. In the case of (34), $p = (0.143, 0.034, 0.004)$ was obtained for $V_a = (10, 15, 20)$, respectively. This shows the importance of the available volume. In the case of the ideal peak cut, $p < 10^{-4}$ was obtained for $V_a = 20$. This result suggests the necessity of fine prediction of flood hydrograph. The exceedance probability $1 - G_f(Q_p)$ which is obtained in this estimation process is also shown with the p.d.f.of forecasted R_{12}.

CONCLUDING REMARKS

The distribution-rate analysis combined with the mathematical model(the random distribution model) analysis is useful to find the statistical properties of the time distribution pattern of heavy rainfalls, such as persistency or degrees of concentration in time, and their differences between regions, levels of magnitude, and meteorological causes of rainfall.

The empirical probability distributions of the maximum distribution-rate of the moving sum type combined with the relation between a peak discharge and its concentration time have the possibility to estimate the conditional and joint probability distributions between rainfall depth in a fixed period and peak discharge in a relatively small river basin.

The estimated conditional and joint probability distributions are useful both for the estimation of the actual safety of flood protection works, and for more effective utilization of a forecasted rainfall depth in regulating the flood at a reservoir.

REFERENCES

Feller, W.(1957) "An Introduction to Probability Theory and Its Application", John Wiley & Sons Inc., Vol.1, 36-37.

Fisher, R.A.(1929) "Tests of significance in harmonic analysis", Proc. of the Royal Society of London, Series A, 125, 796, 54-59.

Ishihara, T., Y. Ishihara, T. Takasao and C. Rai(1962) "A study on the runoff characteristics of Yura river basin", Annuals of Disaster Prevention Res. Inst., 5-A, 147-173 (in Japanese with English abstract and captions).

Ishihara, Y. and K. Tomosugi(1971) "A study on statistical laws of time distribution of a rainfall", Annuals of Disaster Prevention Res. Inst., 14-B, 87-102 (in Japanese with English abstract and captions).

Ishihara, Y. and K. Tomosugi(1974) "A study on flood control by reservoir: reservoir operation based on forecasted rainfall depth", Annuals of Disaster Prevention Res. Inst., 17-B, 479-491 (in Japanese with English abstract and captions).

Ishihara, Y. and M. Kadoya(1960) "A study on time distribution of rainfall based on the probability theory", Lecture note of the 5th hydraulic research meeting (in Japanese).

Tomosugi, K.(1977) "A Study on statistical structure of time distribution of rainfall: characteristics of probability distributions of the distribution rates of rainfall", Annuals of Disaster Prevention Res. Inst., 20-B, 101-118 (in Japanese with English abstract and captions).

Tomosugi, K.(1978) "A study on statistical characterisyics of time distribution of areal average rainfall: on the maximum distribution rates to any sub-period", Annuals of Disaster Prevention Res. Inst., 21-B, 137-152 (in Japanese with English abstract and captions).
Tomosugi, K.(1980) "A study on statistical structures of time distribution of heavy rainfalls: analyses of 10-minutely records by the concept of distribution-rates and runs", Annuals of Disaster Prevention Res. Inst., 23-B, 133-155 (in Japanese with English abstract and captions).

ON ANALYTICALLY DERIVED DISTRIBUTION FUNCTIONS FOR STORM VOLUME

V-T-V. NGUYEN and H. WANG
Dept. of Civil Engineering and Applied Mechanics
McGill University
Montreal, Quebec, Canada H3A 2K6.

A stochastic method is proposed to determine analytically the probability distribution function for storm volume. The volume of a storm is considered as the sum of an unbroken sequence of consecutive hourly rainfalls above a given truncation level. Analytical expressions were then derived for the exact distribution of storm volume for various cases in which successive rainfall exceedances can be assumed to be either independent or correlated, and the cumulative rainfall exceedance can be considered as independent or dependent of the storm duration, respectively. The proposed stochastic method is more general and more flexible than empirical fitting approaches because it can take explicitly into account different stochastic properties of the underlying rainfall process. Results of a numerical application have demonstrated the adequacy of the suggested method .

INTRODUCTION

Knowledge of the probability distribution of total rainfall amount within each individual storm event is essential for successful design and planning of various water resources projects. For instance, when designing storage elements such as reservoirs or overflow structures in urban drainage systems, or when estimating the pollutant concentration from urban storm runoff, one may wish to know, given the occurrence of a storm, what is the total amount of water that could be expected during the total storm duration (see, e.g., Howard, 1976; Etoh, 1992). One common approach is to select a theoretical distribution function which may appear to fit the observed frequencies of storm volume data. Parameters of such empirical distribution functions are properties of the sample and, usually, have no physical interpretation. The alternative approach is to derive analytically a distribution function for storm volume by taking into account some observed stochastic properties of the underlying rainfall series. Such analytical procedure is thus more general and more flexible than the empirical fitting method.

K. W. Hipel (ed.),
Stochastic and Statistical Methods in Hydrology and Environmental Engineering, Vol. 1, 177–183.

Previous investigations, however, have usually relied on simplistic assumptions to derive analytically distribution functions for storm volume. For instance, Eagleson (1972) assumed that both storm duration and average rainfall intensity are exponentially distributed, but the correlation between these two variables was not explicitly considered in the derivation of the storm volume distribution. Cordova and Rodriguez–Iturbe (1985) employed a bivariate exponential distribution for storm duration and average rainfall intensity and demonstrated the significant effect of their correlation on the probabilistic structure of storm surface runoff. Etoh and Murota (1986) have simplified the analytically–derived distribution for storm volume by assuming that the rainfall hyetograph is triangular in shape. Hence, the main objective of the present study is to propose a general stochastic model for determining the exact probability distribution of storm volume. The model suggested in this study can take explicitly into account various correlation properties of rainfall characteristics and does not require the simplistic assumption of triangular storm shape. Further, it was found that the proposed model could provide a very good fit between observed and theoretical results. A similar approach has been sucessfully used in the estimation of flood volume distribution (Nguyen and Chaleeraktrakoon, 1992).

THEORETICAL DEVELOPMENT

For planning and design purposes, it is usually necessary to consider only rainfall depths exceeding a certain base level. The rainfall depths above a given base level will be called *"exceedances"*. A storm event in this study is defined as an unbroken sequence of successive hourly rainfall exceedances. Accordingly, the total storm duration will be considered as the time between an upcrossing of the selected base level and the next downcrossing. Hence, the volume of a storm could be defined as the sum of all exceedances within its total duration. In other words, let ε_i denote the rainfall exceedance above a given base level in the i–th hour, and D represent the duration of a storm. The storm volume V (or the cumulative rainfall exceedances during the storm duration D) can be defined as

$$V = \sum_{i=1}^{D} \varepsilon_i \tag{1}$$

Consider now an interval of time which consists of n hours. In order to consider the volumes of all storms that have durations less than or equal to n hours, all these storms are arranged to begin with the first hour of the n–hour period. According to Eq.1, because the rainfall exceedance ε_i and the storm duration D are random variables, the storm volume V is the sum of a random number of random variables. Hence, the probability distribution of storm volume V can be written as follows (see,e.g., Todorovic, 1970):

$$F_V(x) = P\{V \leq x\} = \sum_{k=1}^{n} P\{X_k \leq x, D = k\} \tag{2}$$

in which $X_k = \sum_{i=1}^{k} \varepsilon_i$ for $k = 1, 2, \ldots, n$. Note that Eq. 2 was restricted to hourly rainfall process. However, it is readily seen that this restriction is not necessary. The general model, Eq. 2, might be used for rainfall data of any time interval. In the following, for hourly rainfall process, we consider two particular cases in which the cumulative rainfall exceedance X_k and the storm duration D are, respectively, independent or dependent.

Cumulative rainfall exceedance X_k and storm duration D are independent

In this case, Eq. 2 becomes

$$F_V(x) = \sum_{k=1}^{n} P\{X_k \leq x\}.P\{D = k\}. \tag{3}$$

Note that, to compute $F_V(x)$, it is necessary to estimate the probabilities $P\{X_k \leq x\}$ and $P\{D = k\}$. Many previous investigations have suggested that the exponential distribution (for continuous variables) (e.g., Eagleson, 1972, Egbuniwe, 1975) or its equivalent geometric distribution (for discrete variables) (e.g., Bogardi and Duckstein, 1993) might successfully be used to describe the distribution of storm duration. Hence, in the present study, the probability $P\{D = k\}$ is assumed to be approximated by a geometric distribution which is consistent to the discrete type of the random variable D considered. That is,

$$P\{D = k\} = p(1-p)^{k-1}. \tag{4}$$

The parameter p is estimated by

$$p = 1/E[D] \tag{5}$$

in which $E[D]$ represents the expected value of the storm duration D.

Previous studies (e.g., Eagleson, 1972; Howard, 1976; Nguyen and Rousselle, 1981) have suggested that the hourly rainfall depths can be roughly described by an exponential distribution. That is, $P\{\varepsilon_i \leq x\} = 1 - e^{-\alpha_i x}$. Further, many previous works (e.g., Chow and Ramaseshan, 1965; Grace and Eagleson, 1966) have pointed out the existence of a significant correlation between successive values of rainfall depths measured at short time intervals. Hence, in the present study, to account for the dependence characteristic inherent in the short–time–increment rainfall process, the successive hourly rainfall exceedances $\varepsilon_1, \varepsilon_2, \ldots, \varepsilon_n$ are assumed to be correlated and non–identically exponentially distributed. Therefore, as in the work by Kotz and Adams (1964), it can be readily shown that the random variable X_k has the same distribution as the weighted sum of k independent and non-identically exponentially distributed random variables, in which the weights represent the eigenvalues λ_i of the correlation matrix of successive hourly rainfall exceedances ε_i. In such cases, the distribution of $P\{X_k \leq x\}$ can be written as:

$$P\{X_k \leq x\} = (-1)^{k-1} \sum_{i=1}^{k} \left[\frac{\prod_{j \neq i} \gamma_j}{\prod_{j \neq i}(\gamma_i - \gamma_j)} \right] .(1 - e^{-\gamma_i . x}) \tag{6}$$

where $\prod$ denotes the product, and $\gamma_i = \alpha_i / \lambda_i$ representing the ratio of the exponential parameter α_i and the eigenvalue λ_i. The proof of this result can be found in Nguyen (1979) (see also Rényi, 1966; Nguyen and Chaleeraktrakoon, 1990).

From Eq. (6), it can be noted that $\lambda_1 = \lambda_2 = \ldots = \lambda_k = 1$ for the case where the successive rainfall exceedances are mutually uncorrelated. In this study, the exponential parameter α_i is determined from the historical data by $\alpha_i = 1/\overline{\varepsilon_i}$ in which $\overline{\varepsilon_i}$ is the mean of the hourly rainfall exceedance ε_i, and the eigenvalues λ_i will be estimated from the observed correlation matrix of hourly rainfall exceedances (see, e.g., Seber, 1984).

Cumulative rainfall exceedance X_k and storm duration D are dependent

Under this hypothesis, Eq. 2 becomes

$$F_V(x) = \sum_{k=1}^{n} P\{X_k \leq x | D = k\}.P\{D = k\}. \tag{7}$$

Further, many previous investigations (e.g., Grace and Eagleson, 1966; Egbuniwe, 1975; Rao and Chenchayya, 1984) have indicated a definite correlation between storm depth and duration, and have suggested that the relationships between X_k and D can be described by

$$X_k = (aD + b)r_k/100. \tag{8}$$

Hence, Eq. 6 can be written as follows

$$F_V(x) = \sum_{k=1}^{n} P\{r_k \leq \frac{100x}{ak + b}\}.P\{D = k\}. \tag{9}$$

in which a and b are constant coefficients, and r_k is the random transformed percent residuals. The parameters a and b of the linear relation between X_k and D will be estimated from the observed data by the least-square method. The distribution of r_k will be approximated by a two–parameter gamma distribution since it's more flexible and could provide a better fit to the observations than the one–parameter exponential distribution, as suggested in the previous work by Nguyen and Huynh (1987).

NUMERICAL APPLICATION

In the following, to test the descriptive capabilities of the theoretical models described above, an illustrative application is presented using hourly rainfall data observed over a period of 32 years (1943-1974) at Dorval Airport in Quebec (Canada). For the moment, to eliminate the effects of small storm events, a threshold value for hourly rainfall depths is set at 0.20 in/hr. According to the definition of a storm in this study, a total of 430 storm events having duration from 1 to 5 hours are selected for this illustrative example.

Figure 1 shows a comparison between observed and theoretical distribution functions for storm volume. It can be readily seen that the theoretical model which can take into account the dependence between cumulative rainfall exceedances X_k and storm durations D (Eq. 9) provides the best fit to the observations. The model which can account for the correlation between successive hourly rainfall exceedances (Eq. 6) gives a slightly better performance as compared with the one for uncorrelated rainfalls, probably due to the weak correlation inherent in the hourly rainfall process considered. Hence, in the estimation of storm volume distribution, the effect of serial correlation of hourly rainfalls is not significant as compared to that due to the correlation between storm depth and duration.

CONCLUSIONS

A stochastic method for determining analytically the probability distribution of storm volume has been proposed in this study. It can be concluded that the suggested method is more general and more flexible than those commonly used in most previous investigations, especially the empirical fitting approach, because it can take explicitly into account various observed stochastic properties of the underlying rainfall process. Results of an illustrative example have indicated the adequacy of the general stochastic model proposed. Finally, it can be seen that the simplistic assumption of triangular shape for storm hyetographs, as usually appeared in previous studies, is not necessary for the estimation of storm volume distribution.

REFERENCES

Bogardi, J.J., and Duckstein, L. (1993) "Event-based analysis of the dry spell phenomenon", Journal of Water Science 6, 23–45.

Cordova, J.R., and Rodriguez-Iturbe, I. (1985) "On the probabilistic structure of storm surface runoff", Water Resources Research 21, 755–763.

Chow, V.T. and Ramaseshan, S. (1965) "Sequential generation of rainfall and runoff data", ASCE J. of the Hydraulics Division 91, 205–223.

Eagleson, P.S. (1972) "Dynamics of flood frequency", Water Resources Research 8, 878–898.

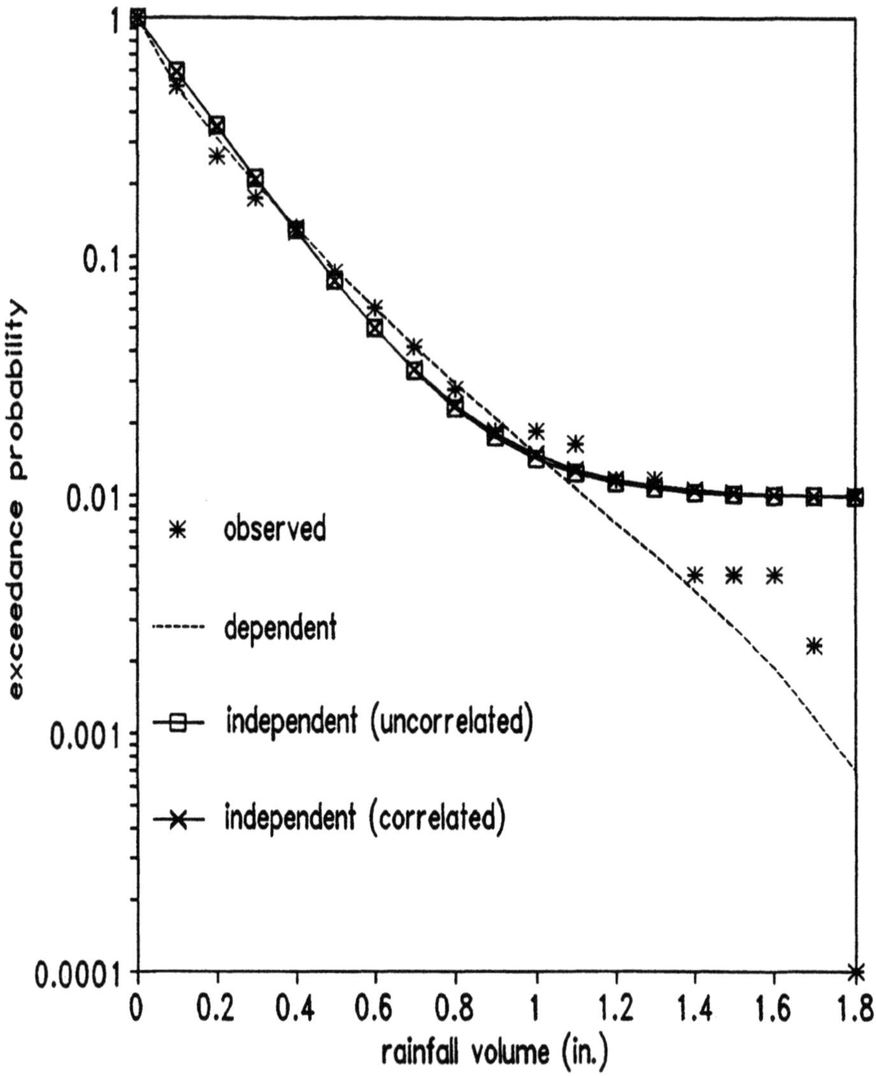

Figure 1. Observed and Theoretical Distributions of Storm Volume.

Egbuniwe, N. (1975) "Generation of hourly rainfall", Water Resources Bulletin 11, 706–713.
Etoh, T. (1992) "Expressions of efficiency of urban runoff storages based on probabilistic models of rainfall series", in Proc. of the Sixth IAHR International Symposium on Stochastic Hydraulics, May 18-20, Taipei, Taiwan, pp. 37–52.
Etoh, T. and Murota, A. (1986) "Probabilistic model of rainfall of a single storm", J. of Hydroscience and Hydraulic Engineering 4, 65–77.
Grace, G.A. and Eagleson, P.S. (1966) The synthesis of short–time increment rainfall sequences, Report 91, Hydrodynamics Laboratory, Massachusetts Institute of Technology, Cambridge.
Howard, C.D.D. (1976) "Theory of storage and treatment plant overflows", ASCE J. of the Environmental Division 102, 709–722.
Kotz, S. and Adams, J.W. (1964) "Distribution of sum of identically distributed exponentially correlated gamma–variables, Annals of Mathematical Statistics 35, 1, 277–283.
Nguyen, V–T–V. (1979) "Caractérisation hydrologique de la pluie pour le contrôle et la planification des systèmes de drainage urbain", Doctoral thesis, Ecole Polytechnique, Montreal, Quebec, Canada, 71 pages.
Nguyen, V–T–V. and Chaleeraktrakoon, C. (1990) "Distribution of sum of correlated exponential random variables and its applications in hydrology", Proc. Int. Conference on Computer Applications in Water Resources, Vol.1: Hydraulics, Hydrology and Water Resources Applications, Tamsui, Taiwan, pp. 398–403.
Nguyen, V–T–V. and Chaleeraktrakoon, C. (1992) "A stochastic method for estimating flood volume distribution", in Proc. of the Sixth IAHR International Symposium on Stochastic Hydraulics, May 18-20, Taipei, Taiwan, pp. 793–798.
Nguyen, V–T–V. and Huynh, N–P. (1987) "Analytically derived distribution for storm volume", Proc. of the Fourth International Conference in Urban Storm Drainage, August 31–September 9, Lausanne, Switzerland, pp. 31–34.
Nguyen, V–T–V. and Rousselle, J. (1981) "A stochastic model for the time distribution of hourly rainfall depth", Water Resources Research 17, 399–409.
Rao, A.R. and Chenchayya, B.T. (1984) "Depth-duration models of short time increment rainfall", Water Science and Technology 16, 109–130.
Rényi, A. (1966) Calcul des probabilités, Dunod, Paris.
Seber, G.A.F. (1984) Multivariate Observations, Wiley, NewYork.
Todorovic,P. (1970). "On some problems involving random number of random variables." Annals of Mathematical Statistics 41, 1059–1063.

PART V

FREQUENCY ANALYSIS AND FITTING DISTRIBUTIONS

ROBUSTNESS OF MODELS AND ESTIMATION METHODS IN FLOOD ANALYSIS

J. BERNIER
invited professor INRS-EAU
Université du Québec

The robustness of estimation methods of quantiles with respect to the modeling uncertainties can be analysed by bayesian methods. Deviations from the studied reference model are described as extra-parameters characterizing a larger class of models. Log-normal distribution is used as reference model and Box-Cox transformations is used to generate the larger class in this paper. Estimation methods of moments, probability weighted moments and maximum likelihood are compared by means of posterior expectations of squared errors of quantiles.

INTRODUCTION

Recently (Bernier 1991, 1993), we have discussed the use of Bayes' methods for studying the robustness of several statistical procedures in hydrology with respect to various kinds of uncertainties. The first ones are the measurement uncertainties appearing in data. We did not deal with them in the following paper. The second ones are sampling uncertainties generated on estimates by the randomness of data as observations of the studied random process and directly linked to the limited size of the available data sample. The latter are the traditionaly and often only uncertainties taken in account in classical studies. The third ones are modeling uncertainties; the use of a given class of probability distributions or a given stochastic model for fitting data asks for a previous choice among a larger set of models. In hydrology this choice is mainly empirical and can be mistaken; the resulting errors are often larger than the sampling ones. The last ones are representativeness uncertainties which appear when some doubt exists in the permanence of the present model, fitted to present and past data, as a good model for the future period of management of the water resources. The inherent uncertainty of the answer is an important ingredient of the problem of management of resources in assumed non steady conditions. If, strictly speaking, representativeness uncertainties belong to the class of modeling uncertainties it is important to consider them as distinct because of their different operational consequences.

Concerning flood problems robustness is of major importance to a number of decisional situations such as estimation of quantiles and their confidence intervals (Bernier 1991), and evaluation and interpretations ot failure probability in a predictive context (Bernier 1993),etc...

K. W. Hipel (ed.),
Stochastic and Statistical Methods in Hydrology and Environmental Engineering, Vol. 1, 187–197.

Here we apply the same Bayesian methodology to analyse modeling uncertainty in flood problems and compare (with the help of flood examples) three popular methods of estimation : maximum likelihood, moments and probability weighted moments).

BAYESIAN ANALYSIS OF MODEL UNCERTAINTIES.

The outline of this analysis are shown in Table I.

Let us consider the way Bayes method determines sampling errors inside the reference model. Each distribution of a studied random variable X_0 belongs to a particular family such as, for instance, the Log - Normal one with a probability density :

$$f(x_0) = \frac{1}{\sqrt{2\pi}.\sigma.x_0}.e^{-\frac{(\mathrm{Log}\,x_0 - m)^2}{2.\sigma^2}} \qquad (3)$$

The parameters m and σ characterize the generic parameter θ inside the reference family. Without data, uncertainty about θ is represented with the prior distribution $\pi_0(\theta)$. Taking into account a sample x of observations from X_0, uncertainty is characterized by the posterior density distribution given by Bayes rule :

$$P(\theta / x) = \frac{f(x / \theta).\pi_0(\theta)}{\int_\theta f(x / \theta).\pi_0(\theta).d\theta} \qquad (4)$$

From this posterior density it is fairly easy to calculate the posterior distribution of any function of parameters θ defining, inside the reference family, a particular characteristic Q such as a quantile (100 year flood for instance)

To take model choice uncertainty into account a possible way to generalize the previous process is to consider the reference family as belonging to a larger class with more parameters θ'.

It is the same idea used in classical non-parametric statistics where involved classes include all continuous densities. However such classes are too large for prediction purposes and it is necessary to limit ourselves to parametric classes. This parametric class is also chosen according the principle of parsimony so that data bring significant information about the position of the reference model and its representative value θ'_0 inside this class. To measure the information into the data relative to θ' or to a given Q we can determine their posterior probability densities using formulae (1) and (2) of Table I which are determined by using Bayes' rule.

Please note: when choosing the larger family it is is necessary to remember that it must represent the uncertainty which affects the prediction purposes. This choice must follow the principle of parsimony to a lesser degree than the usual one when choosing the reference model.

In the following we shall see that the procedure can be applied to larger classes representing model uncertainty as well as representativeness uncertainty.

Sources of uncertainty
- sampling errors
- distribution model **
- representativeness model

<u>Analysis of uncertainties</u>

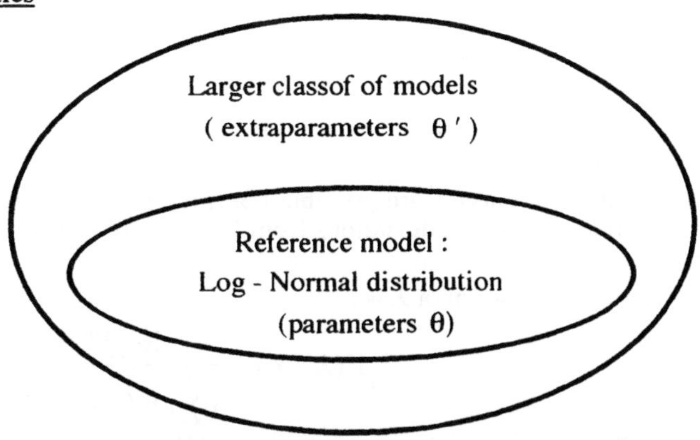

<u>Choice of a Characteristic Q as a function of model parameters, estimate</u> $\hat{Q}$ <u>and Loss function</u> $L(\hat{Q},Q)$ (for instance quadratic $(\hat{Q}-Q)^2$) :

Inside Reference Class :

$$Q = Q(\theta) = Q'(\theta, \theta'_o)$$

$$\text{Prob.}[Q/x] = \int_{\Theta} P(Q/x,\theta).P(\theta/x).d\theta \qquad (1)$$

$$\text{Posterior Loss}: E^x(L) = \int_Q L(\hat{Q},Q).P(Q/x).dQ$$

Inside Larger Class :

$$Q = Q'(\theta, \theta')$$

$$\text{Prob.}(Q/x)) = \int_{\theta'} P(Q/x,\theta').P(\theta'/x).d\theta' \qquad (2)$$

$$\text{Posterior Loss } E^x(L) = \int_Q L(\hat{Q},Q).P(Q/x).dQ$$

θ'_o is the value of extraparameter corresponding to the reference model. The "distance" between θ' and θ'_o characterizes the departure from this reference model.

TABLE I : Framework of bayesian study of uncertainties.

In order to study the model choice uncertainty along the lines above we define the "extended family of distributions" as follows :

let the so called Box - Cox transform :

$$y^{(\lambda)} = \frac{(x_0)^{(\lambda)} - 1}{\lambda} \qquad \text{if } \lambda \neq 0$$
$$y^{(\lambda)} = \text{Log } x_0 \qquad \text{if } \lambda = 0 \qquad (5)$$

$y^{(\lambda)}$ is assumed distributed according a normal distribution with expectation m and standard deviation σ.

Other extended classes are possible (Bernier 1991) but we argue that Box-Cox models are useful to represent deviations from Log-Normal which can be expected in the studied examples.

Now the posterior distribution of λ is :

$$P(\lambda / x) = \frac{K. g^{(n-1).(\lambda-1)}}{[\sum_{i=1}^{n} (y_i^{(\lambda)} - \bar{y}^{(\lambda)})]^{(n-1)}} \qquad (6)$$

where K is a constant (with respect to λ), n the size of the sample and g the geometric mean of this sample. This posterior distribution takes in account non-informative priors proposed by Box and Tiao (1973).

Figures I to III show some fittings to a few French and Canadian examples. It is usually difficult enough to discriminate between models but in some cases log-normal distribution seems not to have the best behaviour for large values compared with the behaviour of Box-Cox models. Sometimes also Log - normal distribution estimated with the help of maximum likelihood, two first moments and PW moments seem better. The posterior distributions of λ shown by the Figure IV enlights the value of the reference model compared with the larger one. In some cases the value $\lambda = 0$ is near by the modal value. In other case the value 0 is not so "credible". Let us note the differences in variability which are due to the different sample sizes and also to the uncertainty of the extreme behaviour of the distribution inside the chosen class. The expectation and standard deviation of λ well describe the behaviour of the posterior distribution due to the gaussian bell shaped density.

ESTIMATION IN APPROXIMATE MODELING

In the study of flood statistical distributions many estimation methods have been proposed : some are popular, others are fashionable during limited periods such as the probability weighted and L moments. The arguments supporting a particular method are either practical or theoretical such as :

- (1) - ease of computation but this argument has limited value when desktop or pocket computers can handle most of the methods used in hydrology now.

(2) - minimum bias and / (or) sample variance compared with others methods. These properties always are studied with respect to a given model. My opinion is that no hydrological or statistical reason can ensure that any distribution model is the true one. In the case of floods deviations are always possible, especially for extremes values. Even if

we use a reference model considered as the best choice, it is often necessary to guard against possible errors and their effects on the estimation of quantiles for instance. To do this we can estimate the parameters of the reference model in such a way that the feared errors decrease. We call this problem estimation in approximate modeling. In such a case the best estimation method is not the one which insures mimimum bias or variance assuming a given distribution but the more robust one with respect to the deviations to be feared.

To illustrate this case we present results concerning Log - Normal (L N) distribution (formula 3) as the reference model and Box - Cox model (B C) as the larger one representing possible errors with respect to the reference model. The approach is described in Table I and can be applied to any reference model and any method of estimation.

Here three methods are compared :

- the maximum likelihood method using mean $\mathrm{Log}(g)$ (g is the geometric mean) and variance s^2_{Log} of logarithms of observations and demonstrated as the theoretically best methods for L N distribution :

$$m_1 = \mathrm{Log}(g) \qquad \sigma_1^2 = s^2_{\mathrm{Log}} \tag{7}$$

-the method of moments using mean $\overline{x}$ and variance s^2 of observations :

$$m_2 = \mathrm{Log}(\overline{x}) - \frac{\mathrm{Log}(cv^2+1)}{2} \qquad \sigma_2^2 = \mathrm{Log}(cv^2+1) \quad \text{where } cv = \frac{s}{\overline{x}} \tag{8}$$

- the method of probability weighted moments using $\overline{x}$ and $\overline{x.F}$, weighted mean with weight $F = \frac{i}{N+1}$:

$$m_3 = \mathrm{Log}(\overline{x}) - \frac{\sigma_3^2}{2}$$

$$\sigma_3 \text{ as solution of } H(\sigma) = \int_{-\infty}^{+\infty} e^{\sigma.u}.g(u).G(u).du = \frac{\overline{x.F}}{\overline{x}} \tag{9}$$

where $g(u)$ and $G(u)$ are density and distribution function of reduced normal variate .

This equation is fairly easy to solve numerically.

In the following Table II the three methods are compared in terms of results concerning the 100 years flood quantile Q_{100}. We give the Q_{100} estimates together with the corresponding optimal value $\hat{Q}_{100}$ in the larger B.C.class with respect to the posterior expected quadratic loss of estimation :

$$W^x_{\mathrm{estimate}} = \int_Q (Q_{\mathrm{estimate}} - Q)^2.P(Q/x).dQ \tag{10}$$

The three estimates are compared with respect to the so called Regret R which is the increase of loss due to the use of the particular estimate instead of optimum $\hat{Q}_{100}$

$$R^{x}_{estimate} = W^{x}_{estimate} - W^{x}_{\hat{Q}_{100}} \qquad (11)$$

Because R has the dimension of square of discharge (in M3/S) we use square root of Regret as scaled R in Table II.

Rivers :	Maximum likelihood (ML)	Moments (M)	P W Moments	Optimum
Chaudiere				
Q100	2155	2081	1989	2016
Scaled Regret	138	64.5	27.4	
Garonne				
Q100	8672	7976	7748	7437
Scaled Regret	1235	539	311	
de Pontois				
Q100	1505	1455	1255	1387
Scaled Regret	118	67.5	132	
Harricana				
Q100	3300	3281	3086	3258
Scaled Regret	41.6	23.1	172	
Chicoutimi				
Q100	539	525	496	513
Scaled Regret	26.6	12.6	17.0	
Vermillon				
Q100	501	488	463	499
Scaled Regret	2.3	11.7	36.5	

TABLE II : Comparison of methods of estimation with respect to Scaled Regret (units : M3/S)

It can be seen that M L method is the best (according our criterion) in one case, PWM method in two cases and M methods in the three others. Some regrets are not large, some others are significant.

SOME CONCLUSIONS

Here the Bayes analysis has been made conditional to the available samples. Similarly to usual simulation studies based on different synthetic sample sizes, the effects of various sample schemes could be treated in bayesian methodology using the so called preposterior analysis.

The purpose was not to make a complete robustness study of estimation methods for any realistic distribution but to show that usual criteria often are not sufficient for comparing methods of estimation in flood frequency analysis.

Still let us remember that Log - Normal reference distribution was chosen only for illustration. However the result concerning PWM method is interesting because it is it is the best when deviations from reference model appear in small values of the studied variable and the behaviour of the large estimate (Q_{100}) then is influenced by these deviations. This kind of behaviour often is met in statistical hydrology. Contrarily, PWM method is dominated (with respect to the given expected loss) when deviations occur for large extreme observations. It seems that this result corresponds to a general feature of PWM methods. As a matter of fact the PW moments give to upper sample values a smaller weight than the usual moments of order 1 or more than 1. This property often is presented as a robustness quality with respect to large values considered as "outliers". However the decision to consider extreme values as "outliers" can be fallacious enough when inefficient tests are used. When no physical reason exist to reject extreme values as "outliers", their use is conservative. Finally we rather prefer the reverse point of view concerning the robustness of PWM method.

REFERENCES

Berger J.O.(1985) "Statistical Decision Theory and Bayesian Analysis", Springer Verlag.

Bernier J.(1991) "Bayesian Analysis of Robustness of Models in Water and Environmental Sciences" in NATO ASI, Water Resources Engineering Risk Assessment - Series G Vol.29 , Springer Verlag.

Bernier J.(1993) "Quantitative Analysis of Uncertainties in Water Resources. Application to Changes Effects Prediction" - NATO ASI on Engineering Risk and Reliability in a Physical Changing Environment, Deauville, France : May 24-June 03.

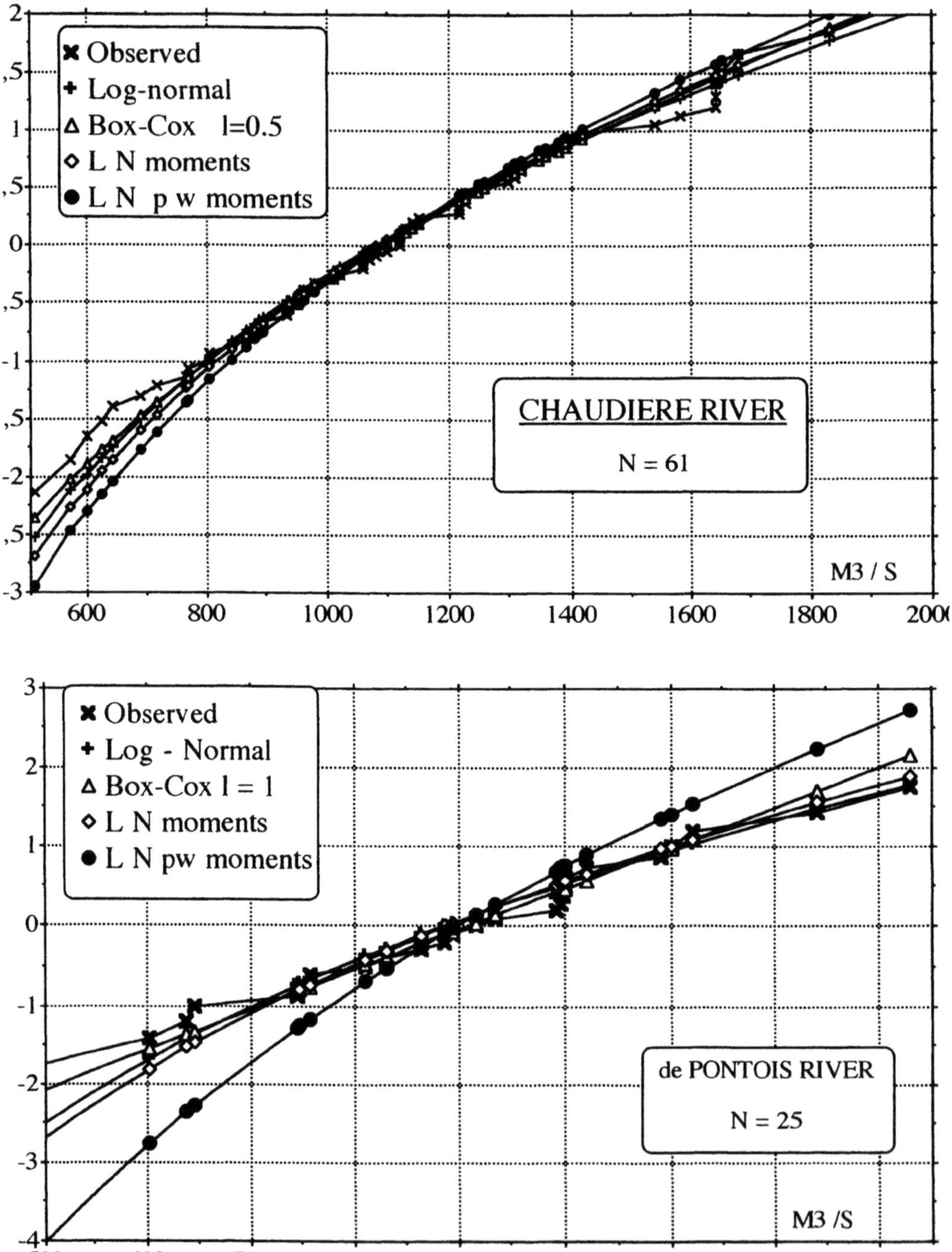

Figure 1 : Examples of fittings to floods samples

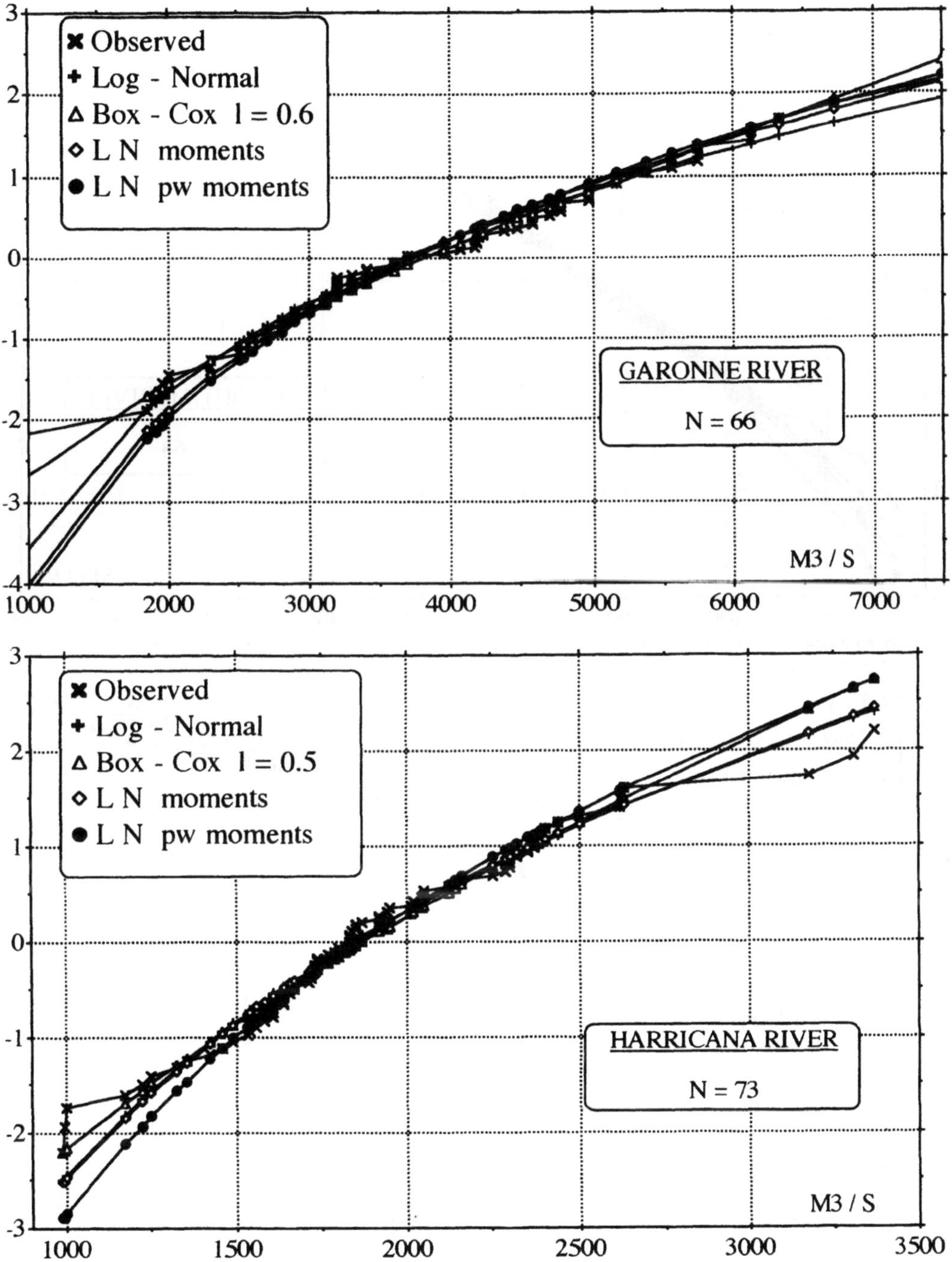

Figure 2 : Examples of fittings to floods samples (continued)

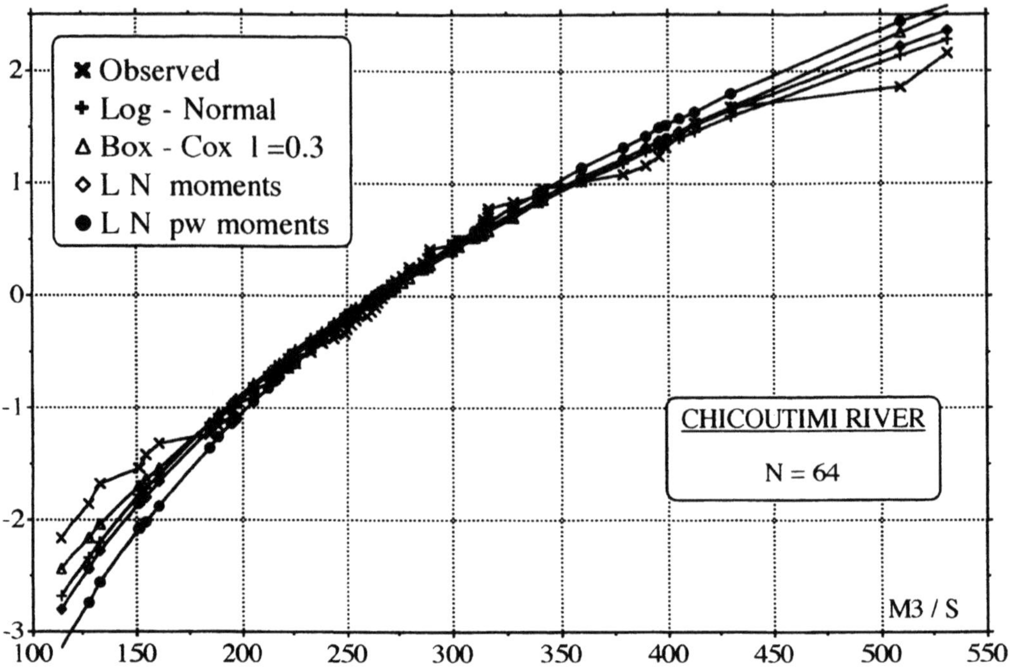

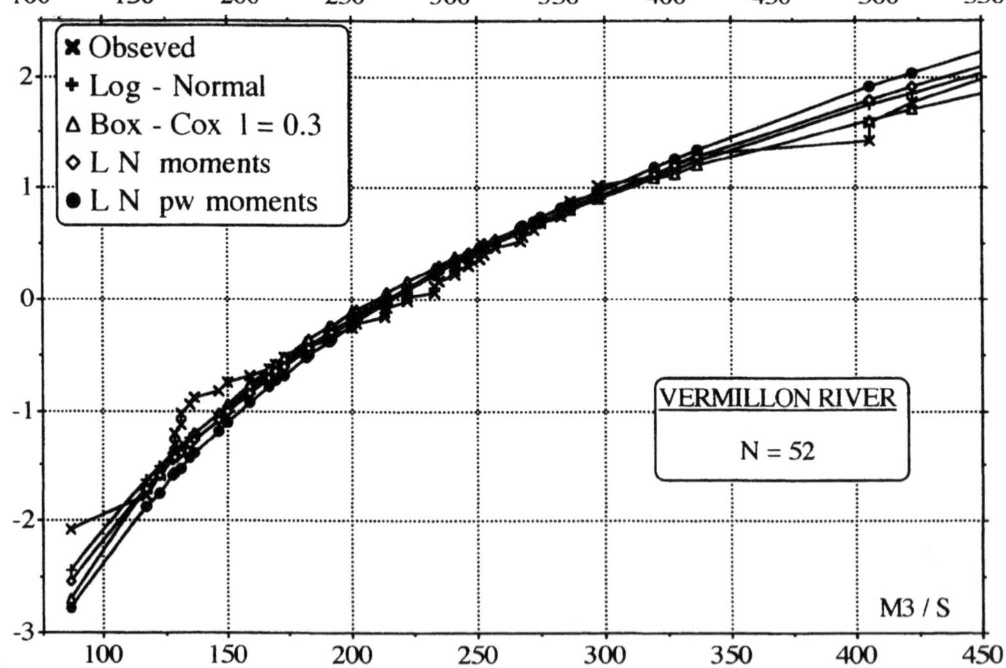

Fligure 3 : Examples of fittings to floods samples (continued)

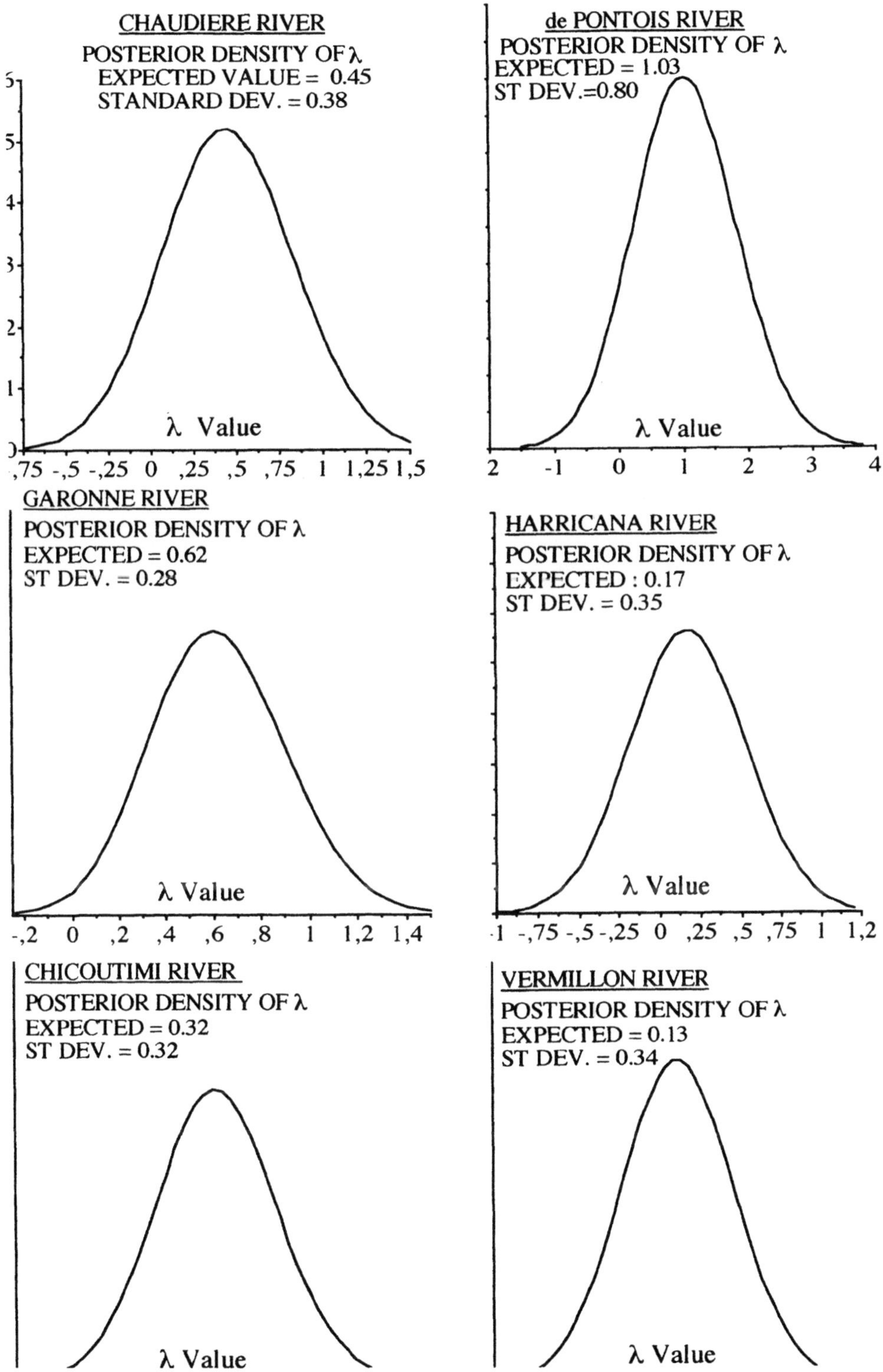
CHAUDIERE RIVER
POSTERIOR DENSITY OF λ
EXPECTED VALUE = 0.45
STANDARD DEV. = 0.38
λ Value
5
4
3
2
1
-,75 -,5 -,25 0 ,25 ,5 ,75 1 1,25 1,5
de PONTOIS RIVER
POSTERIOR DENSITY OF λ
EXPECTED = 1.03
ST DEV.=0.80
λ Value
-1 0 1 2 3 4
GARONNE RIVER
POSTERIOR DENSITY OF λ
EXPECTED = 0.62
ST DEV. = 0.28
λ Value
-,2 0 ,2 ,4 ,6 ,8 1 1,2 1,4
HARRICANA RIVER
POSTERIOR DENSITY OF λ
EXPECTED : 0.17
ST DEV. = 0.35
λ Value
-1 -,75 -,5 -,25 0 ,25 ,5 ,75 1 1,2
CHICOUTIMI RIVER
POSTERIOR DENSITY OF λ
EXPECTED = 0.32
ST DEV. = 0.32
λ Value
VERMILLON RIVER
POSTERIOR DENSITY OF λ
EXPECTED = 0.13
ST DEV. = 0.34
λ Value

UNDERSTANDING FLOOD PROBABILITIES

I. MUZIK
Department of Civil Engineering
The University of Calgary
Calgary, Alberta T2N 1N4
CANADA

Flood frequency analysis is routinely used for estimation of design peak discharges. The procedure adopted usually relies on extrapolation of a probability distribution fitted to a sample of observed flows, without any recourse to physical laws governing flood generation processes. This is not a scientifically sound procedure. The aim of this paper is to illustrate how physical laws applicable to runoff also affect the probability distribution of peak flows. From an elemental runoff plane to a complex watershed, such a physically-based approach to the estimation of flood frequencies should supplement traditional, purely statistical types of flood frequency analysis.

INTRODUCTION

"With respect to flood frequency analysis, it's all in the tails, so to speak" (Greis, 1983). Indeed, often small sample of annual floods can be fitted equally well by several probability distributions, such as Gumbel, log-normal and log Pearson type III; but outside the sample range each tail of the fitted distributions tells a different story, so to speak, of the extreme floods. Which story, or tail should a hydrologist believe? This question, as yet, remains largely unanswered.

The problem of estimating probabilities of flood occurrences is not just an exercise in statistics. As Klemes (1987) put it, "Flood frequency theorists often cannot comprehend that the difference between the occurrence of small and large floods is more fundamental than the difference between drawing a card marked with a small number and with a large one. Unable to see beyond a sample of random numbers, each of which contains the same amount of information about the parent distribution, they cannot appreciate that the smallest floods from a given historic period may contain little or no information relevant to the likelihood of a very large flood, i.e. that the processes governing the formation of small and large floods may be vastly different." Chow (1964) pointed out that "frequency analysis is only a procedure to fit hydrologic data to a mathematical model of distribution," and that "it is only experience and verification of data that decide the use of a certain distribution." He recognized the need for some theoretical interpretation or reasoning for the preference of one distribution to another by stating that "Such interpretations would describe a physical process of the hydrologic phenomena and thus help understand the procedure of frequency analysis and the significance of the results ...". Chow's well-known justification for the use of the lognormal distribution (Chow, 1955) was based on the reasoning that a hydrologic event is a result of the cumulative

K. W. Hipel (ed.),
Stochastic and Statistical Methods in Hydrology and Environmental Engineering, Vol. 1, 199–207.

action of many independent factors, thus leading by means of the central limit theorem, to a lognormal distribution.

Chow's (1955) argument in favor of the lognormal distribution was, however, more probabilistic than physical. The first attempt to use a physically- based mathematical model of a catchment to derive the probability distribution of peak streamflow from a given distribution of rainfall was made by Eagleson (1971, 1972) by means of a kinematic wave catchment model. More recently, Rodriguez-Iturbe et al (1982) analytically derived the probability density function of the peak and time to peak of the instantaneous unit hydrograph (IUH) as functions of rainfall probability characteristics and basin geomorphological parameters. Coupling of the IUH and rainfall characteristics was again achieved, as in Eagleson's approach, by means of the analytical form of the kinematic wave equations.

It is the intention of this paper to describe the linkage between probabilities of rainfall intensity and peak discharge for the case of runoff from an impervious plane. Effects of the parent distribution of rainfall intensity and physical laws of fluid motion on the derived distribution of peak discharge are clearly demonstrated. Next, the discussion is extended to flood probabilities on a catchment scale. Arguments are presented to support the view that floods of various magnitudes may not necessarily conform to a single probabilistic model. Monte Carlo simulation is used to illustrate the generation of a physically-based synthetic flood frequency curve for a test watershed, and this result is compared to a log-Pearson type III distribution fitted to observed annual flood series.

ELEMENTAL RELATIONS

The analytical solution of the kinematic wave equations of overland flow from an impervious runoff plane due to uniformly distributed rainfall (Henderson and Wooding, 1964) is used here to study the transformation of a probability density function (pdf) of rainfall intensity f(i), all rainfalls having a constant duration t_o, into a pdf of peak discharge at the end of the runoff plane, $f(q_p)$. The discharge per unit width of the plane at any time t since the beginning of rain, and shorter than the time to steady state t_s, is given by

$$q = \alpha(it)^m \tag{1}$$

where i is the rainfall intensity. If the flow is turbulent and Manning's equation is used to solve for the velocity of flow, then the exponent m is equal to 5/3 and α is equal to

$$\alpha = S^{1/2}/n \tag{2}$$

where S and n are the slope and Manning's roughness coefficient of the plane, respectively. With increasing α the flow velocity also increases. If the rain continues long enough a steady state is reached at time t_s, after which the following equation applies

$$q = i \tag{3}$$

Figure 1 shows overland flow hydrographs for three values of α, resulting from such rainfall intensities that the steady state is reached at t_s = 60 s in all three cases. The runoff plane length L equals 10 m.

Since the interest is in peak discharge only it is convenient to draw the relationship between peak discharge q_p and i (t_o = 60 s), as shown in Figure 2. The i-q_p curves are in general given by

$$q_p = \alpha(it_o)^{5/3}/L \qquad t_o < t_s \tag{4}$$

$$q_p = i \qquad t_o \geq t_s \tag{5}$$

where

$$t_s = (L/\alpha i^{2/3})^{3/5} \tag{6}$$

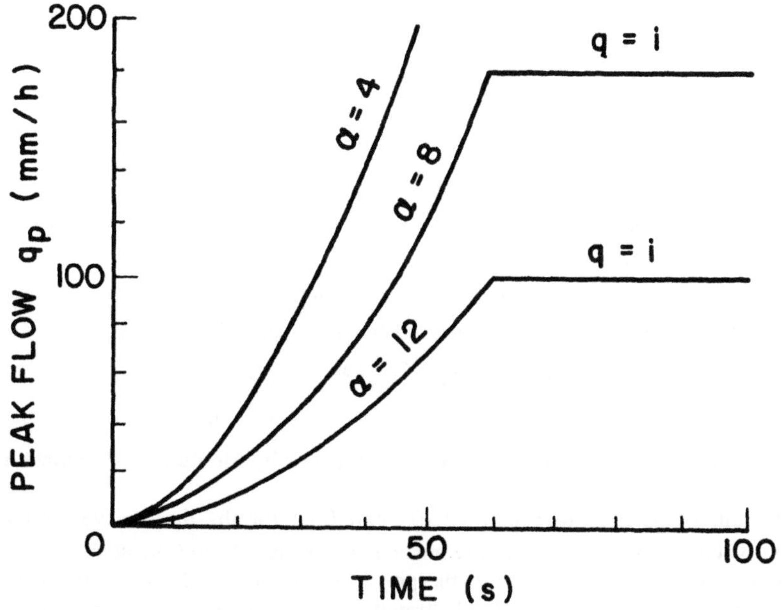

Figure 1. Overland flow kinematic wave hydrographs.

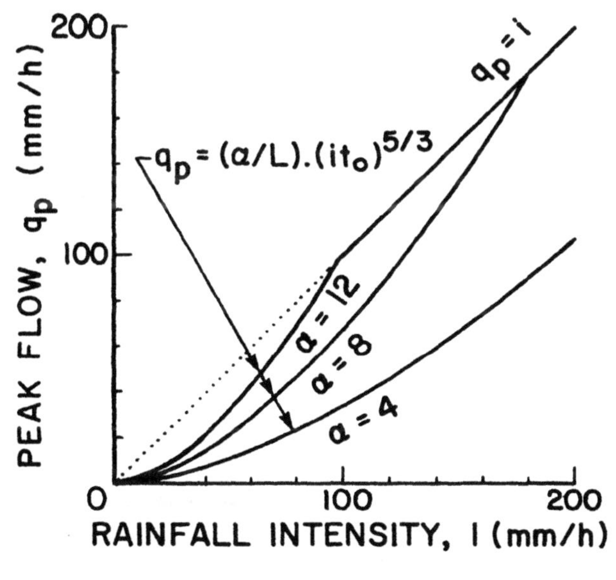

Figure 2. Example of i-q_p curves for L = 10 m and t_o = 60 s.

The i-q_p relationship for kinematic flow over a plane of length L thus consists of two parts: a power curve for nonequilibrium flow, and a linear part for equilibrium flow. The former is governed by the physical parameters of the plane, S and n, as expressed by the coefficient α. The changeover from a power curve to a straight line relationship occurs abruptly when $t_o = t_s$. The corresponding rainfall intensity i_s can be computed by (6) rearranged into

$$i_s = (L/(\alpha\ t_0^{5/3}))^{3/2} \tag{7}$$

When i is a random variable with a known pdf f(i), then according to the theory of derived distributions (Haan, 1982), the pdf of q_p is given by

$$f(q_p) = f(i)\ \ |di/dq_p| \tag{8}$$

Since the derivative di/dq_p can be expressed from (4) and (5) exactly as

$$di/dq_p = 3L/5\alpha\ t_0^{5/3}\ i^{2/3} \qquad i < i_s \tag{9}$$

$$di/dq_p = 1 \qquad i > i_s \tag{10}$$

the derived distribution $f(q_p)$ can be obtained by eq. (8) or by Monte Carlo simulation (Muzik and Beersing, 1989).

Figure 3 shows the transformation of f(i) into $f(q_p)$ for the three α values considered previously, when f(i) is normal, with the mean $\bar{i}$ = 100 mm/h and the standard deviation s_i = 20 mm/h. The effect of α, i.e. of the physical parameters of the runoff plane, on the distribution of peak discharge is clearly illustrated. As α increases (faster response to rainfall), the $f(q_p)$ approaches the parent distribution f(i). This results in a mixed distribution of q_p, even though there has been no change in the physical parameters of the runoff plane.

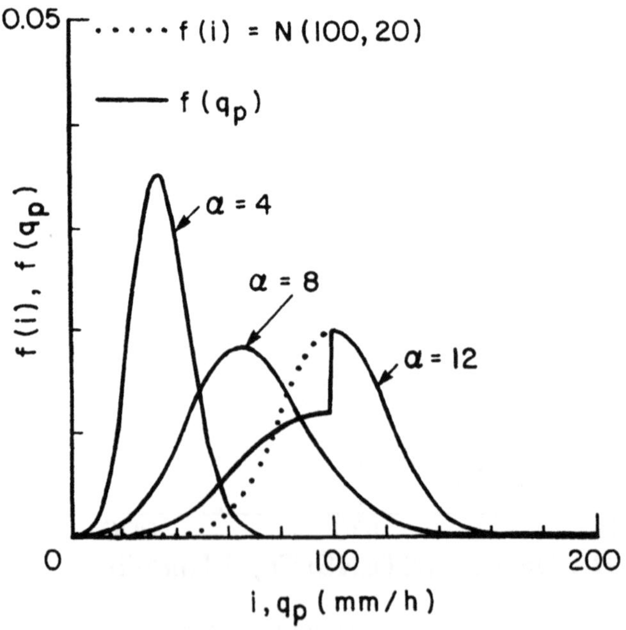

Figure 3. Sensitivity of $f(q_p)$ to α.

Muzik and Beersing (1989) showed, by numerical and laboratory experiments, that the relationship between the coefficient of variation, C_v (q_p), and skewness, C_s (q_p), of $f(q_p)$ is a curve of the type as shown in Figure 4. In the present example the curve applies to overland flow when the rainfall intensity is normally distributed with the coefficient of variation C_v (i) equal to 0.3.

The following observations are made:

1. The curve is valid for any value of α and t_o.

2. The coefficients C_v (q_p) and C_s (q_p) have maximum values, and remain constant and independent of α and t_o for all transformations which do not involve a significant number of i values greater than i_s (equation (7)). In other words, as long as primarily eq.(4) applies in the transformation process, the derived distributions of q_p will have the same maximum values of C_v (q_p) and C_s (q_p) (point M in Fig. 4). Analytical proof can be found in Muzik and Beersing (1989). Could this result be used in search for hydrologically homogeneous regions?

3. As $\bar{i}$ increases, the proportion of i values being greater than $\bar{i}$ also increases. This results in the decrease of C_v (q_p) and C_s (q_p) values. When $\bar{i}$ becomes greater than a certain limit, depending on α and t_o, the coefficient C_s (q_p) becomes negative.

4. Eventually, when $\bar{i}$ is large enough so that primarily eq. (5) applies in the transformation process, the derived distributions approach f(i), C_v (q_p) approaches C_v (i) and C_s (q_p) approaches zero.

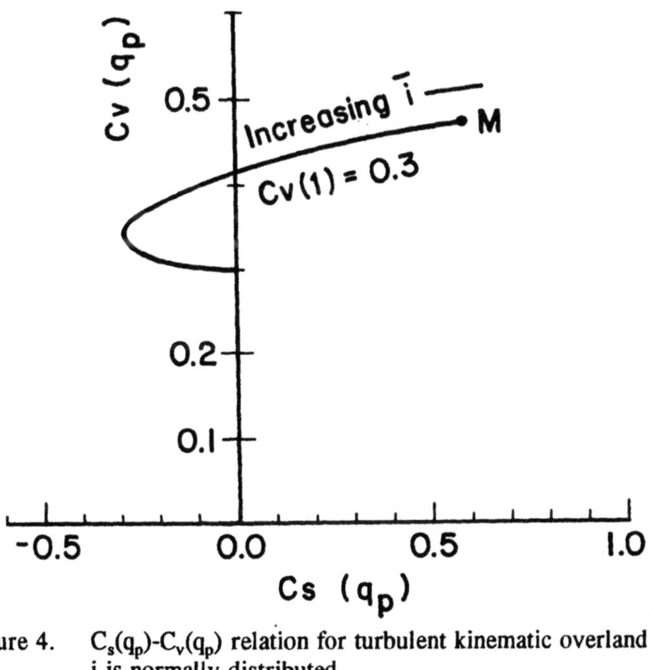

Figure 4. $C_s(q_p)$-$C_v(q_p)$ relation for turbulent kinematic overland flow when i is normally distributed.

Similar results were obtained when i follows the two-parameter gamma distribution. This distribution has a fixed relationship between its coefficients such that C_s (i) = 2 C_v (i). The curves shown in Figure 5 correspond to C_v (i) equal to 0.3, 0.2, and 0.1, respectively. The results are similar to those obtained for normally distributed i, and observations 1 through 4 apply to each curve.

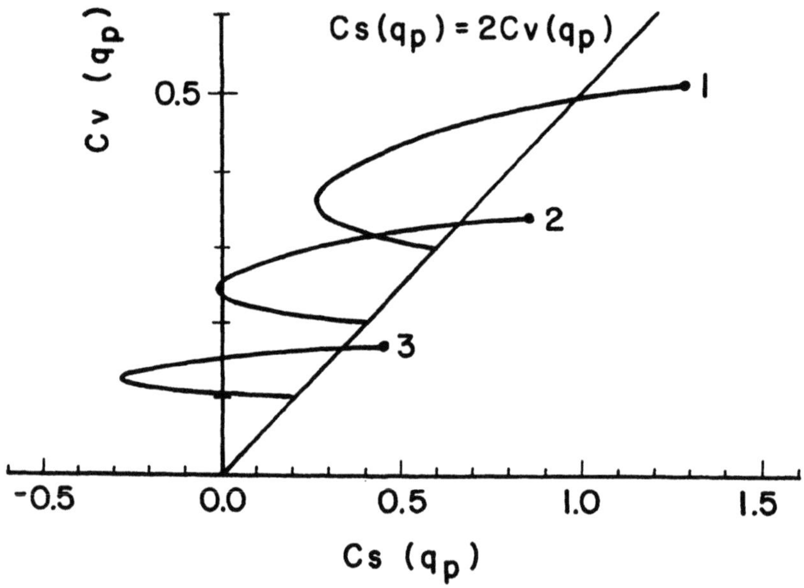

Figure 5. $C_s(q_p)$-$C_v(q_p)$ relations for turbulent kinematic overland flow when i is two-parameter gamma distributed and (1) $C_v(i)$ = 0.3, (2) $C_v(i)$ = 0.2, (3) $C_v(i)$ = 0.1.

NATURAL FLOODS

The preceding results reveal that on an elementary scale (uniform rainfall of constant duration and impervious runoff plane) the probability distribution of peak discharge depends on hydraulic parameters of the runoff plane and the probability distribution of rainfall intensity. In view of this, what can be said about the distribution of an annual flood series? On a catchment scale, the rainfall-runoff relationship and associated probabilities certainly get complex. It seems reasonable to say, that annual flood series will more likely contain discharges generated by precipitation inputs which cannot be described by a single probability distribution, and by overland and subsurface flow processes which are nonstationary and of various types, than by rainfalls of a single probability distribution and one runoff law. It follows then that, in general, the annual flood distribution will be a "mixed" distribution, which may have very different forms for various ranges of discharge. The following discussion elaborates on the major points of this proposition.

First, there is the question of precipitation input. An example given frequently is that of annual floods arising from two sources, snowmelt and rainfall, each process producing floods having their own characteristic frequency distribution (Jarrett and Costa, 1982; Waylen and

Woo, 1982). Hirschboeck (1987) attempted to separate events in a flood series according to various synoptic atmospheric circulation mechanisms and patterns that generate each flood event in a series. The flood events in a study region were hydroclimatically classified into eight categories using surface and upper air weather maps and precipitation data. The analysis demonstrated that floods in the study basin originated from a variety of atmospheric processes and suggested that the parameters of the flood frequency distributions of different hydroclimatic subgroups may have a physical basis. One could further reason that besides the storm type (convectional, frontal, cyclonic), the storm location and movement over the basin and the storm hyetograph have a bearing on the frequency distribution of peak discharge.

Secondly, consider the land portion of flood generation processes. A mixed population of annual floods will arise due to various antecedent moisture conditions, seasonal variations in land use and cover, and nonstationarity of any other significant parameter. Channel hydraulics is another important factor. Singh (1987) gave an example of a plot of observed annual floods on lognormal probability paper showing a change in curvature which may be attributed to a change from predominantly channel flow to a predominantly floodplain flow.

Finally, a mention should be made about measurement errors. In flood frequency analysis it is generally assumed that peak flood discharges are measured without significant errors. River stage, continuously recorded at a gaging station, is converted to discharge with a rating curve (or stage versus discharge curve) developed from 10 to 20 flow measurements made each year and kept current to minimize errors. It is the large flood discharges, which exceed the upper limits of the established rating curve, that are susceptible to significant errors because of the extrapolation involved. Potter and Walker (1985) define gaging ratio as the ratio of the largest directly measured discharge to the largest observed discharge. Low values of gaging ratio, found mostly for small drainage areas, cause higher values of standard deviation and skew for some flood series with high outliers which may be caused by errors in extrapolation. Characteristics of the storm causing such an outlier can be analyzed to determine if the perceived outlier is a genuine outlier or is caused by extrapolation. Knowledge of the gaging ratio can be used in assigning weights to flood estimates from various gaging stations to develop regional flood frequency curves (Singh, 1987).

Synthetic Flood Frequency Curves

Since a multitude of stochastic variables influences the probability of occurrence of a flood of a certain magnitude, Monte Carlo simulation is a good method for estimation of the frequency distribution of an annual flood series. Muzik and Chang (1992, 1993) used a GIS-assisted flood simulation model to generate synthetic flood frequency curves for a number of watersheds in the Alberta Rocky Mountain foothills region. The model was based on a modified SCS runoff curve number method and a regional dimensionless unit hydrograph. Input parameters included the five-day antecedent precipitation, initial abstraction, duration, depth and time distribution of rainfall. Values of the input parameters were drawn randomly from frequency distributions derived from regional data. Figure 6 shows an example of a synthetic flood frequency curve for the Little Red Deer River, based on 5000 generated peak discharges. Also shown are the observed flood peaks and the fitted Log-Pearson III distribution. The LPIII and the synthetic distribution differ significantly in their tail sections. By way of this example, the following general observations can be made.

1. While fitting of LP III or any other theoretical distribution to observed data is a strictly mathematical procedure, the generation of a synthetic distribution has the advantage

of being physically based, thus enabling the modeller to make an educated judgement of its validity through examination of the inputs and the capabilities of the deterministic model used.

2. The synthetic distribution is non-parametric, thus the upper end of the distribution corresponding to extreme floods is not unduly influenced by extreme low flows, as is often the case when fitting theoretical distributions to a sample.

3. Finally, it was shown (Fig. 3) that the distribution of peak discharge of elemental runoff approaches the parent distribution of rainfall intensity when the physical parameters of the runoff plane exceed certain critical values. By induction, it may be reasoned that the distribution of floods on a natural watershed, when runoff conditions are maximized, should similarly approach a limiting distribution in its upper tail governed by probability distributions of rainfall parameters in the region. This convergence will be automatically assured in the generating process of synthetic flood distributions, if a realistic deterministic simulation is maintained.

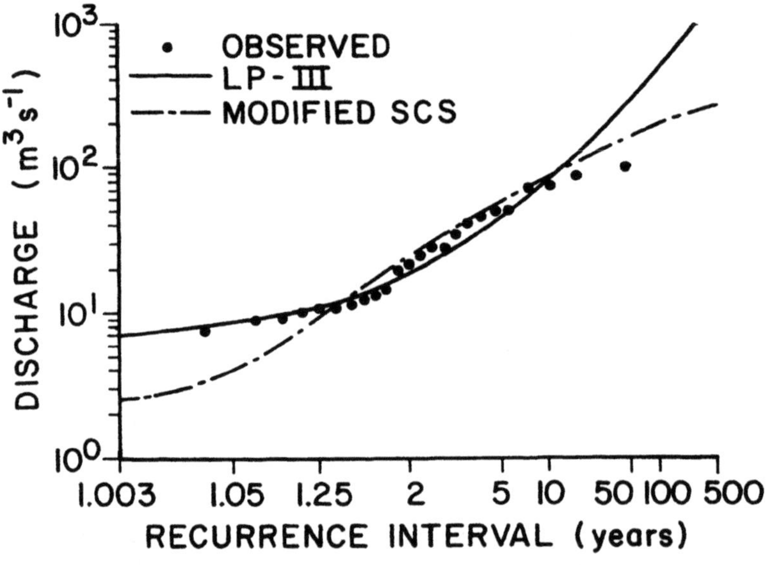

Figure 6. Comparison of flood frequency curves for Little Red Deer River.

DISCUSSION

It has been shown that the probability distribution of peak discharge of overland flow from an impervious plane, due to uniform rainfall, depends on the probability distribution of rainfall intensity and the hydraulic parameters of the plane. Thus, if the rainfall intensity distribution is known, the peak discharge distribution can be derived more or less exactly.

There is evidence, documented in literature, that frequency distributions of annual flood series are similarly dependent on climatic and watershed physical parameters. Because of the runoff process complexity and the large number of parameters involved in generation of floods,

Monte Carlo simulation, using stochastic inputs and deterministic modelling, provides an excellent approach to study flood probabilities. In other words, the way to understanding flood probabilities is through understanding flood physics.

There is no physical rationale for assuming a single, underlying distribution for a flood series. Variations in storm types, antecedent soil moisture conditions, and channel and flood plain storage cause variations in the basin response and flood peak. Therefore, the flood series can hardly be expected to follow any particular theoretical distribution in its entirety.

REFERENCES

Chow, V.T. (1985) "On the determination of frequency factor in log-probability plotting", Eos Trans. AGU, 36, 481-486.

Chow, V.T. (1964) "Statistical and probability analysis of hydrologic data", in V.T. Chow (ed), Handbook of Applied Hydrology, McGraw-Hill, New York, pp. 1-42.

Eagleson, P.S. (1971) "The stochastic kinematic wave," in Systems Approach to Hydrology, Water Resources Publication, Fort Collins, Colorado, pp. 202-222.

Eagleson, P.S. (1972) "Dynamics of flood frequency," Water Resources Res., 8(4), 878-897.

Greis, N.P. (1983) "Flood frequency analysis: A review of 1979-1982", Review of Geophysics and Space Analysis, 21(3), 699-706.

Haan, C.T. (1982) Statistical Methods in Hydrology, Iowa State University Press, Ames.

Henderson, F.M. and Wooding, R.A. (1964) "Overland flow and groundwater flow from a steady rainfall of finite duration," J. Geophys. Res., 69(8), 1531-1540.

Hirschboeck, K.K. (1987) "Hydroclimatically-defined mixed distributions in partial duration flood series", in V.J. Singh (ed), Hydrologic Frequency Modeling, D. Reidel Publishing Co., Dordrecht, pp. 199-212.

Jarrett, R.D. and Costa, J.E. (1982) "Multidisciplinary approach to the flood hydrology of foothills streams in Colorado", in C. Johnson (ed), International Symposium on Hydrometeorology, American Water Resources Association, Bethesda, pp. 565-569.

Klemes, V. (1987) "Hydrological and engineering relevance of flood frequency analysis", in V.J. Singh (ed), Hydrologic Frequency Modeling, D. Reidel Publ. Co., Dordrecht, pp. 1-18.

Muzik, I. and Beersing, A.K. (1989) "Stochastic-deterministic nature of an elemental rainfall runoff process," Water Resources Research, 25(8), 1805-1814.

Muzik, I. and Chang, C. (1992) "Physically based estimates of flood probabilities", in D. Shrubsole (ed), Resolving Conflicts and Uncertainty in Water Management, Canadian Water Resources Association, Cambridge, Ontario, pp. 18.1-18.10.

Muzik, I. and Chang, C. (1993) "Flood simulation assisted by GIS", in K. Kovar and H.P. Nachtnebel (eds), Application of Geographic Information Systems in Hydrology and Water Resources Management, IAHS Publication No. 211, IHHS Press, Institute of Hydrology, Wallingford, UK, pp. 531-540.

Potter, K.W. and Walker, J.F. (1985) "An Empirical study of flood measurement error". Water Resources Research, Vol. 21, No. 3, pp. 403-406.

Rodriguez-Iturbe, I., Gonzalez-Sanabria, M. and Bras, R.L. (1982) "A geomorphoclimatic theory of the instantaneous unit hydrograph," Water Resources Research, 18, 877-886.

Singh, K.P. (1987) "Flood data, underlying distribution, analysis, and refinement", in V.J. Singh (ed), Hydrologic Frequency Modeling, D. Reidel Publ. Co., Dordrecht, pp. 227-241.

Waylen, P. and Woo, M. (1982) "Prediction of annual floods generated by mixed processes", Water Resources Research, 18, 1283-1286.

COMPARISON OF REGIONAL FLOOD FREQUENCY PROCEDURES FOR CANADIAN RIVERS: AN OVERVIEW

B. BOBÉE[1], K. ADAMOWSKI[2], F. ASHKAR[3], V. NGUYEN[4], J. ROUSSELLE[5], and P. F. RASMUSSEN[1]

[1] INRS-Eau, University of Quebec, 2800 rue Einstein, C.P. 7500, Sainte-Foy (PQ), Canada, G1V 4C7

[2] Dept. of Civil Engineering, University of Ottawa, 161 Louis Pasteur, Ottawa (ONT), Canada K1N 6N5

[3] Dept. of Mathematics, University of Moncton, Moncton (N-B), Canada, E1A 3E9

[4] Dept. of Civil Engineering, McGill University, 817 Sherbrooke West, Montreal (PQ), Canada H3A 2K6

[5] Dept. of Civil Engineering, Ecole Polytechnique, C.P. 6079 Succ. A, Montreal (PQ), Canada H3C 3A7

An initiative has been taken at five Canadian universities and at Hydro-Québec to compare a number of regional flood frequency estimation methods (RFM). Supported by the NSERC Strategic Program, the research project focuses on flood data from Quebec and Ontario, but the results should be applicable to other parts of the country. The aim is to recommend the most adequate RFM's for these flood data.

INTRODUCTION

Hydraulic design works often require the estimation of a flood quantile X_T (flood corresponding to the return period T) at sites where streamflow records are limited or not available. At such locations, it is necessary to estimate flood quantiles using either regionalization methods or rainfall-runoff modeling. A current research project aims at comparing some of the most frequently used regional flood frequency estimation methods by applying them to Canadian flood data. The focus of the project will be on data from Quebec and Ontario, but the results will be applicable to other parts of the country. The time frame of the project is 1991-1994, during which two workshops are planned: one in 1993 to present the objectives and the methodologies of the project (Waterloo International Conference on Stochastic and Statistical Methods in Hydrology and Environmental Engineering), and another at the end of the project in 1994 to present the results.

K. W. Hipel (ed.),
Stochastic and Statistical Methods in Hydrology and Environmental Engineering, Vol. 1, 209–215.

STEPS IN REGIONAL FLOOD FREQUENCY ANALYSIS

Regional flood frequency estimation involves the pooling of data from sites within a region considered hydrologically homogeneous with respect to certain flood characteristics. The regional pooling of information can be regarded as a means of substituting spatial data for temporal data at sites where there is a lack of sufficient information. Generally, regional techniques are used to improve both the at-site estimation of flood quantiles at gauged sites with short record, or to predict such quantiles at sites with no flood record at all (ungauged sites).

The regional approach followed in the Strategic Research Project is the classical one involving three steps: (1) at-site flood frequency analysis; (2) homogeneous region delineation (grouping of catchments into homogeneous groups); and (3) flood-quantile estimation from regional information. The research project intends to compare three principal methodologies for delineation of homogeneous regions ($[MHG]_i$, i=1, ..., 3):

$[MHG]_1$ = canonical correlation (Cavadias, 1990);
$[MHG]_2$ = correspondence analysis (Benzecri *et al.*, 1973) combined with hierarchical ascending classification (Jambu, 1976);
$[MHG]_3$ = "region of influence" approach (Burn, 1990);

Based on homogeneous regions delineated according to the above methods $[MHG]_i$, the application of three general methodologies to estimate flood quantiles ($[MEQ]_j$, j=1, 2, 3) will be examined in the project:

$[MEQ]_1$ = parametric distributions applied to annual flood series (AFS);
$[MEQ]_2$ = nonparametric distributions applied to AFS;
$[MEQ]_3$ = peaks-over-threshold (POT) methodology.

The various combinations of $[MHG]_i$ with $[MEQ]_j$ will be compared in the final stage of the project.

SPECIAL INTEREST OF REGRESSION METHODS IN REGIONAL ESTIMATION

Regionalization techniques based on the regression of flood quantiles on catchment characteristics are among the most frequently used regional methods for flood frequency estimation. Due to the importance of such methods, an initial phase of the project has dealt with that topic. The main objectives of the regression research are: (1) to assess the performance of various regression methods in estimating flood quantiles X_T based on hydrometeorologic and physiographic characteristics; and (2) to analyze the causal relationships between flood quantiles and various hydrometeorologic and physiographic characteristics in order to identify the most important explanatory variables. Various regression models will be tested and compared, and the best method evolving from this comparison will be employed in other phases of the project.

DELINEATION OF HOMOGENEOUS REGIONS

Various approaches to delineating homogeneous groups of basins for the purpose of regional flood frequency analysis have been proposed in the literature, with criteria ranging from geographical proximity to proximity in some multidimensional space of basin and/or flood characteristics. Many classical statistical techniques have been used for this purpose, for example factor analysis, cluster analysis, and canonical correlations analysis. One common characteristic of regional delineation methods is that they all require a proximity measure in order to quantify similarity. Potential attributes for identifying similar stations include flood statistics (e.g. coefficient of variation or skewness of the annual flood series), physical basin characteristics (e.g. drainage area, lake percentage), and hydrometeorologic characteristics (e.g. mean annual precipitation). It is generally necessary to standardize the attributes in order to reduce unit and scale effects.

One major objective of the project is to identify which regional techniques among a set of candidates allow for the most efficient transfer of spatial information to the site of interest. The delineation methods to be compared are outlined in the following sections.

Canonical correlation analysis: $[MHG]_1$

Cavadias (1990) introduced a technique for delineating homogeneous regions based on canonical correlation analysis. The proximity is defined in the space of canonical variables based on flood statistics, and the regions defined in this space are subsequently transferred to the space of canonical variables based on catchment characteristics. An ungauged site can be assigned to a homogeneous region by determining its position in the space of catchment characteristics.

Correspondence analysis: $[MHG]_2$

Essentially, the factorial correspondence analysis (Benzecri *et al.*, 1973) is a principal components analysis applied to data transformed into frequencies. In this method, the classical Euclidian distance is replaced by a distance measure analogue to chi square. The correspondence analysis has several advantages over principal components analysis, in particular that, due to the transformation of the data, the proximity in the factor space is a clear indication of similarity. Based on distances in a reduced factor space (i.e. two or three factors), regions can be objectively delineated by cluster analysis. The so-called hierarchical ascending classification analysis is used (Jambu, 1976).

Region of influence approach: $[MHG]_3$

Many of the classical statistical techniques used in regional flood frequency analysis have produced regions with fixed boundaries. However, Burn (1990) proposed a methodology, referred to as the "region of influence" approach, in which each site is allowed to have a unique set of stations which constitute the region for this site. The Euclidian distance in a p-dimensional attribute space is used as proximity measure. Attributes are standardized by dividing them with the standard deviation of values from all the stations in the study.

The proximity of sites is used to establish a weighting function that reflects the relative importance of each station in the region of influence for the estimation of quantiles at the target station.

The region of influence methodology is currently based on certain subjective inputs, for example (1) the selection of a distance threshold, (2) the choice of attributes to use as a measure of similarity, and (3) the definition of a weighting function that reflects the relative importance of stations.

Test of homogeneity using L-moments

L-moments techniques are not only being used for the delineation of homogeneous regions, but also for parameter estimation and distribution identification. L-moments seem to possess interesting statistical properties which are being investigated in more detail in the project. The L-moment ratio diagram is considered by some investigators to provide an effective tool for identifying the distribution of floods (Pilon and Adamowski, 1992).

Three statistics are employed in regional analysis based on L-moments: a heterogeneity measure (for testing whether the identified regions are homogenous), a discordancy measure (for identifying unusual sites in the region), and a goodness-of-fit measure (for determining whether a candidate distribution gives an adequate fit to the data).

The purpose of this phase of the project is to investigate the value of L-moments in regional flood frequency.

FLOOD-QUANTILE ESTIMATION FROM REGIONAL INFORMATION

Upon the delineation of homogeneous regions by the methods outlined above, three principal methodologies for estimating quantiles are considered in the project. These are briefly discussed in the following sections.

Parametric distributions applied to AFS: $[MEQ]_1$

The classical index flood method is used here. Particular attention is given to the question of which properties should be averaged when computing a regional dimensionless probability distribution. The GEV distribution will be used as regional growth curve. The family of Halphen distributions (Halphen, 1941) is revisited and its suitability for regional flood frequency analysis is examined in the project.

Nonparametric distributions applied to AFS: $[MEQ]_2$

Nonparametric methods can be employed at all three steps of the regional flood analysis procedure: at-site flood frequency analysis, homogeneous region delineation, and regional estimation. Nonparametric frequency analysis is found particularly suitable for the analysis of flood data with a multimodal distribution.

Probability density shape, as determined from nonparametric methods, is related to flood generation mechanisms. For example, unimodal distributions correspond to flood series dominated by snowmelt floods, while bimodal distributions arise from a combination of snowmelt and rainfall floods. Sites can be grouped on a geographical basis to form regions which are homogeneous with respect to distribution shape. A statistical test based on L-moments is used to verify the hypothesis that a delineated homogenous bimodal region contains stations whose data are drawn from the same statistical population. Nonparametric regression is being investigated for the development of regional relationships. The research has led to a regional estimation methodology involving only nonparametric methods.

Peaks-over-threshold (POT) methods: [MEQ]3

The generalized Pareto distribution (GPD), which contains the exponential as a special case, is widely used for modeling flood exceedances in peaks-over-threshold (POT) series. It can be shown that if the exceedances arrive according to a Poisson process, and if they are distributed according to a GPD, then the annual flood maxima will follow a GEV distribution.

Two main approaches have been identified in the project as possible ways to apply the POT approach in a regional context:

1. Model based on direct regression, which seeks to establish a relationship between the at-site estimates of flood quantiles and catchment characteristics. One of the advantages of this method is the flexibility in choosing the type of distribution to represent the exceedances (GPD, exponential, Weibull, etc.)
2. Model based on regionalization of model parameters. By assuming that a common distribution of exceedances applies to all stations in a given region (GPD, for example), and that the shape parameter of the distribution is constant within the region, this approach resembles the AFS index flood method.

COMPARISON OF THE REGIONAL ESTIMATION PROCEDURES

The ultimate objective of the project is to compare the various regional statistical estimation procedures in terms of their capacity to estimate flood quantiles. All combinations of $[MHG]_i$ and $[MEQ]_j$ developed in the first phase of the project will be compared. The comparison will be done using: (1) a computer intensive resampling approach (bootstrap); and (2) a Bayesian approach.

Comparison using the bootstrap resampling method

A major advantage of computer-intensive methods such as the bootstrap is that they allow us to extract stastistical information that cannot be obtained by analytical means. The idea of the bootstrap method is to mimic the flood generating process by drawing with replacement N samples of size n from the observed flood data set. The distribution of the N X_T-flood estimates provides an indication of the statistical accuracy of the estimate obtained from the historical sample. A regional bootstrap procedure, which explicitly preserves the dependence structure between flood samples at different sites in a region, has been developed. At-site and regional bootstrap procedures will be used to compare the different RFM's.

Bayesian comparison

A model comparison methodology based on Bayesian analysis has been conceived (Bernier, 1992), in which the bias and variance of the regional models are quantified either by empirical Bayes or by hierarchical Bayes analysis. At-site quantile estimators are assumed to be normally distributed with negligible bias and a variance that can be easily quantified (for instance by asymptotic theory). From the at-site model, one can obtain unbiased estimates of X_T (with known variances) at M key sites. A given regional model can provide alternative estimates of the same quantities at the M sites, but these estimates have unknown biases and variances. The objective of the empirical and hierarchical Bayes analyses is to quantify the bias and variance of regional models and thus to provide a basis for comparing them.

DATA BASE FOR THE PROVINCES OF QUEBEC AND ONTARIO

The data base used in this study is composed of three distinct data sets containing respectively: physiographical, hydrological and meteorological information for 189 gauging stations in the provinces of Quebec and Ontario.

ACKNOWLEDGMENTS

The financial support provided by the Natural Sciences and Engineering Research Council of Canada is gratefully acknowledged (Strategic Grant, no. STR0118482). The following researchers and students participate in the project: **INRS-Eau:** Dr. L. Mathier, H. Perron, H. Messaudi, I. Abi-Zeid; **McGill:** G. Pandey; **University of Moncton:** Dr. T. Ouarda, S. Hebabi; **University of Ottawa:** Y. Alila, Dr. D. Gingras, J.L. Daviau; **Ecole Polytechnique (Montreal):** Dr. S. Birikundavy, J. Correa; **Hydro-Quebec:** R. Roy. Finally, we owe thanks to Prof. J. Bernier for many simulating inputs to the project.

REFERENCES

Benzecri J.P. et al. (1973) "Analyse des données", Vol. 3, Dunod, Paris.

Bernier, J. (1992) "Modèle régionale à deux niveaux d'aléa", working paper, INRS-Eau, Quebec.

Burn, D.H. (1990) "An appraisal of the region of influence approach to flood frequency analysis", Hydrologic Science J. 35, 2, 149-165.

Cavadias, G. (1990) "The canonical correlation approach to regional flood estimation", Regionalisation in Hydrology. Proc. of the Ljubiana Symposium, IAHS Publ. No 191, pp. 171-178.

Halphen, E. (1941). "Sur un nouveau type de courbe de fréquence", Compte rendu de l'Académie des Sciences, tome 213, pp. 634-5. (Published under the name of D. Dugué).

Hosking, J.R.M. and Wallis, J.R. (1993) "Some statistics useful in regional frequency analysis", Water Resources Research 29, 2, 271-281.

Jambu, M. (1976) "Sur l'interprétation mutuelle d'une classification hiérarchique et d'une analyse des correspondances", Revue de la Statistique Appliquée 24, 2, 45-73.

Pilon, P. and Adamowski, K. (1992) "The value of information to flood frequency analysis using the method of L-moments", Canadian J. Civil Engineering 19, 137-147.

RECENT JAPANESE CONTRIBUTIONS TO FREQUENCY ANALYSIS AND QUANTILE LOWER BOUND ESTIMATORS

K.T. TAKARA[1] and J.R. STEDINGER[2]
[1]Disaster Prevention Research Institute, Kyoto University,
Uji, Kyoto 611, Japan
[2]School of Civil & Environmental Engineering, Cornell University,
Ithaca, NY 14853-3501, U.S.A.

This paper addresses frequency analysis of extreme events such as storms and floods. In particular, it considers two problems: choice of a distribution, and choice of fitting (parameter estimation) method. The authors review recent Japanese contributions to these problems and summarize their results. For distribution selection, they discuss a model selection procedure that uses quantile estimation error estimated using the jackknife method, as well as four goodness-of-fit criteria: SLSC, COR, MLL, and AIC (Takara and Takasao, 1988ab). The procedure evaluated eleven distributions with twelve datasets. Use of two-parameter distributions is recommended from the viewpoint of quantile estimation accuracy for datasets having sample skews greater than 0.38 and less than 1.8. The square-root-exponential distribution for maxima proposed by Etoh et al. (1987) could be useful when the Gumbel distribution does not fit well. As for the choice of fitting method, the authors summarize Monte Carlo studies that evaluate various fitting methods for the lognormal, Pearson type III, log-Pearson type III, Gumbel, and GEV distributions conducted in Japan as well as those in other countries. Recent Japanese research suggests that the quantile lower bound estimators proposed by Iwai (1947), Stedinger (1980), and Hoshi and Leevayanija (1986) are likely to provide more accurate quantile estimators than other procedures.

INTRODUCTION

Hydrologic frequency analysis plays an important role in water resources planning and disaster prevention. Significant issues that need to be addressed include:

- data characteristics (homogeneity, independence),
- sample size (effect of years of records on accuracy and appropriate estimation method),
- parameter estimation (selection of parameter values),
- model evaluation (selection of a distribution), and
- accuracy of quantile estimates (unbiasedness, estimation error).

When dealing with samples of extreme events (datasets of storms, floods and droughts), we generally assume homogeneity and serial independence to make frequency analysis procedures straightforward. This assumption is usually accepted since neither heterogeneity nor dependence is evidently recognized. Even if these assumptions are reasonable, the other four problems are still serious.

Many researchers have worked on these problems (Stedinger et al., 1993). As a result, there are several options for estimation methods and choice of distribution. Practitioners now wonder which estimation method they should use and which is the best probability distribution to employ.

K. W. Hipel (ed.),
Stochastic and Statistical Methods in Hydrology and Environmental Engineering, Vol. 1, 217–234.

The aims of this paper are:

- to describe Japanese research giving some informative results for quantile estimation,
- to give some answers to the above problems by synthesizing the Japanese results with those obtained in other countries, and
- to suggest direction for future research.

CHOICE OF DISTRIBUTION

There are many reasonable probability distributions (frequency analysis models) for modeling extreme events. Important families include the normal, extreme-value type I and type II, and Pearson type III. If we had sufficient knowledge of a hydrologic phenomenon, one could derive its population probability distribution without depending on observed data. If we had sufficiently long record of the phenomenon, we could determine its frequency distribution precisely, as long as the distribution did not change over time. But how should we select a model that approximates the population distribution based on the limited knowledge and data generally available?

For a set of hydrologic data, hydrologists often try several distributions and select one to calculate quantile estimates. When fitting distributions to a set of data, we often find that several distributions achieve almost the same goodness of fit, but can generate different estimates of extreme quantiles. In this situation, how should we select the "best" distribution?

Takara and Takasao (1988ab, 1990) propose an attractive model selection procedure with six steps:

1. Evaluate homogeneity and independence of data;
2. Enumerate several distributions as candidates for quantile estimation;
3. Estimate parameters for each distribution;
4. Screen the distributions to assess goodness of fit (excluding those distributions that give worse fit);
5. Analyze the variability of quantile estimates for distributions that have not been excluded in the prior step, by using a resampling method such as the jackknife or the bootstrap; and
6. Select a distribution that fits data well and exhibits the smallest variability for quantile estimators.

Conventional model selection procedures have generally used goodness-of-fit criteria and/or tests in step 4, and not included steps 5 and 6. When two or more distributions achieve almost the same good results in fitting, conventional procedures have not been able to decide which is most appropriate. The proposed procedure will select the final distribution in terms of both the goodness of fit and the stability of quantile estimators, which is preferable from a practical viewpoint. While it is not difficult to estimate quantile variability, quantile accuracy (which is really of greater concern) is more difficult to determine because the true distribution of the events and its associated parameters are unknown.

Parameter estimation

The maximum likelihood (ML) method is used to obtain the estimates of the parameters for all the distributions. Suppose that X is a hydrologic random variate and its realizations are $x_1, ..., x_N$, where N is the number of observations. Let $f(x;\theta)$ be a probability density function (pdf) corresponding to a cumulative distribution function $F(x;\theta)$, for the variate X, where θ is a parameter vector consisting of k parameters. The method selects the estimates θ that maximizes the likelihood function:

$$L(x_1,\cdots,x_N;\theta)=\prod_{i=1}^{N} f(x_i;\theta) \tag{1}$$

Numerical optimization techniques are employed for distributions having no simple analytical expressions for the maximum likelihood estimators.

Goodness-of-fit criteria

To screen candidate distributions, Takara and Takasao use four criteria to evaluate goodness of fit quantitatively.

Suppose that S is the reduced or standardized variate for X: $S=g(X)$. Let q be a non-exceedance probability. For q^*, a specific value of q, define s^*:

$$s^*=g(F^{-1}(q^*)) \tag{2}$$

Let $y_1, \ldots, y_N$ be the order statistics for the original observations $x_1, \ldots, x_N$, and q_i be the non-exceedance probability assigned to y_i. Note that y_1 is the smallest value. Using the tranformation function g, we obtain

$$s_i=g(y_i) \tag{3}$$

and

$$r_i=g(F^{-1}(q_i)). \tag{4}$$

a) SLSC (standard least-squares criterion): The SLSC was originally introduced for evaluating linearity of the data (order statistics) plotted on a probability paper (Takasao et al., 1986). It is calculated as:

$$\mathrm{SLSC}=\frac{\sqrt{\delta^2_{\min}}}{\left|s^*_{1-q}-s^*_q\right|} \tag{5}$$

in which s^*_q is a specific value of the reduced variate s corresponding to the non-exceedance probability q^*, and $\delta^2_{\min}$ is obtained by minimizing:

$$\delta^2=\frac{1}{N}\sum_{i=1}^{N}(s_i-r_i)^2. \tag{6}$$

This minimization operation corresponds to the so-called least-squares method (or one of the graphical fitting methods) based on a plotting position formula such as:

$$q_i=\frac{i-\alpha}{N+1-2\alpha} \tag{7}$$

in which α is constant. They recommend the use of Hazen's formula ($\alpha=0.5$) to give q_i. The denominator in (5) is introduced to standardize the square root of $\delta^2_{\min}$. Thus the SLSC can be used to compare goodness of fit across distributions. For the non-exceedance probability q in (5), $q=0.99$ is used because most hydrologic samples have less than 100 observations. Smaller SLSC values imply better fits. The SLSC is useful for absolute goodness-of-fit evaluation as well as comparison (relative evaluation). When using the ML method instead of the least-squares method, we substitute δ^2 obtained by the ML method into (5).

Takasao et al. (1986) fitted five two-parameter distributions (the normal, log-normal, exponential, Gumbel, and log-Gumbel distributions) by the least-squares method to samples in the Lake Biwa basin: the monthly and yearly precipitations and inflows, and the annual maximum m-day precipitations (m=1, 2, and 3). They concluded that SLSC $\approx$ 0.02 corresponds to a good fit; if SLSC>0.03, other distributions should be tried. They also compared six plotting positions formula: the Weibull ($\alpha=0.0$), Adamowski ($\alpha=0.25$),

Blom ($\alpha = 0.375$), Cunnane ($\alpha = 0.4$), Gringorten ($\alpha = 0.44$), and Hazen ($\alpha = 0.5$). For the 70-year datasets of annual maximum m-day precipitations (m=1, 2, and 3), Hazen's formula gave better quantile estimates than the other five formulas for the log-normal and Gumbel distributions (Here, "better" means "nearest to those obtained by the ML method".), so it was recommended.

b) COR (correlation coefficient): Another goodness-of-fit criterion is the correlation coefficient between the ordered statistics y_i and r_i:

$$\mathrm{COR} = \frac{\sum_{i=1}^{N}(y_i - \bar{y})(r_i - \bar{r})}{\left[\sum_{i=1}^{N}(y_i - \bar{y})^2 \sum_{i=1}^{N}(r_i - \bar{r})^2\right]^{1/2}} \tag{8}$$

in which $\bar{y}$ and $\bar{r}$ are the means of y_i and r_i, respectively. Values of COR closer to unity corresponds to better fits. This probability plot correlation coefficient test has been applied to the normal, lognormal, and Gumbel distributions (Vogel, 1986), and the Pearson type III distribution (Vogel and McMartin, 1991).

c) MLL (maximum log-likelihood): For the ML method, for reasons of computational tractability, we usually maximize the log-likelihood function obtained by taking logarithms of the likelihood function. The maximized log-likelihood is given by

$$\mathrm{MLL} = \sum_{i=1}^{N} \log f(x_i;\hat{\theta}) \tag{9}$$

where $\hat{\theta}$ is the maximum likelihood estimator of θ. When several distributions are fitted to a sample, the distribution that gives the greatest MLL value can be regarded as fitting the best.

The MLL is not only the maximum of the log-likelihood; it has some interpretation from a viewpoint of information theory. If the population distribution is known, the Kullback-Leibler (KL) information is used as an evaluation criterion for the models that approximate the population distribution. In general, it is unknown; then the MLL can be used as an alternative criterion instead of the KL information (Sakamoto et al., 1983).

d) AIC (Akaike information criterion): In general, distributions with three free parameters fit better than those having two free parameters. As the number of parameters increases, goodness of fit should appear to improve: the SLSC values decrease and the COR and MLL values increase. Consequently, as long as the SLSC, COR, or MLL is used, distributions having more parameters tend to be evaluated as "better". In the evaluation of models, we must consider model simplicity as well as the goodness of fit. The AIC proposed by Akaike (1974) balances the number of parameters k and the quality of fit using:

$$\mathrm{AIC} = -2 \log(\text{maximum likelihood}) + 2k = -2\mathrm{MLL} + 2k \tag{10}$$

As k increases, the second term of (10) increases, while the first term decreases because the goodness of fit becomes better (the MLL increases). Akaike (1974) suggests that the model that minimizes the AIC is best. The AIC has been applied to hydrology, for example, in determination of the optimal order of time series models (Hipel et al., 1977) and in evaluation of runoff models (Takasao et al., 1984).

Resampling methods for evaluating variability of quantile estimates

Two resampling methods, the jacknife and the bootstrap, are employed to quantify the variability (estimation error) of quantile estimators. The resampling methods correct the bias of the statistics obtained from the original dataset and estimate the variance of the statistics, producing many datasets by repeatedly sampling a part of the data from the original set or by repeatedly drawing samples of the same size as the original set with replacement. Datasets produced in this way are relatively easily generated using computers. Examples of these resampling methods are provided by Bardsley (1977), Tung and Mays (1981), Cover and Unny (1986), and Potter and Lettenmaier (1990).

Analytical expressions are available for the variance of quantile estimators in many cases (Stedinger at al., 1993). Namely, the variance is given by a function of the non-exceedance probability q, sample size N, and parameter vector θ:

$$\mathrm{Var}(\hat{x}_q) \approx \mathrm{func}(q, N, \theta) \tag{11}$$

Equation (11) provides accurate variances for the normal distribution and some other two-parameter distributions. For some distributions (in particular three-parameter distributions), equation (11) provides approximate values that may be inacurate (Hoshi et al, 1984). In addition, this kind of equation has not been developed for some distributions. On the other hand, resampling methods can be used for any distributions. They directly provide the bias-corrected quantile estimators and their variance, as well as their empirical distributions from observed data. If there is good coincidence between the variance analytically provided by (11) and that obtaind by resampling methods, then we can use either one; though this kind of research has not been done. Takara and Takasao used the jackknife on all distributions thereby being consistent.

Let $\psi(x_1, \ldots, x_N)$ be a statistic representing some characters of the parent dataset. It is obtained by using N data: $x_1, \cdots, x_N$. Then the jackknife algorithm is:

i) For a sample of size N, using all the data, we obtain the estimate of ψ

$$\hat{\psi} = \psi(x_1, \cdots, x_N) \tag{12}$$

ii) Using the N- 1 data excluded the ith datum ($i = 1, \ldots, N$), we obtain

$$\hat{\psi}_{(i)} = \psi(x_1, \cdots, x_{i-1}, x_{i+1}, \cdots, x_N) \tag{13}$$

iii) The average of $\hat{\psi}_{(i)}$ is given by

$$\hat{\psi}_{(\cdot)} = \frac{1}{N}\sum_{i=1}^{N}\hat{\psi}_{(i)} \tag{14}$$

iv) Quenoulli's estimate of bias is

$$\mathrm{Bias} = (N-1)(\hat{\psi}_{(\cdot)} - \hat{\psi}) \tag{15}$$

Consequently, the jackknife estimate $\hat{\psi}_J$ corrected for bias is

$$\hat{\psi}_J = \hat{\psi} - \mathrm{Bias} = N\hat{\psi} - (N-1)\hat{\psi}_{(\cdot)} \tag{16}$$

The jacknife estimate of variance of the statistic ψ is

$$\hat{s}_J^2 = \frac{N-1}{N}\sum_{i=1}^{N}(\hat{\psi}_{(i)} - \hat{\psi}_{(\cdot)})^2 \tag{17}$$

It is known that $\hat{\psi}_J$ has a bias of order N^{-2} (Efron, 1982).

A cumulative distribution function $F(x)$ is given by

$$F(x) = \int_{-\infty}^{x} f(t;\theta)\,dt \tag{18}$$

Using F, we obtain the quantile x_q for a non-exceedance probability q:

$$x_q = F^{-1}(q) \tag{19}$$

where the return period (recurrence interval) T is

$$T = \frac{1}{1-q} \tag{20}$$

Regarding the quantile x_q (T-year event) as the statistic ψ in the jackknife algorithm, we can easily quantify the variability of quantile estimators. Namely, fitting a distribition to the sub-datasets constructed in the manner of resampling through the ML method, we obtain the quantile estimators corresponding to $\hat{\psi}_{(i)}$ in (13) and then obtain the jackknife estimate of the quantile and its variance given in (16) and (17), respectively. The proposed procedure for selection of a quantile estimation method, in the step 5, compares the variability of the quantile estimate for each distribution obtained by this method.

Distributions compared and data used

Takara and Takasao (1988b, 1990) considered eleven distributions: the normal, N(2); lognormal, LN(2, 3); Pearson type III, PIII(2, 3); log-Pearson type III, LPIII(3); Gumbel, G(2); log-Gumbel, LG(2, 3); generalized extreme-value, GEV(3) distributions; and the square-root-exponential distribution for maximum, SQET(2) proposed by Etoh et al. (1987). The SQET(2) distribution has a cumulative distribution function:

$$F(x) = \begin{cases} 0 & (x < 0) \\ \exp\left[-\lambda\left\{\left(1+\sqrt{\beta x}\right)\exp\left(-\sqrt{\beta x}\right)\right\}\right] & (x \geq 0) \end{cases} \tag{21}$$

where β and λ are parameters and corresponds to modeling the square-root of x by a Gumbel distribution.

The abbreviations for the distributions introduced here will be used hereafter for simplicity. Note that the number in the parentheses denotes the number of free parameters. Though the LG(3) distribution is essentially identical to the GEV(3) distribution, the expressions are different and different algorithms were employed to obtain parameter estimates; thus different quantile estimates are obtained in most cases.

The model selection procedure was applied to twelve datasets of annual maxima of precipitation and flood discharge (Takara and Takasao, 1988ab, 1990b):

- daily precipitation at Osaka (1889-1980);
- m-day precipitation at Otsu (1912-1985, m =1, 2, 3);
- m-day precipitation at Hikone (1912-1985, m =1, 2, 3);
- m-day areal precipitation in the Lake Biwa basin (1912-1985, m =1, 2, 3);
- daily discharge of the St. Marys River at Stillwater (1916-1974) (Kite,1977); and
- peak discharge of the Ganga River at Hardwar (1885-1971) (Raghunath, 1985).

The respective data included in the datasets are assumed to be serially independent and homogeneous.

Results and discussions

This section introduces results of the analysis of the annual maximum precipitation at Osaka (Figure 1, Tables 1 and 2). A summary of the results for all the datasets is given in Table 3.

(a) Evaluation in terms of goodness of fit for Osaka:

Figure 1 shows the result of fitting the LN(3), PIII(3), and SQET(2) distributions to the annual maximum daily precipitations at Osaka. Table 1 shows the values of thegoodness-of- fit criteria for the eleven distributions with this dataset. The SLSC, COR, and MLL gave good rankings to distributions with three adjustable parameters such as the LN(3), PIII(3), LPIII(3), and LG(3), while the best distribution for the AIC was the SQET(2) distribution with only two parameters. From the results shown in Figure 1 and Table 1, we could not identify which distribution should be used to estimate quantiles.

(b) Evaluation in terms of quantile estimator variability for Osaka:
For the Osaka dataset, Table 2 shows the estimates of three *T*-year events (*T*= 50, 100, 200), and the estimates of their standard errors using the jackknife method, for the models which not excluded in the model screening step. Table 2 indicates that:
1) The SQET(2) distribution was ranked as the best for the dataset of annual maximum precipitation at Osaka in terms of the smallest variation of the quantile estimates.
2) The variations (standard errors) of the *T*-year precipitations for *T*= 50, 100, and 200 for the SQET(2) distribution were about 6.2, 6.7, and 7.1%, respectively.
3) Three-parameter distributions tend to give larger variability of quantile estimates. In general, three-parameter distributions fit each dataset well; but larger quantile variation is a penalty that results from such flexibility.

TABLE 1. Comparison of the goodness of fit for the annual maximum daily precipitation at Osaka (92 years, 1989-1980)

Distributions	(*k*)	SLSC	COR	MLL	AIC
Normal	(2)	0.07937	0.9312	-450.15	904.30
Lognormal	(2)	0.02996	0.9902	-434.91	873.83
Lognormal	(3)	0.01666***	0.9970***	-432.82***	871.64**
Pearson III	(2)	0.06116	0.9685	-438.17	880.34
Pearson III	(3)	0.03765	0.9865	-432.90**	871.80*
Log-Pearson III	(3)	0.01749**	0.9967**	-432.91*	871.82
SQET	(2)	0.02423	0.9932	-433.09	870.18***
Gumbel	(2)	0.04769	0.9846	-434.41	872.83
GEV	(3)	0.02124	0.9944	-433.17	872.34
Log-Gumbel	(2)	0.03496	0.9895	-434.53	873.06
Log-Gumbel	(3)	0.01858*	0.9960*	-433.17	872.34

The number in the parentheses is the number of parameters. *** indicates the best distribution for each criterion; ** and * the second and the third, respectively.

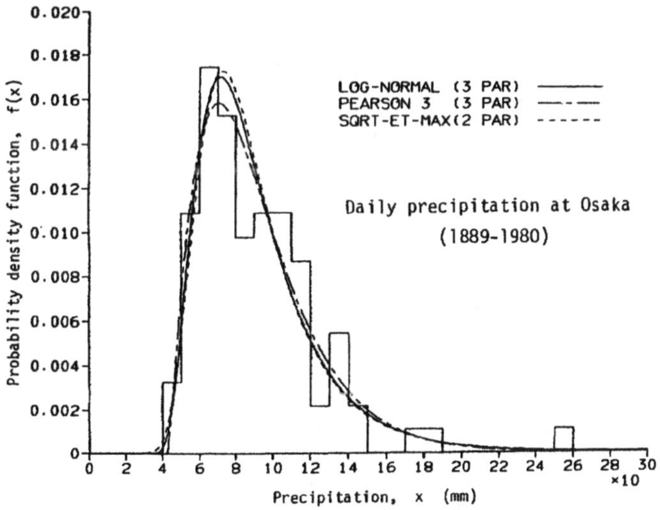

Figure 1. Three distributions fitted to the annual maximum precipitations at Osaka (1889-1980) by the maximum likelihood method.

TABLE 2. The jacknife estimates and standard error for T-year daily precipitation at Osaka (in mm, T = 50, 100, and 200)

Distributions	(k)	T = 50	T = 100	T = 200
SQET	(2)	180.46 [11.25]	203.56 [13.58]	227.85 [16.08]
Pearson III	(3)	172.82 [14.68]	189.34 [17.34]	205.51 [20.36]
Lognormal	(3)	179.94 [17.83]	201.66 [23.69]	224.17 [30.44]
Log-Pearson III	(3)	181.99 [19.56]	205.69 [27.06]	230.87 [36.13]
Log-Gumbel	(3)	182.95 [21.07]	207.62 [30.35]	233.86 [43.03]
GEV	(3)	183.10 [21.19]	207.83 [30.42]	234.15 [42.13]

The value in the square brackets is the standard error obtainded by the jackknife.

TABLE 3. Best distributions for each criterion for each datasets

Dataset	Skew-ness	* Goodness-of-fit criteria SLSC	COR	MLL	AIC	*Quantile variability	Rank by G-o-F **
Precipitation							
Osaka (m=1)	1.785	LN(3)	LN(3)	LN(3)	SQET(2)	SQET(2)	5, 5, 4, **1**
Ohtsu (m=1)	0.841	LN(3)	LN(3)	PIII(3)	PIII(3)	G(2)	4, 4, 7, **3**
Ohtsu (m=2)	0.382	LN(3)	LN(3)	PIII(3)	PIII(2)	PIII(2)	4, 4, 2, **1**
Ohtsu (m=3)	0.607	LN(3)	LN(3)	PIII(3)	PIII(2)	PIII(2)	4, 2, 2, **1**
Hikone (m=1)	0.931	LPIII(3)	LPIII(3)	PIII(3)	LN(2)	G(2)	6, 5, 7, **2**
Hikone (m=2)	1.362	LN(3)	LN(3)	PIII(3)	LN(2)	G(2)	6, 6, 7, **2**
Hikone (m=3)	1.244	LN(3)	LN(3)	LN(3), PIII(3)	LN(2)	G(2)	5, 5, 6, **2**
Lake Biwa (m=1)	1.071	LN(3)	LN(3)	LN(3), PIII(3)	SQET(2)	G(2)	8, 4, 7, **2**
Lake Biwa (m=2)	0.990	PIII(3)	PIII(3)	PIII(3)	PIII(3)	G(2)	2, 2, 8, **4**
Lake Biwa (m=3)	0.860	LN(3)	LN(3)	PIII(3)	LN(2)	G(2)	5, 5, 8, **4**
Flood							
St Marys	1.369	LG(3)	LG(3)	PIII(3)	LN(2)	G(2)	8, 7, 7, **3**
Ganga	1.461	LN(3)	LN(3)	PIII(3)	SQET(2)	SQET(2)	6, 5, 4, **1**

* The number in the parentheses denotes the number of parameters included in the distributions, which are denoted by the abbreviations:
LN, lognormal; PIII, Pearson type III; LPIII, log-Pearson type III; G, Gumbel; LG, log-Gumbel; and SQET, SQRT-ET-max.

** The numbers in the "Rank by G-o-F" column indicate the ranking of the best model based on quantile variability in terms of SLSC, COR, MLL, and AIC, respectively.

(c) Summary of the results for all the datasets:

Table 3 shows the best distributions for each criterion for all twelve datasets. This table indicates that:

1) The SLSC, COR, and MLL goodness-of-fit criteria always select three-parameter distributions. The MLL tends to select PIII(3), while the SLSC and the COR tend to select LN(3). As for the relationship between SLSC and COR, Takara and Takasao (1989) verified the almost one-to-one correspondence between COR and SLSC using the ML method. SLSC≈0.02 corresponds to COR≈0.995, and SLSC >0.03 corresponds to COR<0.990. Since the SLSC and COR criteria give almost the same rankings, COR is to be recommended because it is simpler than SLSC.
2) The AIC tends to select two-parameter distributions: LN(2), PIII(2), and SQET(2). In only two cases, was the PIII(3) distribution selected.
3) The quantile variability criterion always selected two-parameter distributions. The

Gumbel distribution was selected for eight cases; each of SQET(2) and PIII(2) was selected for two cases.

4) The four numbers in the "Rank by G-o-F" column show the ranking of the best model based on the quantile variability, for the four goodness-of-fit criteria: SLSC, COR, MLL, and AIC, respectively. AIC picks well fitting distributions which also provide less variable quantile estimators.

Paying attention to the sample skew of the datasets and the distribution selected by the quantile variability criterion, we observed that:

5) The square-root-exponential (SQET(2)) distribution proposed by Etoh et al. (1987) is selected by the AIC as the best distribution for the precipitation at Osaka (sample skew is 1.785) and the flood discharge in the Ganga River in India (sample skew is1.461). These are the two largest skews. Since the SQET(2) has a longer upper tail than the PIII(2) and G(2) distributions, it is able to fit data having a skew larger than 1.1396, the popuration skew of the Gumbel distribution.
6) The two-parameter Pearson type III (PIII(2)) distribution is selected for the datasets of 2- and 3-day precipitation at Otsu (sample skews are 0.382 and 0.607, respectively). The PIII(2) apppears to be a good alternative when the sample skew is much less than 1.1396.
7) For datasets with the sample skew between 0.841 and 1.369, the Gumbel distribution was selected.

The proposed evaluation procedure using the variability of quantiles as a criterion tends to select two-parameter models rather than three-parameter ones, because in general two-parameter models have smaller variability. In fact, all of the best models for the above twelve sets of extreme-value hydrologic data have two parameters. Three-parameter models fit better to each dataset; but fitting better means that the quantile estimates change much more; the improvement over two-parameter distribution may reflect overfitting. Only when none of the two-parameter distributions provide a good fit, we would select one of the three-parameter models.

From the above discussion, we recommend the following procedure. When fitting distributions to extreme maximum data, try the Gumbel distribution first. If it fits well (SLSC<about 0.02 or COR>about 0.995), then adopt it as the best one without trying other models. However, datasets of flood discharges often have higher skewness such as 4.0 or 5.0; the Gumbel distribution can not fit such data well. When the Gumbel distribution does not provide a good fit, try other two-parameter distributions using Takara and Takasao's procedure. Based on the results obtaind for the twelve datasets, the PIII(2) could be the best for two samples with the skew less than 0.7; and the SQET(2) for two samples with the skew greater than 1.4 and less than 1.8.

In some sense by using the sample skew to select among two parameter models, the procedure is a three parameter estimation method. The jackknife analysis does not include the variability in quantile estimators introduced by that model selection step. This deficiency can be overcome if the analysis of several (independent) datasets for a region (defined in some reasonable way) can identify a statistically reasonable and physically appropriate two-parameter model to employ with each of the datasets. This procedure should be successful in many cases. If none of the two-parameter models fit well, apply the proposed procedure to three-parameter models and adopt the model giving the smallest quantile estimate variability.

Recently, Lu and Stedinger (1992) compared the two- and three-parameter GEV/PWM quantile estimators in terms of mean square error (MSE) for the 100-year flood estimation. They found that the two-parameter estimators still have smaller MSE than the three-parameter estimator despite substantial errors in the shape parameter. They described the combinations of sample size and the shape parameter for which the Gumbel/PWM estimator (i.e. the two-parameter GEV/PWM estimator with the shape parameter fixed to zero) is superior to the three-parameter GEV/PWM estimator. For reasonable values of the

GEV shape parameter κ corresponding to skews on the order of 1, use of the Gumbel distribution for quantile estimation is recommended rather than the GEV (3) distribution for samples of size $N < 1/(1-P)$.

For twelve sets of annual maximum hydrologic data, the goodness of fit of eleven distributions and the variability (estimation error) of the quantile estimates obtained with those models have been quantified. Since the variability of quantiles is very important in practice, adopting it as a model evaluation criterion is reasonable and understandable. The resampling methods are useful for quantifying it in a consistent manner accross distributions. With the availability of computers and statistical data processing techniques, the proposed model evaluation procedure can be easily implemented.

CHOICE OF PARAMETER ESTIMATION METHOD

There are several philosophies upon which fitting methods (parameter estimation methods) can be based. The most common are methods of moments, probability weighted moments, quantiles, maximum likelihood, maximum entropy, and least squares. It is difficult to definitely say which is the most appropriate method for a particular model and dataset. However, Monte Carlo experiments are able to evaluate fitting methods for a range of distributions. The experiments basically include the following procedure:

1. Assume a population distribution (including population parameters);
2. Generate many samples of a specified size, whose observations are drawn from the assumed popuration by using random number generation techniques;
3. Fit the distribution model to each sample by some method and obtain estimators of parameters and quantiles; and
4. Evaluate the fitting method in terms of accuracy of the estimators.

This paper reviews Monte Carlo experiments reported in the literature with an emphasis on recent Japanese contributions from the viewpoint of quantile estimation and on quantile lower-bound estimators. We discuss evaluation of fitting methods for the lognormal (LN), Pearson type III (PIII), log-Pearson type III (LPIII), Gumbel (G), and generalized extreme-value (GEV) distributions.

Table 4 summarizes recent Japanese papers and related papers in other countries. It shows what distributions and methods were addressed, as well as sample sizes considered. Abbreviations for fitting methods are used for simplicity (e.g., ML, the method of maximum likelihood; MoM, the method of moments). See the explanation given below the table. The methods with asterisk in the column of fitting methods were generally best for estimating quantiles with small exceedance probabilities. This
preference may depend on population skew and sample size.

From Table 4, we notice that:

1) The maximum likelihood (ML) method gave accurate quantile estimators for two-parameter distributions: LN(2), G(2).
2) For three-parameter distributions such as the LN(3), PIII(3), and LPIII(3), the quantile lower bound estimation method coupled with the methods of moments or maximum likelihood method (Qt&MoM or Qt&ML) gives more accurate quantile estimates.

The following paragraphs summarize the results for choice of fitting method for the distributions in Table 4, including the concept of a quantile lower bound estimator.

Lognormal distribution

The lognormal distribution has a long history. It was first used by Gulton in 1875 (Chow, 1954). In hydrology, Hazen (1914) used the lognormal probability paper for estimating flood quantiles. During the 1930s, Grassberger in Germany, Gibrat in France, and Slade in the U. S. A. discussed applications in these countries. Iwai (1947) introduced it into hydrology in Japan.

There are alternative methods for fitting the lognormal distribution to data. Chow (1954), Sangal and Biswas (1970), Burges et al. (1975) and others have discussed a range of fitting methods for the LN(3) distribution. In Japan, Iwai (1947, 1949, 1950), Ishihara

TABLE 4. Japanese and related Monte Carlo studies comparing various fitting methods

P.D.F.	Paper	Fitting methods	Sample sizes
LN(2)	Stedinger (1980)	5: ML*, MoM(b), modified ML, MoM(u), MoM(W)	10, 25, 50
LN(3)	Stedinger (1980)	4: Qt(S)+MoM(real)*,Qt(S)+ML,MoM(W), MoM(u)	25, 50
	Hoshi et al. (1984)	7: Qt(S)+MoM(real)*, MoM(u), Qt(I)+MoM(real), ML, Qt(S)+MoM(log),Qt(I)+MoM(log), MoM(BR)	20, 40, 80
	Takeuchi & Tsuchiya (1988)	7: Qt(S)+MoM(real)*, MoM(u), Qt(S)+MoM(log), PWM, Qt(S)+PWM(log,real), Sx	10 - 100
	Takara & Takasao (1990a)	10: Qt(I)+MoM*, MoM(b)*, MoM(u), PWM, LS(H), ML, MoM(IT), MoM(BR), LS(W), Sx	10 - 500
PIII(3)	Hoshi & Leeyavanija (1986)	5: Qt(S)+MoM*, Qt(S)+ML*, Sx*, MoM(u, BR)	20, 40, 80
LPIII(3)	Hoshi and Leeyavanija (1986)	6: Qt(S)+MoM(real)*, MoM(real)*,Qt(S)+MoM(log), Sx, Qt(S)+ML, MoM(BR)	20, 40, 80
G(2)	Landwehr et al. (1979)	3: ML*, PWM, MoM(u)	5 - 999
	Lettenmaier & Burges (1982)	4: ML*, MoM(b), MoM(u), Gum	10 - 100
	Raynal & Salas (1986)	7: ML*, BLCOS, PWM, MoM(b, u), LS(W), MIR	9, 19, 49, 99
	Phien (1987)	4: ML*, ME, PWM, MoM(u)	30, 40, 50
	Takara et al. (1989)	6: ML*, ME, PWM, MoM(u), LS(H), LS(W)	10 - 1000
GEV(3)	Hosking et al. (1985)	3: PWM*, Sx, ML	15, 25,50,100
	Takara et al. (1989)	4: PWM*, ML, MoM(u), MoM(b)	10 - 1000

Distributions considered are: **LN**, lognormal; **PIII**, Pearson type III; **LPIII**, log-Pearson type III; **G**, Gumbel; **GEV**, generalized extreme-value; the number of parameters is given in the parentheses after the abbreviation.

Fitting methods are: **MoM**, method of moments, where the abbreviations (b, u) for two-parameter distributions indicates that each uses the biased and unbiased variances, respectively, whereas the abbreviations (b, u, W, BR, and IT) for three-parameter distributions indicate (b) the skew based on the biased variance, (u) the skew based on unbiased variance, (W) the standard deviation and skew corrected by Wallis et al. (1974), (BR) the skew corrected by Bobée and Robitaille (1975), and (IT) the skew corrected by Ishihara and Takase, respectively ('log' and 'real' indicate that the variable is considered in log and real space, respectively); **PWM**, probability weighted moments; **ML**, maximum likelihood; **Qt**, quantile method using lower-bound-parameter estimation proposed by Iwai (I) or Stedinger (S); **Sx**, sextile method; **LS**, least-squares (the H and W indicate the use of plotting formula by Hazen and Weibull, respectively); **Gum**, Gumbel's regression method; **BLCOS**, best linear combination of order statistics method; **MIR**, mode and interquartile range method; **ME**, maximum entropy method.

Note: In the fitting method column, * indicates generally best. However, best methods may change with the population skew, the sample size, and the exceedance probability for quantiles.

and Takase (1957), Kadoya (1962) proposed their own methods. Here, we review several Monte Carlo studies which evaluated various fitting methods for the lognormal distribution.

For the LN(2) distribution, Stedinger (1980) evaluated the five fitting methods shown in Table 4. Four population skews of 0.377, 0.766, 1.63, and 4.00 were considered with sample sizes of 10, 25, and 50. The maximum likelihood (ML) method (i.e. use of the mean and sample variance of $\log x_i$) is generally best for estimating $x_{0.99}$, the quantile with a non-exceedance of 0.99, in terms of root mean square error (RMSE). The ML is inferior to MoM(b) and MoM(u) only for a case where the population skew is largest (4.00) and the sample size smallest ($N=10$).

The LN(3) distribution has three parameters: μ_y, σ_y, and ξ, where $y=\log(x-\xi)$. It is well known that the lower bound parameter (i.e. location parameter) ξ can be difficult to estimate accurately. Greater variablity of ξ affects the estimation accuracy of quantiles. Therefore, a good procedure for fitting three-parameter models to data and obtaining less variable quantile estimates is to combine the ML or MoM, which are good for two-parameter models, with some techniques which provides a reasonable estimate of the lower bound parameter ξ.

The method of quantiles can provide good estimates of ξ. Stedinger (1980) evaluated four fitting methods for the LN(3) distribution and showed that using the method of quantiles to estimate ξ is likely to give generally good estimates for both $x_{0.01}$ and $x_{0.99}$. Hoshi et al. (1984) verified the superiority of a method of quantiles by comparing seven methods.

The method of quantiles uses the relation among the median $x_{0.5}$, the q and $1-q$ quantiles x_q and x_{1-q}, and the lower bound parameter ξ:

$$\xi = \frac{x_q x_{1-q} - x_{0.5}^2}{x_q + x_{1-q} - 2x_{0.5}} \tag{22}$$

Given the lower bound estimate, one can estimate the other parameters included in the distribution considered.

Iwai (1947) proposed use of this method with $x_{0.5}=x_{\text{median}}$ (the sample median) and a pair of $x_q=x_{(N+i-1)}$ and $x_{1-q}=x_{(i)}$ for large samples, where $x_{(i)}$ is the i-th smallest order statistic of the N observations:

$$\xi_i = \frac{x_{(N+i-1)}x_{(i)} - x_{\text{median}}^2}{x_{(N+i-1)} + x_{(i)} - 2x_{\text{median}}} \tag{23}$$

For small samples, he recommended use of the geometric mean of the observations rather than the sample median, and employed the averaged lower bound:

$$\bar{\xi} = \frac{1}{M}\sum_{i=1}^{M}\xi_i \tag{24}$$

where M is the nearest integer to $N/10$. This estimator focuses upon the extreme observations corresponding to probabilities less than 10 per cent and greater than 90 per cent. Such an idea was applied in a ratio-test for goodness of fit presented by Kimball (1936). This procedure using (23) and (24) was called Iwai's method (Iwai, 1949), and has been widely used in Japan even with large samples.

Stedinger (1980) recomended use of ξ_1 (M=1)which is based on the largest and smallest observed values. Hoshi et al. (1984) compared Stedinger's recommendation (Qt(S) in Table 4), Iwai's method (Qt(I)), and other methods. Their results showed that Stedinger's quantile method generally provides better quantile estimators. The estimator in (23) has been called the quantile-lower-bound estimator (Hoshi et al., 1984; Stedinger et al., 1993).

Takeuchi and Tsuchiya (1988) also conducted Monte Carlo experiments on LN(3) distribution to examine the performance of PWM method, Qt(S)&PWM, and the sextile method proposed by Jenkinson (1969) originally for the GEV distribution, in comparison with MoM and Qt(S)&MoM. According to their results, the PWM method had no significant advantage for the LN(3) in terms of bias or RMSE for the quantiles with non-exceedance probabilities of 0.99 and 0.998. The sextile method was not very good. The MoM and Qt(S)&MoM(RS) gave the smallest RMSE. In particular, the MoM gave the smallest RMSE for samples of 40 or less when the population skew is 0.766 and 1.625.

Takara and Takasao (1990) reported another Monte Carlo analysis to evaluate ten fitting methods including Iwai's quantile method (Table 4). They evaluated the methods in terms of both the bias and RMSE of estimators of quantiles with non-exceedance probabilities of 0.98, 0.99, and 0.995. Results were as follows.

1) The MoMs did well for small samples with sizes of 50 or less, which is consistent with results obtained by Hoshi et al. (1984) and Takeuchi and Tsuchiya (1988). Especially, the MoMs using the skew estimators by Ishihara and Takase (1957), which has been often used in Japan, and by Bobée and Robitaille (1975) always (i.e. for sample sizes of 10 to 500) gave small bias, though they were not good in terms of RMSE.
2) Takeuchi and Tsuchiya's PWM method gave unbiased quantile estimates when the sample is large (100 or more).
3) For large samples with sizes of 70 or greater, the LS(H) was good; the LS(W) was always worse than the LS(H). The least-squares method can be regarded as an objective graphical analysis using probability papers. The use of the Weibull formula on the lognormal probability paper gave very biased estimators, which was also verified by Takasao et al. (1986).
4) The ML gave almost the same RMSE and bias as the LS(H) for large samples with size greater than 70; it was not good for small samples.
5) The sextile method did not do well.
6) Iwai's quantile method Qt(I)&MoM showed good performance in terms of RMSE. It gave the smallest RMSE for the samples with the sizes of 30 to 200, as well as unbiased quantile estimators for small sample with size of 20. In terms of bias, however, it gave relatively large negative bias (i.e. often underestimated the quantiles).

From the results obtained by Stedinger (1980), Hoshi et al. (1984), Takeuchi and Tsuchiya (1988) and Takara and Takasao (1990), use of the quantile-lower-bound estimators combined with the MoM and ML is likely to be a good method to obtain accurate quantile estimators for the LN(3).

Pearson type III and log Pearson type III distributions

Gamma-type distributions have been often used in hydrologic frequency analysis especially in the United States since 1960s. Bobée and Ashkar (1991) have completed a comprehensive text summarizing various aspects of gamma-type distributions, as well as computer programs which are available for small computers.

Hoshi and Leeyavanija (1986) developed the quantile-lower-bound estimator for the PIII(3) distribution:

$$\hat{\xi} = \frac{(G_1 - 3E^2/4) \pm \sqrt{(G_1 - 3E^2/4)^2 - 4(G_2 - 3E/2)(G_3 - E^3/8)}}{2(G_2 - 3E/2)} \tag{25}$$

where

$$G_1 = x_{0.5}x_{1-q} + x_{1-q}x_q + x_q x_{0.5} \quad ; \quad G_2 = x_{0.5} + x_{1-q} + x_q$$
$$G_3 = x_{0.5}x_{1-q}x_q \quad ; \quad 3E = 8x_{0.5} - x_{1-q} - x_q \tag{26}$$

As was done by Stedinger (1980) for the LN(3) distribution, Hoshi and Leeyavanija (1986) employed the sample median, largest, and smallest values of observetions as the best choices of $x_{0.5}$, x_q, and x_{1-q}, respectively.

a) Pearson type III distribution: They compared five methods (see Table 4) for the PIII(3) distributions to samples with sizes of 20, 40, and 80 drawn from a PIII(2) distribution with population skews of 0.5, 1.0, and 2.0 (which correspond the coefficients of variation (CV) of 0.25, 0.50, and 1.00). Qt(S)&ML, Qt(S)&MoM, and sextile methods yielded a very good all-around quantile estimate and generally peformed best or about best, in terms of the RMSE of estimates of quantiles with non-exceedance probabilities 0.01, 0.99, and 0.998. When fitting the PIII(3) distribution, there is little advantage to using unbiased skew estimates in the MoM for estimating the upper quantiles with samples from a population with skew of 1.0 or less.

b) Log Pearson type III distribution: Hoshi and Leeyavanija (1986) employed the quantile-lower-bound estimator with the LPIII(3) distribution. They conducted Monte Carlo experiments using two population distributions: one had the skew of 3.95 (CV was 0.66) and thus had a lower bound, and the other had a skew of 1.0 (CV was 0.5) and thus had an upper bound. Six methods (Table 4) were fitted to samples of 20, 40, and 80 observations drawn from the secified population sets. Replacing $x_{0.5}$, x_q, and x_{1-q} with $\ln x_{0.5}$, $\ln x_q$, and $\ln x_{1-q}$, respectively, they used (25) and (26) to obtain the lower bound estimate, which is used in Qt(S)&MoM(real), Qt(S)&MoM(log), and Qt(S)&ML.

For the LPIII(3) distribution having the lower bound, they conclued as follows. The most accurate estimates of high extreme quantiles with non-exceedance probabilities of 0.99 or greater were obtained by Qt(S)&MoM(real). In terms of RMSE for upper quantiles with the non-exceedance probabilities of 0.99 and 0.998, the Qt(S)&MoM(real). Other quantile estimators Qt(S)&ML and Qt&MoM(log) yielded the least RMSE values for low extreme quantiles with non-exceedance probabilities of 0.01 and 0.1. The sextile method also became competitive. The two MoMs were not very good.

For the LPIII(3) distribution with an upper bound, however, overall performances of MoM(real) were the best for estimating quantiles with either small or large non-exceedance probability, while MoM(log) performed poorly. Qt(S)&MoM(real) was capable of estimating the quantiles with non-exceedance probabilitiy of 0.99 and 0.998 with acceptable accuracy. Other quantile methods were not very good for this distribution with an upper bound.

Gumbel and generalized extreme-value distributions

In the 1920s, Fréchet, Fisher, Tippett and others derived the theoretical distributions of extremes. Since Gumbel applied the extreme-value type I (EV1) distribution (the double exponential distribution) to annual maximum floods in the 1940s, it has been often used in hydrologic frequency anlysis all over the world. It is called the Gumbel distribution.

Jenkinson (1955) obtained the generalized extreme-value (GEV) distribution as a generalization of three extreme-value distributions and showed that this distribution is able to describe well various meteorological observations. It is often used in the United Kingdom since the National Environment Research Council recommendation in 1975. In Japan, the GEV distribution has been used in a different formulation. Kadoya (1956) called it the log extreme-value distributions type A and type B, corrsponding to the Fisher-Tippett type II ("Fréchet" or log-Gumbel distribution) and type III ("Weibull" distribution), resectively. Independently of Jenkinson (1955), Kadoya (1956) showed that the Gumbel distribution is a special case of the log extreme-value distribution. The log extreme-value distribution type A is often used in Japan as well as the Gumbel and lognormal distributions.

a) Gumbel distribution: Lowery and Nash (1970) compared four methods: MoM, LS(W), Gumbel's method, and ML. Based on two experiments (one for 500 samples with size of 10 and the other for 100 samples of size 50), they conlude that MoM is the best and ML is slightly more efficient than MoM but slightly biased. They recommended use of MoM rather than ML, which requires repetitive computation to obtain the ML estimators.

Table 4 summarizes Monte Carlo experiments conducted by Landwehr et al. (1979), Lettenmaier and Burges (1982), Raynal and Salas (1986), Phien (1987), and Takara et al. (1989). Synthesizing their results, the ML method is generally recommended, though it provides slightly larger bias in estimation of parameters and quantiles than the method of PWM.

b) Generalized Extreme-Value (GEV) Distribution: Hosking et al. (1985) applied the PWM method to the GEV distribution and compared it with the ML and the method of sextile proposed by Jenkinson (1969). Results in terms of accuracy and ease of use supported adoption of the PWM method. Takara et al. (1989) compared four methods: ML, MoM(b), MoM(u), and the PWM. They found that the PWM method is always the best in terms of both RMSE and bias, except for the sample with the size of 1000 when the ML is superior.

Since the GEV(3) distribution has a lower bound parameter, a method of quantiles could be developed to estimate that parameter. If a quantile-lower-bound estimator for GEV(3) is developed, it might give more accurate quantile estimators.

CONCLUSIONS

For choice of distribution:

1) The four criteria, SLSC, COR, MLL, and AIC, are able to assess goodness of fit. The SLSC and COR are useful for both absolute and relative evaluation, while MLL and AIC are only for relative evaluation. The one-to-one relationship between the SLSC and the COR is empirically documented. The use of the COR is to be recommended from the viewpoint of tractability.
2) Takara and Takasao's model evaluation procedure, which considers both the goodness of fit and the variability (estimation accuracy) of quantile estimtors, could be a good solution for selecting a distribution. This procedure tends to select distributions with two parameters because two-parameter distributions provide less variable quantile estimators.
3) To save time and effort, it is recommended that one might evaluate the Gumbel distribution first to see if it fits well. If it does not, then one could try other two-parameter models before fitting distributions with three free parameters. If a two-parameter model provides a good fit with COR>0.995 for the dataset considered, the model can be accepted. According to the evaluation results, the Gumbel distribution was good for samples with the skew from 0.7 to 1.4; the PIII(2) for samples with the skew from 0.3 to 0.7; and the SQET(2) for samples with the skew greater than 1.4 and less than 1.8. The SQET(2) distribution proposed by Etoh (1987) could be a good alternative when the Gumbel distribution does not fit datasets having larger skew than about 1.4.
4) The discussion in the previous item 3) is based on the evaluation results obtained by using the twelve datasets having sample skews greater than 0.38 and less than 1.8. In fact, datasets of flood discharges can have higher skewness, such as 4.0 or 5.0. Since the Gumbel distribution, which has fixed skewness, is not able to fit such data, Takara and Takasao's model evaluation procedure could give a good selection among three-parameter distributions.
5) The resampling methods such as the jackknife and the bootstrap are useful for evaluating the variability of quantile estimates based on available samples. The quantile estimation accuracy obtained by resampling methods have not been assessed; it should

be compared with that obtained by analytical formula given for each distribution.

For choice of fitting method:

6) For two-parameter distributions, the maximum likelihood estimation method is to be recommended.
7) For three-parameter distributions, the methods employing the quantile lower-bound estimator Qt+MoM and Qt+ML seem to perform better than other methods with the LN(3), P(3), and LP(3) distributions having lower-bound parameter. This suggests that the quantile methods proposed by Iwai (1947, 1949, 1950), Stedinger (1980), and Hoshi and Leeyavanija (1986) deserve additional study.
8) The limitation or drawback of some traditional methods that rely on plotting positions, such as Gumbel's, were revealed in terms of quantile estimation accuracy. Plotting position formulas should not be used for precise estimation of quantiles, though they are a useful tool for visual evaluation of goodness of fit.

In general,

9) This paper has described the advantages and disadvantages of various distributions and fitting methods for estimating upper quantiles (i.e. T-year storms or floods). Based on that discussion, one direction of future research is to find a good quantile estimator, namely a good combination of a distribution and a fitting method that gives accurate quantile estimates for hydrological datasets.
10) Takara and Takasao's evaluation procedure used the maximum likelihood method for all distributions. However, better methods exist for some distributions as seen here. Therefore, their evaluation procedure can be improved by introducing such method (it may be a quantile method) for each distribution.
11) The lower-bound parameter essentially has physical meaning. Therefore, if the parameter is able to be defined (or estimated) by some physical consideration, three-paramer distributions can be regarded as "two-free-parameter models", and then will give more stable quantile estimates.

REFERENCES

Akaike, H. (1974) "A new look at the statistical model identification", IEEE Transactions on Automatic Control, AC-19, 716-723.

Birdsley, W. E. (1977) "A test for distinguishing between extreme value distributions", Journal of Hydrology, 34, 377-381.

Bobée, B., and Ashkar, F. (1991) The Gamma Family and Derived Distributions Applied in Hydrology, Water Resources Pub., Litteleton, Colorado, 203 pp.

Bobée, B., and Robitaille, R. (1975) "Correction of bias in the estimation of the coefficient of skewness", Water Resources Research 11(6), 851-854.

Burges, S. J., Lettenmaier, D. P., and Bates, C. L. (1975) "Properties of the three-parameter log normal probability distribution", Water Resources Research 11(2), 229-235.

Chow, V. T. (1954) "The log-probability low and its engineering applications", Proc. of ASCE, Pap. No. 536, Vol. 80, 25 pp.

Cover, K. A. and Unny, T. E. (1986) "Application of computer intensive statistics to parameter uncertainty in stream flow synthesis", Water Resources Research, 22(3), 495-507.

Efron, B. (1982) The Jackknife, the Bootstrap and Other Resampling Plans, SIAM Monograph, No. 38, 92 pp.

Etoh, T., Murota A., and Nakanishi M. (1987) "SQRT-exponential type distribution of maximum", in V. P. Singh (ed.), Hydrologic Frequency Modeling, D. Reidel Pub. Co., Dordrecgt, Holland, 253-264.

Hipel, K. W., McLead, A. I., and Lennox, W. C. (1977) "Advances in Box-Jenkins modeling, 1. Model construction", Water Resources Research, 13(3), 567-575.

Hoshi, K. and Leeyavanija, U. (1986) "A new approach to parameter estimations of

gamma-type distributions", Journal of Hydroscience and Hydraulic Engineering, JSCE, 4(2), 79-95.
Hoshi, K., Stedinger, J. R., and Burges, S. J. (1984) "Estimation of log-normal quantiles—Monte Carlo results and first-order approximations", Journal of Hydrology, 71, 1-30.
Hosking, J. R. M., Wallis, J. R., and Wood, E. F. (1985) "Estimation of the generalized extreme-value distribution by the method of probability-weighted moments", Technometrics, 27(3), 251-261.
Ishihara, T. and Takase, N. (1957) "The logarithmic-normal distribution and its solution based on moment method", Trans. JSCE, 47, 18-23 (in Japanese).
Iwai, S. (1947) "On the asynmetric distribution in hydrology", Collection of Treatises, No. 1, 2, Journal of the Civil Engineering Society, 93-116 (in Japanese).
Iwai, S. (1949) "Studies on nonsymmetrics of Slade-type distribution and several new solutions", Proc. JSCE, 4, 84-104 (in Japanese).
Iwai, S. (1950) "Duration curves of logarithmic distribution type and their approximations", Memoirs of Engineering, Kyoto University, 12(3), 1-19.
Jenkinson, A. F. (1955) "The frequency of the annual maximum (or minimum) values of meteorological elements", Quarterly J. of the Royal Meteorological Society, 81, 158-171.
Jenkinson, A. F. (1969) "Statistics of extremes", Technical Note 98, WMO, Geneva.
Kadoya (1956) "The extreme-value law and its some solutions for engineering problems", Nougyou Doboku Kenkyuu, 23(6), 350-357 (in Japanese).
Kadoya (1962) "On the applicable ranges and parameters of logarithmic normal distributions of the Slade-type", Nougyou Doboku Kenkyuu, Extra publication 3, 12-27 (in Japanese).
Kimball, B. F. (1936) "Probability-distribution-curve for flood-control studies", Trans. American Geophysical Union, Part I., 460-483 (with discussions).
Kite, G. W. (1977) Frequency and risk analyses in hydrology, Water Resources Pub., Fort Collins, U.S.A., 224 pp.
Landwehr, J. M., Matalas, N. C., and Wallis, J. R. (1979) "Probability weighted moments compared with some traditional techniques in estimating Gumbel parameters and quantiles", Water Resources Research, 15(5), 1055-1064.
Lettenmaier, D. P., and Burges, S. J. (1982) "Gumbel's extreme value I distribution: a new look", Journal of Hydraulic Div., Proc. ASCE, 108, HY4, 502-514.
Lowery, M. D. and Nash, J. E. (1970) "A comparison of methods of fitting the double exponential distribution", Journal of Hydrology, 10, 259-275.
Lu, L.-H., and Stedinger, J. R. (1992) "Variance of two- and three-parameter GEV/PWM quantile estimators: formulae, confidence intervals, and a comparison", Journal of Hydrology, 138, 247-267.
Phien, H. N. (1987) "A review of methods of parameter estimation for the extreme value type-1 distribution", Journal of Hydrology, 90, 251-268.
Potter, K. W. and Lettenmaier, D. P. (1990) "A comparison of regional flood frequency estimation methods using a resampling method", Water Resour. Res., 26(3), 415-424.
Raghunath, H. M. (1985) Hydrology—Principles, Analysis and Design, Wiley Eastern Ltd., New Delhi.
Raynal, J. A., and Salas, J. D. (1986) "Estimation procedures for the type-1 extreme value distribution", Journal of Hydrology, 87, 315-336.
Sakamoto, H., Ishiguro, M., and Kitagawa, G. (1983) Information Statistics, Kyoritsu, Tokyo, 27-64.
Sangal, B. P. and Biswas, A. K. (1970) "The 3-parameter lognormal distribution and its applications in hydrology", Water Resour. Res., 6(2), 505-515.
Stedinger, J. R. (1980) "Fitting log normal distributions to hydrologic data", Water Resources Research, 16(3), 481-490.
Stedinger, J. R., Vogel, R. M., and Foufoula-Georgiou, E. (1993) "Frequency analysis of extreme events", in D.R. Maidment (ed.), Handbook of Hydrology, McGraw-Hill, Chap.

18, 66 pp.
Takara, K. and Takasao, T. (1988a) "Criteria for evaluating probability distribution models in hydrologic frequency analysis", Proc. JSCE, 393/II-9, 151-160 (in Japanese).
Takara, K. and Takasao, T. (1988b) "Evaluation criteria for probability distribution models in hydrologic frequency analysis", Preprint, Fifth IAHR Int'l Symp. on Stochastic Hydraulics, Univ. of Birmingham, U.K., Paper A5, 10pp.
Takara, K. and Takasao, T. (1989) Closure for Y. Goda's Discussion on "Criteria for evaluating probability distribution models in hydrologic frequency analysis", Proc. JSCE, 405/II-11, 267-272 (in Japanese).
Takara, K., Takasao, T., and Shimizu, A. (1989) "Comparison of parameter estimation methods for extreme value distributions", Annuals, Disas. Prev. Res. Inst., Kyoto University, 32 B-2, 455-469 (in Japanese).
Takara, K. and Takasao, T. (1990a) "Comparison of parameter estimation methods for hydrologic frequency analysis models", Proc. Hydraulic Engineering, JSCE, 34, 7-12, (in Japanese).
Takara, K. and Takasao, T. (1990b) "Evaluation of hydrologic frequency analysis models based on quantile variability obtained by resampling mrthods", Proc. Fifth International Conference on Urban Storm Drainage, Suita, Osaka, Japan, 2, 587-592.
Takasao, T., Shiiba, M., and Takara, K. (1984) "Evaluation of runoff models by an information criterion", Annuals, Disas. Prev. Res. Inst., Kyoto University, 27 B-2, 275-290 (in Japanese).
Takasao, T., Takara, K., and Shimizu, A. (1986) "A basic study on frequency analysis of hydrologic data in the Lake Biwa basin", Annuals, Disas. Prev. Res. Inst., Kyoto University, 29 B-2, 157-171 (in Japanese).
Takeuchi, K. and Tsuchiya, K. (1988) "On relative accuracy of PWM estimates of normal and 3-parameter lognormal distributions", Proc. JSCE, 393/II-9, 103-112 (in Japanese).
Tung, Y.-K. and Mays, L. W. (1981) "Generalized skew coefficients for flood frequency analysis", Water Resources Bulletin, 17(2), 262-269.
Vogel, R. M. (1986) "The probability plot correlation coefficient test for the normal, lognormal, and Gumbel distributional hypothesis", Water Resources Reserach, 22(4), 587-590 (with correction, Water Resour. Res., 23(10), 2013.
Vogel, R. M. and McMartin, D. E. (1991) "Probability plot goodness-of-fit and skewness estimation procedures for the Pearson type III distribution", Water Resources Research, 27(12), 3149-3158.
Wallis, J. R., Matalas, N. C., and Slack, J. R. (1974) "Just a moment!", Water Resources Research, 10(2), 211-219.

ESTIMATION OF QUANTILES OF THE GENERALIZED PARETO DISTRIBUTION

F. ASHKAR[1], T. B.M.J. OUARDA[1], B. BOBÉE[2] and M. HACHÉ[2]
[1] Dept. of Mathematics, University of Moncton
Moncton, N.B. E1A 3E9, Canada
[2] INRS-EAU, University of Québec
2800 Einstein, Ste Foy, Québec, G1X 4N8, Canada

This paper reviews the properties of the two-parameter Generalized Pareto Distribution (GPD) and investigates the use of the Generalized Method of Moments (GMM) for its fitting. A general formula, easily programmed on computer, is derived for the calculation of the asymptotic variance of the T-year event X_T obtained by combining any two moments of the GPD. Monte Carlo simulation is used to study the effect of the order of the two moments to be used in applying the GMM for various sample sizes and population characteristics. Six GMM pairs of moment orders are used to estimate parameters and quantiles X_T for five different values of the return period T. Results indicate that the adequate choice of the order of the two GMM moments can lead to a considerable reduction in the variance of the estimate of X_T in comparison to estimation by the traditional method of moments. Moments of order (0,-1) and (2,-1) are shown to generate best results for positive values of the shape parameter α (GPD-2 form), and the traditional method of moments (moments 1 and 2) leads to superior results for negative values of α (GPD-3 form). The GMM with moments of order (0,1) resulted generally in a satisfactory and stable behaviour in all conditions.

INTRODUCTION

The two-parameter generalized Pareto distribution (GPD) is a special case of the log-Pearson Type 3 (LP3) distribution, and has the uniform, exponential, and Pareto distributions as special cases. The GPD has been recommended for the analysis of natural extreme phenomena (Van Montfort and Witter, 1986; Smith,1987; Hosking and Wallis, 1987; Jin and Stedinger, 1989), and has mainly been used in hydrology in association with the Peaks Over Threshold (POT) approach.

Fitting of the GPD

The fitting the GPD has been traditionally carried out by using such methods as the method of maximum likelihood, the classical method of moments and the method of

K. W. Hipel (ed.),
Stochastic and Statistical Methods in Hydrology and Environmental Engineering, Vol. 1, 235–248.

probability weighted moments (PWM). Hosking and Wallis (1987) compared the performance of these methods in a simulation study and demonstrated that unless the sample size is larger than 500, estimators derived by the method of moments or PWM were more reliable than those obtained by maximum likelihood. Ashkar and Bobée (1987) indicated that there is no unique way of applying the GMM to a given distribution because the order of the moments to be used in estimating the parameters of the distribution is not unique. These authors have stressed, however, that this flexibility of the method can be considered as an advantage, because it provides the user with the possibility of choosing a version of the method which gives larger weight to the larger elements of the sample, or another version which assigns more weight to the smaller elements. Thus, the user can choose the method that best fits the problem at hand, and departures from the traditional method of moments, which uses the mean and variance of the distribution, are perfectly justified in most practical applications. It is expected therefore, that the shape of the distribution being considered determines to a large degree which method performs best for that given distribution.

A general formula, easily programmed on computer, is derived in the present paper for the calculation of the asymptotic variance of the T-year event X_T obtained by combining any two moments of the GPD. This method which fits the distribution by combining any two of its moments in real space shall be called "Generalized Method of Moments" (Bobée and Ashkar, 1991) or GMM. It is shown, in this work, that by carefully choosing the order of the two moments to be used in applying this method, a significant reduction in the variance of the estimate of the T-year event X_T can be achieved, in some cases, in comparison to estimation by the traditional method of moments which uses only moments of order 1 and 2.

SOME PROPERTIES OF THE GPD

The cumulative distribution function (cdf) of the GPD is given by:

$$\begin{aligned} F(x) &= 1-(1-kx/\beta)^{1/k} && k\neq 0 \\ &= 1-e^{-x/\beta} && k=0 \end{aligned} \tag{1}$$

The range of x is $0 \leq x < \infty$ for $k \leq 0$ and $0 \leq x \leq \beta/k$ for $k > 0$. β is a scale parameter and k is a shape parameter. This form of the cdf of the GPD was used by Hosking and Wallis (1987) who define the GPD as the distribution of a random variable X obtained from the standard exponential distribution, Y, by a transformation of the form:

$$X = \beta(1 - e^{-kY})/k$$

The probability density function (pdf) of this distribution is

$$f(x) = \beta^{-1}(1-kx/\beta)^{1/k-1} \qquad k\neq 0$$
$$\quad = \beta^{-1}e^{-x/\beta} \qquad k=0 \tag{2}$$

The simple mathematical form of this distribution is one of its attractive features. The other interesting property of the GPD is that it is closed under the operation of truncation from below: If X has the generalized Pareto pdf (2), and X is truncated from below at t, then X-t, given X $\geq$ t, has also a generalized Pareto pdf with the same shape parameter k (Ashkar et al., 1992). In other words, for this distribution, truncation is equivalent to scaling.

A reparameterized form of the GPD (Johnson and Kotz, 1970) is more often used in hydrology, and allows for the application of the commonly used versions of the GMM. The reparameterization on the random variable X changes its pdf (2) into one that is a special case of the widely used log-Pearson Type 3 (LP3) distribution (Ashkar et al., 1992). Let X be a random variable with pdf (2), k $\neq$ 0, and let

$$U = e^{m}(1-kX/\beta) \tag{3}$$

where m is any real number. It is shown (Ashkar et al., 1992) that the pdf of U is

$$f(u) = \alpha u^{-(\alpha+1)}e^{\alpha m} \qquad \alpha>0$$
$$\quad = |\alpha| u^{-(\alpha+1)}e^{\alpha m} \qquad \alpha<0 \tag{4}$$

and its cdf is

$$F(u) = 1-(ue^{-m})^{-\alpha} \qquad \alpha>0$$
$$\quad = (ue^{-m})^{-\alpha} \qquad \alpha<0 \tag{5}$$

The range of u is $e^{m} \leq u < \infty$ for $\alpha > 0$ and $0 \leq u \leq e^{m}$ for $\alpha < 0$. The parameter m acts as a parameter of origin and of scale at the same time, whereas α is a parameter of shape. The parameter α in (4) is related to the parameter k in (2) by the relationship $\alpha = -1/k$. It is seen that for positive values of k, the right tail of pdf (2) is transformed into the right tail of pdf (4), but when k is negative, the right tail of (2) is transformed into the left tail of (4).

It is easy to show (Ashkar et al., 1992) that if α goes to ∞ in absolute value (for example $|\alpha| > 10$), then the generalized Pareto cdf in its standardized form tends to the standardized exponential cdf (when α goes to $-\infty$ an "inverted" exponential is obtained). On the other hand, the uniform distribution is obtained as a special case of the GPD (4) when $\alpha = -1$. It is useful to note that the reparameterization (3) conserves, in a certain sense, the property of the GPD of closure under the operation of truncation from below (Ashkar et al., 1992).

Three forms or "Types" of generalized Pareto pdf may be defined, depending on the value of the shape parameter α:

(1) GPD-2: $\alpha > 0$

(2) GPD-3: $\alpha < 0$

(3) GPD-1: limiting form of both GPD-2 and GPD-3, obtained when α is very large in absolute value ($\alpha \rightarrow \mp \infty$: exponential distribution).

This notation is analogous to the one given by Van Montfort and Witter (1986) for the GPD in its initial parameterized form (2). Ashkar et al. (1992) show that for the GPD in its reparameterized form, the following interesting observations can be made:

1. if U follows GPD-2 then 1/U follows GPD-3 (and vice versa). GPD-3 has sometimes been called the "power-function distribution";
2. GPD-2 is a special case of the Pearson Type VI distribution, while GPD-3 is a special case of the Pearson Type I;
3. if U follows GPD-2 then ln(U) follows a 2-parameter exponential distribution and if U follows GPD-3 then ln(U) follows an "inverted" 2-parameter exponential (i.e. one with negative skewness). Therefore, one advantage of expressing the GPD in its reparameterized form (cdf 5) is that exponential probability paper can now be used to examine the fit of the GPD to a given set of data.

GMM FOR FITTING THE GPD

The rth moment about the origin (or "noncentral moment") of the GPD has the general form:

$$\mu_r'(x) = e^{mr}\left(1-\frac{r}{\alpha}\right)^{-1} \tag{6}$$

which exists for all r (notably r can be negative, and not necessarily an integer), as long as the following condition is satisfied:

$$1-\frac{r}{\alpha} > 0; \quad \text{i.e. } r<\alpha \quad if \quad \alpha>0 \quad \text{and} \quad r>\alpha \quad if \quad \alpha<0 \tag{7}$$

Whenever necessary, we assume that when (6) is used, (7) is satisfied. The application of the GMM to a two-parameter distribution such as the GPD was described by Ashkar and Bobée (1987), Bobée and Ashkar (1988), and Ashkar et al. (1988) and can be defined as follows: For a sample $x_1,\ldots,x_N$ of size N from the GPD; define its rth noncentral moment:

$$m_r' = \frac{1}{N} \sum_{i=1}^{N} x_i^r \tag{8}$$

and choose any two distinct real numbers s and t, different from zero (in practice two integers small in absolute value are used). The next step is to write the following two equations relating two moments of the sample to the corresponding two moments of the population:

$$m_s' = \mu_s' \tag{9a}$$

$$m_t' = \mu_t' \tag{9b}$$

These two equations can be solved for the parameters m and α, when such a solution exists, to give the desired parameter estimates $\hat{m}$ and $\hat{\alpha}$. In the present work, the method of bisection (Burden and Faires, 1985) was used and found to be efficient for solving the system of Equations (9a) and (9b). The GMM which uses the system of Equations (9a) and (9b) is denoted by GMM(s,t).

Four population moments, μ_r', are considered in this study, with the first three representing special cases of (6) (corresponding to moments of order r=-1, 1, 2):

(1) r = 1 arithmetic mean, $\mu_1'(X)$

(2) r = -1 inverse of the harmonic mean (1/h), where by definition 1/h = $\mu_{-1}'(X)$

(3) r = 2 quadratic mean, $\mu_2'(X)$

The fourth moment is derived from the geometric mean, g, of X, which can be formally considered as the moment of order "quasi-zero" (Bobée and Ashkar, 1991) because it corresponds to the noncentral moment of order r, as r tends to 0. It is evident, however, that the logarithm of the geometric mean of a random variable X, is the arithmetic mean of Y=log X, so it is often convenient in applying the GMM to define the moment of order quasi-zero (μ_r', "r = **0**") as follows:

(4) r = **0** $\mu_0'(X) = \log g = \mu_1'(Y)$ where Y = log X.

In the case of the GPD, the moments of order quasi-zero for the sample and for the population, are respectively as follows:

$$m_0'(x) = \bar{y} = \frac{1}{N}\sum_{i=1}^{N} \ln(x_i) \tag{10a}$$

$$\mu_0'(x) = \mu_1'(y) = m + 1/\alpha \tag{10b}$$

where it is readily seen that $\mu_0'(X)$ does not have the same functional form as $\mu_r'(X)$ in Equation (6). This point is treated later in the study.

ESTIMATION AND VARIANCE OF T-YEAR EVENTS

If X follows an LP3 distribution with shape parameter $\lambda = 1$ (i.e. GPD), then $Y = \ln X$ follows a P3 distribution with $\lambda = 1$ (i.e. 2-parameter exponential distribution with positive skewness if $\alpha > 0$, and with negative skewness if $\alpha < 0$). Following the same procedure proposed by Bobée and Ashkar (1988) for the LP3/P3 case, and posing $\lambda = 1$, leads to the variance of the T-year event for the GPD case, in a straightforward manner.

The relationship between the estimators $\hat{X}_T$ (GPD) and $\hat{Y}_T$ (2-parameter exponential) is

$$\ln(\hat{X}_T) = \hat{Y}_T \quad i.e. \quad \hat{X}_T = e^{\hat{Y}_T} \tag{11}$$

For large samples we have

$$var\ \hat{X}_T = \left(\frac{d\hat{X}_T}{d\hat{Y}_T}\right)^2 var\ \hat{Y}_T \tag{12}$$

which together with (11) gives

$$var\ \hat{X}_T = \hat{X}_T^2\ var\ \hat{Y}_T \tag{13}$$

Thus $var(\hat{Y}_T)$ can be calculated first, and then (13) can be used to calculate $var(\hat{X}_T)$. Now Y_T is a function of the parameters of the 2-parameter exponential distribution, which has the following form:

$$Y_T = \xi(\alpha,m) = m + \frac{1}{\alpha}\ln(T) \tag{14}$$

where α and m are respectively the scale and location parameters of the distribution. For large sample sizes we have:

$$var\ Y_T = \sum_{j=1}^{2} \left(\frac{\partial \xi}{\partial \theta_j}\right)^2 var\ \theta_j + 2\left(\frac{\partial \xi}{\partial \theta_i}\right)\left(\frac{\partial \xi}{\partial \theta_j}\right) cov\ (\theta_i, \theta_j) \tag{15}$$

where $\theta_1 = \alpha$ and $\theta_2 = m$. The terms var(θ_j) and cov(θ_i,θ_j) in (15) are related to the variances and covariances of the two moments $\mu_s'(X)$ and $\mu_t'(X)$ that are used in the estimation (Equation 9). The notation is simplified by posing $\mu_s'(X) = M_1$ and $\mu_t'(X) = M_2$, so for large sample sizes we can write:

$$cov\ (M_r, M_q) = \sum_{j=1}^{2} \left(\frac{\partial M_r}{\partial \theta_j}\right)\left(\frac{\partial M_q}{\partial \theta_j}\right) var\ \theta_j + \sum_{j=1}^{2} \sum_{\substack{i=1 \\ i \neq j}}^{2} \left(\frac{\partial M_r}{\partial \theta_j}\right)\left(\frac{\partial M_q}{\partial \theta_i}\right) cov\ (\theta_j, \theta_i) \tag{16}$$

where r,q = 1, 2; $\theta_1 = \alpha$, $\theta_2 = m$. By imposing r = q in (16) we have var(M_r) and in matrix form this equation can be expressed as:

$$\begin{bmatrix} var\ M_1 \\ var\ M_2 \\ cov\ (M_1, M_2) \end{bmatrix} = \begin{bmatrix} V_{11} & V_{12} & V_{13} \\ V_{21} & V_{22} & V_{23} \\ V_{31} & V_{32} & V_{33} \end{bmatrix} \cdot \begin{bmatrix} var\ \alpha \\ var\ m \\ cov\ (\alpha, m) \end{bmatrix} \tag{17}$$

which is of the form $V_M = V.V_P$. The matrix V is given by:

$$V = \begin{bmatrix} A_{11}^2 & A_{12}^2 & 2A_{11}A_{12} \\ A_{21}^2 & A_{22}^2 & 2A_{21}A_{22} \\ A_{11}A_{21} & A_{12}A_{22} & (A_{11}A_{22}+A_{21}A_{12}) \end{bmatrix} \tag{18}$$

where the terms A_{ij} are easily obtained from (6), and in matrix form they give

$$A = A_{ij} = \begin{bmatrix} A_{11} & A_{12} \\ A_{21} & A_{22} \end{bmatrix} = \begin{bmatrix} \dfrac{-s \cdot \mu_s'}{\alpha^2(1-s/\alpha)} & s \cdot \mu_s' \\ \dfrac{-t \cdot \mu_t'}{\alpha^2(1-t/\alpha)} & t \cdot \mu_t' \end{bmatrix} \tag{19}$$

The elements of the vector V_M in (17) can be easily determined by noting that for arbitrary sample size N, and arbitrary orders of moment r and q we have (Kendall and Stuart, 1963):

$$var\ \mu_r' = \frac{1}{N}(\mu_{2r}' - \mu_r'^2) \tag{20}$$

$$cov\ (\mu_r', \mu_q') = \frac{1}{N}(\mu_{q+r}' - \mu_r'\ \mu_q') \tag{21}$$

Once the elements of the vector V_M and the matrix V are completely specified, the vector V_P can be determined from (17). Knowing the variances and covariances of the parameters (vector V_P) and the partial derivatives of Y_T with respect to these parameters (obtained from 14), the calculation of var($\hat{Y}_T$) becomes straightforward from Equation (15), and var($\hat{X}_T$) can be easily deduced by using (13) and (11). The calculation of this variance can be easily programmed on a computer. Table 1 gives the conditions under which GMM(s,t) is applicable, and those for the existance of Var(X_T), as a function of moment orders s and t.

SPECIAL CASE: MOMENT OF ORDER QUASI ZERO

The moment of order quasi-zero, as defined by (10b) does not have the same functional form as μ_r', r different from **0**, (Equation 6). This means that if $\mu_0'(X)$ is to be used in the estimation, then certain terms of the vector V_M in (17), and of the matrix A in (19), will have to be modified. If we take s = **0**, so that $M_1 = \mu_s'(X) = \mu_0'(X)$ in (16) and (17), then all terms involving M_1 or s in (19), (20) and (21) should be modified. These terms are:

(1) all terms involving M_1 in the vector V_M in (17); the new terms are

$$var\ M_1 = \frac{1}{N}\left(\frac{1}{\alpha^2}\right) \tag{22}$$

$$cov\ (M_1, M_2) = \frac{1}{N}\mu_t'(x)\left[\left(1 - \frac{t}{\alpha}\right)^{-1}\left(-\frac{mt}{\alpha} + \mu_y'\right) - \mu_y'\right] \tag{23}$$

(2) the two elements in the first row of the matrix A in (19); these become

$$A_{11} = -\frac{1}{\alpha^2} \tag{24}$$

$$A_{12} = 1 \tag{25}$$

SIMULATION EXPERIMENTS

Monte Carlo simulation was used to study the effect of the choice of the order of the two moments to be used in applying the GMM for various sample sizes and population characteristics. In each experiment, n = 1000 samples of various sizes (N = 15, 25, 50, 100, 200, and 500) were generated from a GPD with a particular set of parameters. Five values of the parent distribution shape parameter α were considered (α = 100, 5, 2.5, -2.5, and -5) and the parameter m was fixed to the value m = 0. Each experiment is therefore characterized by the number of samples generated n, the size of each sample N, and the set of generated population characteristics (α,m).

For each Monte Carlo experiment, 6 versions of the GMM were used to calculate parameter estimates $\hat{\alpha}$ and $\hat{m}$, and design event estimates $\hat{X}_T$ for five different values of the return period (T = 10, 50, 100, 500, and 1000 years). The six two-moment sets used in accordance with the GMM are as follows: (2,0); (2,-1); (1,0); (2,1); (1,-1) and (0,-1). The application of these six versions of the GMM requires putting the GPD in its reparameterized form (pdf 4). The following performance indices were computed for each of the parameter and quantile estimates obtained by the six GMM versions: mean, observed bias, observed standardized bias, observed standard error, observed standardized standard error, standardized standard error based on asymptotic theory, observed root mean square error (RMSE), observed standardized root mean square error, and observed coefficient of skewness. We present in the next section a portion of the final results and figures for illustration purposes. A complete set of results for all moments, parameters and quantiles is presented in Ashkar et al. (1992).

SIMULATION RESULTS

Estimation of parameters

Results show that all six GMM estimation procedures produce parameter estimates that are relatively highly variable and biased, especially for small sample sizes. It was observed that, for all GMM methods, the sign of the bias follows the sign of the parameter α. Moment pair (0,-1) is least biased and results in the lowest values of the standard error and RMSE when the parameter of shape α is positive (GPD-2). However, this moment pair performs poorly for negative values of α (GPD-3), in which case moment pair (2,1) outperforms other methods with respect to all performance indices. Moment pairs (2,-1) and (1,0) tend to have a consistently average performance. For small and moderate sample sizes the standard error of the parameter estimates based on asymptotic theory (SE_{as}) is seen to differ significantly from the observed standard error (SE_{obs}) for both estimation of α and m. Large sample sizes N and large absolute values of α result in higher performances with respect to all statistics.

Estimation of quantiles

The Generalized Pareto pdf (4) could be useful for representing two types of hydrologic variables: When α is positive (GPD-2), the distribution could be useful for representing variables such as floods which are bounded from below but have a long right tail. When α is negative (GPD-3), the distribution could be useful for representing variables such as low flows which are bounded from both sides but with a left tail. The quantiles X_T investigated below refer to return period $T=1/(1-p)$ when α is positive (modeling of flood flows) and to return period $T=1/p$ when α is negative (low-flow modeling). As a whole, moment pair (0,-1) resulted in best performance when $\alpha > 0$ (estimation of flood quantiles) except for the observed standard error and RMSE for quantile estimation, for which method (2,-1) generated superior results, followed closely by method (1,-1) . However, more often than not, method (0,-1) leads to poorest results when $\alpha < 0$ (estimation of low flows), in which case, moment pair (2,1) was shown to be consistently superior, followed by methods (2,0) and (1,0). Unlike other methods, the use of moment orders (1,0) resulted generally in a stable and non-extreme behaviour. Figures 1 to 5 illustrate some of these results.

The difference between SE_{obs} and SE_{as} increases with the decrease of $|\alpha|$, and can be rather significant for some estimation methods and shape-parameter values even when the sample size is as large as 500. The return period seems to have very little effect on the ratio SE_{obs}/SE_{as}. High values of the RMSE seem to be directly caused by immoderate values of the standard error, and in general it is correct to state that SE_{obs} contributes more to the RMSE than does the bias.

SUMMARY AND CONCLUSIONS

The use of the Generalized Method of Moments (GMM) for fitting the Generalized Pareto Distribution (GPD) was investigated and a general formula was derived for the calculation of the asymptotic variance of the T-year event X_T obtained by combining any two moments of the GPD. The use of this formula was easily programmed on computer and its asymptotic results were compared through simulation with results obtained from small samples. The following conclusions can be drawn from the results of this study:

(1) Large absolute values of the shape parameter α, low return periods, and large sample sizes are shown to produce better results and cause a rapprochement of all methods, independantly of the performance index considered.

(2) In general, the methods that employ the moment of order -1 are recommended for $\alpha > 0$ (flood flows), while the methods that make away with this moment are recommended for $\alpha < 0$ (low flows). Pair (1,0) resulted generally in a stable and non-extreme behaviour.

(3) When α is large in absolute value, the GPD tends to the exponential distribution (positive skewness when $\alpha > 0$, and "inverted" exponential for $\alpha < 0$) and in this case consistently better results were obtained in terms of bias, standard error

and RMSE by all GMM methods considered.

The traditional method of moments (2,1) was shown to generally lead to the best results for the estimation of low-flow quantiles using the GPD-3 distribution. Note that this method puts more weight on the larger elements of the sample. However, this does not imply that it is the method of choice for estimating the parameters of the distribution.

REFERENCES

Ashkar, F., Ouarda, T. B.M.J., and Haché, M. (1992). The Generalized method of moments for fitting the generalized Pareto distribution, Scientific Report, Department of Mathematics, University of Moncton, N.B (in French).

Ashkar, F., Bobée, B., Leroux, D., and Morisette, D. (1988). The generalized method of moments as applied to the generalized gamma distribution, Stochastic Hydrol. Hydraul., 2, 161-174.

Ashkar, F., and Bobée, B. (1987). The generalized method of moments as applied to problems of flood frequency analysis: some practical results for the log-Pearson type 3 distribution, J. of Hydrol., 90, 199-217.

Bobée, B., and Ashkar, F. (1991). The Gamma Family and Derived Distributions Applied in Hydrology, Water Resources Publications, Littleton, CO. 203p.

Bobée, B., and Ashkar, F. (1988). Generalized method of moments applied to LP3 distribution, J. Hydraul. Eng. Am. Soc. Civ. Eng., 114(8), 899-909.

Burden, R. L., and Faires, J. D. (1985). Numerical analysis, PWS publishers, 3rd edition, Boston, Mass. 676p.

Hosking, J. R. M., and Wallis, J. R. (1987). Parameter and quantile estimation for the generalized Pareto distribution, Technometrics, 29(3), 339-349

Jin, M., and Stedinger, J. R. (1989). Partial duration series analysis for a GEV annual flood distribution with systematic and historical flood information, Unpublished paper.

Johnson, N. L., and Kotz, S. (1970). Distributions in statistics: continuous univariate distributions, 1, 2. Wiley, New York.

Kendall, M. G., and Stuart, A. (1963). The advanced theory of statistics, Vol. 1. Charles Griffin, London.

Smith, J. A. (1987). Estimating the upper tail of flood frequency distributions, Water Resour. Res., 23(8), 1657-1666.

Van Montfort, M. A. J., and Witter, J. V. (1986). The generalized Pareto distribution applied to rainfall depths, Hydrol. Scienc. J., 31, 2, 6, 151-162.

TABLE 1. Conditions of applicability of GMM(s,t), and existence of $Var(X_T)$

s	t	Applicability of GMM(s,t)	Existence of $Var(X_T)$
-1	0	$\alpha>0$ or $\alpha<-1$	$\alpha>0$ or $\alpha<-2$
-1	1	$\alpha>1$ or $\alpha<-1$	$\alpha>2$ or $\alpha<-2$
-1	2	$\alpha>2$ or $\alpha<-1$	$\alpha>4$ or $\alpha<-2$
0	1	$\alpha>1$ or $\alpha<0$	$\alpha>2$ or $\alpha<0$
0	2	$\alpha>2$ or $\alpha<0$	$\alpha>4$ or $\alpha<0$
1	2	$\alpha>2$ or $\alpha<1$	$\alpha>4$ or $\alpha<1$

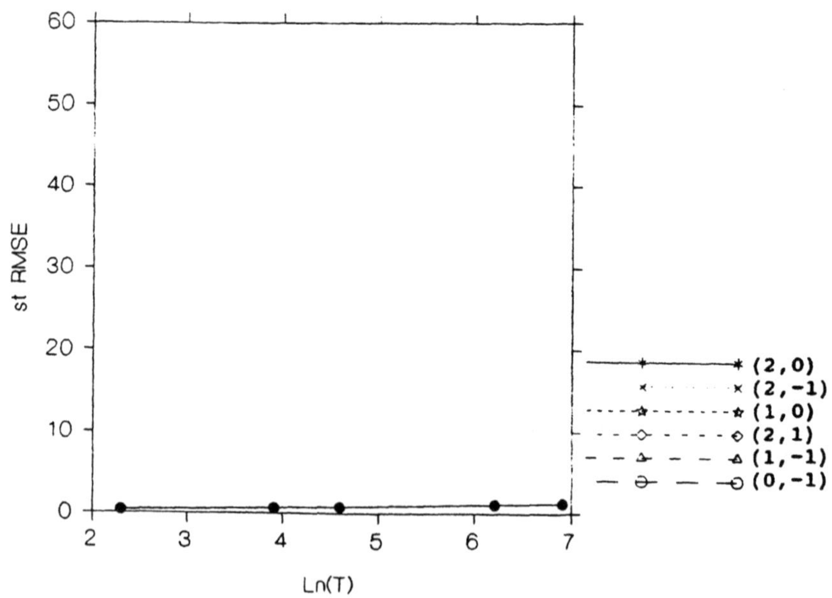

Figure 1. Root mean square error (%) vs. return period for the estimation of quantiles, for N=50 and α = 100.0.

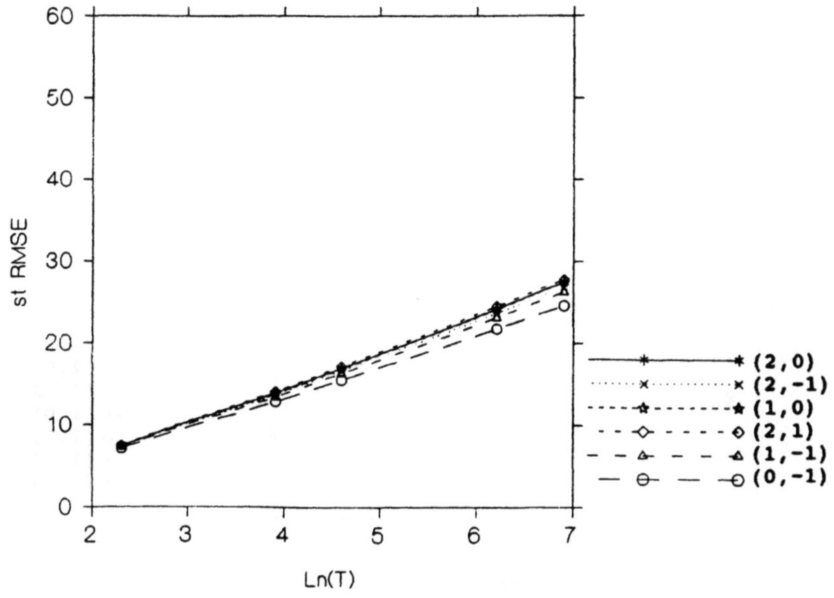

Figure 2. Root mean square error (%) vs. return period for the estimation of quantiles, for N=50 and α = 5.0.

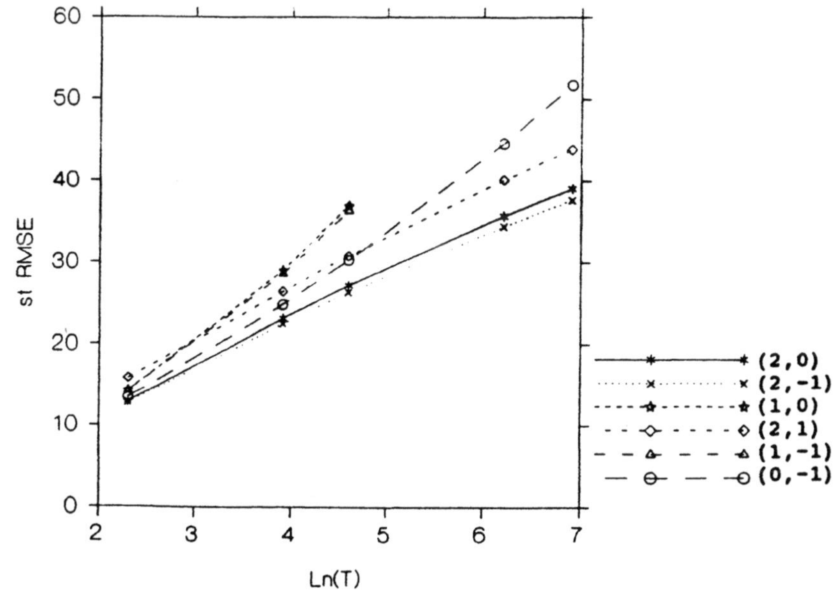

Figure 3. Root mean square error (%) vs. return period for the estimation of quantiles, for N=50 and α = 2.5.

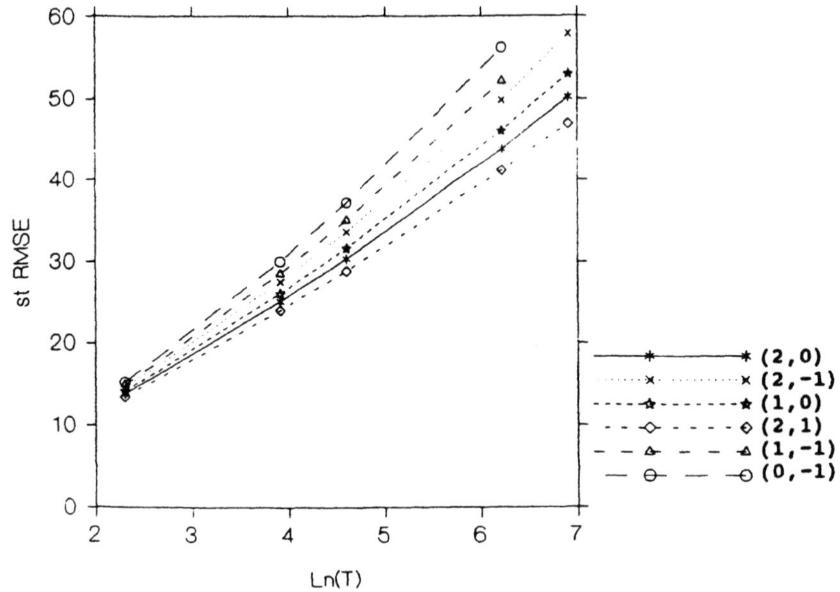

Figure 4. Root mean square error (%) vs. return period for the estimation of quantiles, for N=50 and α = -2.5.

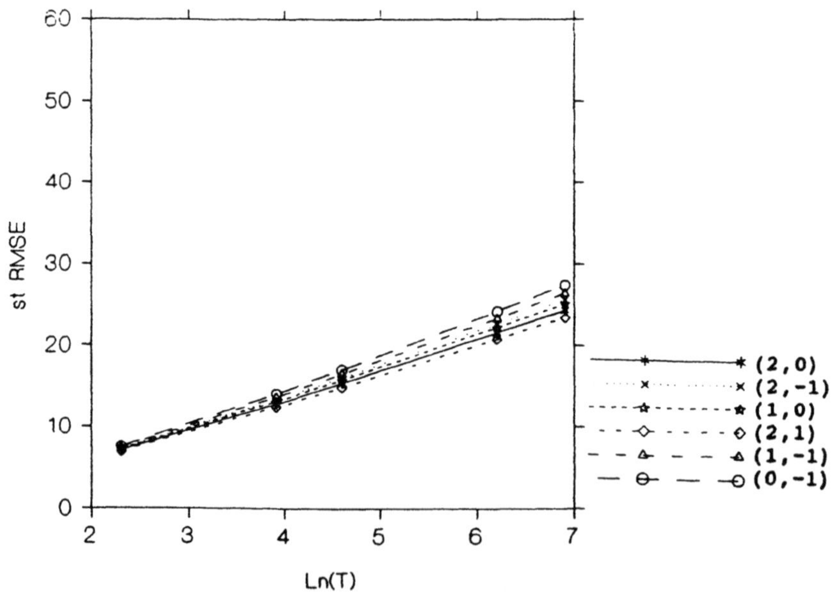

Figure 5. Root mean square error (%) vs. return period for the estimation of quantiles, for N=50 and α = -5.0.

FLOOD HYDROGRAPHS OF GIVEN EXCEEDENCE PROBABILITY

PEGRAM G G S and DEACON M P
Department of Civil Engineering
University of Natal
King George V Avenue
4001 Durban, South Africa

Floods are usually characterised by their peaks and much research has gone into the estimation of exceedence probabilities associated with flood peaks. These efforts include modelling using appropriate probability distributions and regionalization. In some applications the hydrograph as a whole and not just the peak is required. The pioneering work of Hiemstra and others (Hiemstra et al, 1976, Hiemstra and Francis, 1979), on the so-called runhydrograph, attached exceedence probability to the flood hydrograph as a whole when it is defined by the two variables **peak** and **volume**. Hiemstra coined the word 'runhydrograph' to describe the full hydrograph shape; the terminology was based on the theory of runs and crossings - in particular the necessity of selecting those hydrographs independent of each other in time.

We report on two new ideas that derive from the earlier study based on some work with more modern data collected by Deacon, (1993). The first idea is that of defining the joint distribution of the peaks and volumes of the flood hydrograph so that the distribution of the volume is conditioned on that of the corresponding peak selected from the sequence of annual peak maxima. The second idea is the development of a relationship between the probability of exceedence of the causative storm and the resulting flood hydrograph as a whole, defined in terms of its peak-volume pair (rather than the usual approach of confining attention to the peak alone).

INTRODUCTION

The Flood Studies Report (FSR) of the UK Natural Environment Research Council, (NERC, 1975) dealt extensively with the problem of predicting floods in the United Kingdom and Ireland. Besides work done in the United States, little else has been done to enable the practising engineer to estimate design floods at ungauged locations. This

K. W. Hipel (ed.),
Stochastic and Statistical Methods in Hydrology and Environmental Engineering, Vol. 1, 249–258.

estimation process possibly comprises the major work effort in the hydrological context in the profession. It is still practice throughout the engineering hydrology world to use the Unitgraph as the transfer function from rainfall to runoff. This even includes calculation of such economically important floods as the Probable Maximum Flood (to which some ascribe a nominal 10 000 year recurrence interval) and design floods with more conventional return periods of 50 to 1000 years. The procedure involved in Unitgraph-based flood calculation does not in itself guarantee that the recurrence interval of the causative storm and the computed flood have the same recurrence intervals, although it would be convenient if this were so. This is because of the well-known problem of defining the condition of the catchment prior to the flood-producing storm.

There are several aspects of the flood design problem which are under-studied. One of the most difficult ones to resolve (following on the argument of the preceding paragraph) is the relationship between the recurrence interval of the design storm and that of the resulting flood. The FSR commented on some simulation work relating the two (NERC,1975) but their treatment was unsatisfactory in that the relationship was indicated by a lower envelope and not by the more usual regression line, thus obscuring the precision (or lack of precision) in the relationship.

This paper reports on two contributions from a preliminary study. The first contribution, that of conditioning the distribution of the volume on that of the peaks, allows one to use whatever model is desired for estimating the quantiles of the peaks, from lognormal to Two Component Extreme Value (TCEV). The volumes are then computed of those floods which correspond with the annual (peak) flood series. The combined hydrograph can then be defined in terms of the peak and volume via a shape function (most conveniently chosen as a Gamma function).

The second contribution is: that causative storms (be they 1, 2, 3, ..., 7 days long) over the catchments of interest have themselves certain exceedence probabilities and these can be easily computed using an areal integration routine based on multiquadric surface-fitting over polygons. These two ideas show that if the flood hydrograph is taken as a whole (defined by the peak-volume pair) then it has a closer probability relationship to the causative storm than if the peak is taken in isolation.

DEFINITION OF FLOOD HYDROGRAPH RECURRENCE INTERVAL

Following the pioneering work of Hiemstra et al (1976) and culminating in that of Hiemstra & Francis (1979), who defined the Flood Hydrograph jointly in terms of its peak and volume and a resulting shape factor, Deacon (1993) suggested an easier construct to understand. Firstly, he dealt with annual rather than peak-over-threshold series as employed by Hiemstra & Francis (1979) and NERC (1975); secondly, he considered the distribution of the volume of the hydrograph conditioned on that of the annual maximum flood peak; thirdly, he defined the volume of the hydrograph in a simple but meaningful way. To understand the definition, it is to be noted that flow data are conventionally stored in so-called 'break-point' form, the data consisting of the time and flow-rate pairs at the points where the hydrograph trace changes slope by a noticeable amount. Four hydrographs corresponding to a sequence of annual maxima from a record nineteen years long appear in Figure 1. The river gauge from which these hydrographs come is a catchment of 409 km^2 in the hinterland of South Africa in a semi-arid region. They will be seen to be defined in break-point form - an irregular sequence of connected straight lines.

To define flood volume of a hydrograph, one needs to separate it in a sensible way from other hydrographs. In applications of the Unitgraph, several methods are suggested for base-flow separation. By contrast Hiemstra & Francis (1979) defined the flood hydrograph as a distinct flow event bounded by a rising and recession limb with zero flow as the reference datum, not the base-flow as in the Unitgraph. This was to emphasise the fact that the feature of interest is the actual flow in a river in order to be able to compute the frequency that a particular level might be inundated. This can be best determined from the flow record itself not from some ill-defined rainfall-runoff relationship. We take that point of view in this paper. Their treatment of the flood hydrograph was to produce the average slopes of the rising and recession limbs respectively to zero from the closest points adjacent to the peak where a minimum (upturn) in the flow-rate was detected. Integration of the flow under the hydrograph thus defined its volume.

We take the view that the flood hydrograph is delimited from the point where an upturn precedes the peak to where the flow returns to the same level after the peak on the recession limb. This is implicitly shown in Figure 1 where all four hydrographs are defined from the moment of upturn (at the origin in the figures) to the time when the flow returns again to the same level. The argument is based on a linear system one - that the storage in the catchment is linearly related to its outflow - the very argument behind Unitgraph theory. Thus the flow occurring between the time of the upturn prior to the flood peak to the time when the flow returns again to that level must all be due to the flood-causing storm. Note that in the three single-peaked hydrographs in Figure 1, the time between upturn and return to that level of flow lies between 20 and 26 hours, a remarkable consistency given that the peaks range between 50 and 290 m^3/s. This observation lends credence to the linearity assumption underlying Unitgraph theory and supports the argument behind the method used to compute the volume of a hydrograph

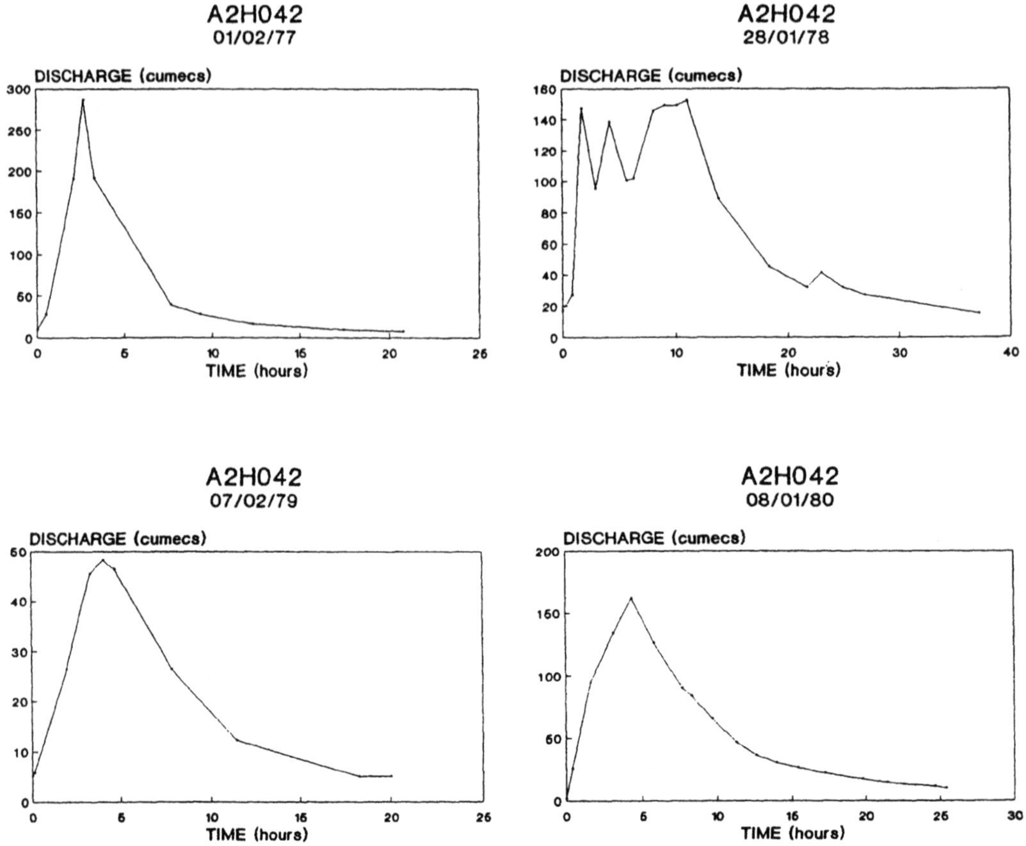

Figure 1. Four hydrographs corresponding to the largest annual peaks in four successive years, taken from a record 19 years long of the flow from a catchment of $409 km^2$ in South Africa, called A2H042. The volumes are computed as the integral of the flow from the upturn (at the origin) to the time the flow returns to the same level of flow as at the upturn - the time where the hydrograph is cut off in the figures. (The data from this gauge do not appear in the comparison between the recurrence intervals of flood and causative storm listed in Table 1 and depicted in Figures 2 & 3 because there were no useful concurrent rainfall records for gauges in or near this catchment during its 19-year life. The hydrographs are included for description purposes.)

as suggested here. Using this technique to delimit a flood makes the determination of flood volume a simple matter of integration of the flow of the hydrograph from the beginning (the upturn) to the end (the moment the flow returns to the level of the upturn).

Once the sequence of annual flood peak and volume pairs have been identified it remains to use this information to define the probability of a hydrograph as a whole when defined by these data-pairs. Hiemstra & Francis' (1979) approach was to compute the joint exceedence probability of the pair from a fitted continuous distribution function. They found that the bivariate lognormal distribution most often fitted the data-sets adequately. Their point of departure was the univariate exceedence distribution of the peak which is not always lognormal. The methodology made it a little difficult for practitioners to understand. This difficulty and objection can be overcome in the following way.

Our proposal is to fit the most appropriate marginal distribution to the annual peaks in the usual way, yielding the conventional interpretation of the exceedence probability of a selected peak. For a given value of peak flow, the conditional distribution of volumes can be defined. Thus there is a probability of 1 that the volumes conditional on the given peak level will exceed zero; one can also conceive of a median flood volume i.e. one which is exceeded 50% of the time that the given flood peak is exceeded. In such a way the conditional distribution can be constructed.

To illustrate the argument by means of an example, let a given peak have a recurrence interval of 100 years (a probability of 0.01 of being exceeded) and associated with it a volume defined as the median conditioned on the 100-year flood peak. The joint probability of both peak and volume being exceeded simultaneously in any one year is evidently 0.005 or 1 in 200, the product of the two probabilities. Again, the joint probability of the 100-year flood peak being associated with any volume is trivially 1 in 100, the product of 0.01 and 1.0, the latter being the conditional probability of a peak having a volume in excess of zero. In such a way is the probability of a flood hydrograph defined in terms of its peak-volume pair - the bigger the volume associated with a given peak, the rarer the event.

The next step is to find the distribution of the flood volume conditioned on the peak. In our experience with the South African data-set the conditional distribution of the volumes was adequately modelled by a lognormal distribution function judging by a Chi-square test at the 95% level of significance. Nonetheless, whatever marginal distribution functions most satisfactorily describe the peaks and volumes of the floods, it is a relatively straightforward matter to transform them into a bivariate normal distribution function by matching quantiles. This is a particularly useful step because the algorithms for computing the joint exceedence probabilities of a bivariate normal distribution are available (e.g. Tihansky, 1970, as quoted in Johnson & Kotz, 1972) whereas for other bivariate distribution functions, such computational algorithms may be hard to come by.

In such a way one can ascribe to any sequence of flood-peak volume pairs describing hydrographs, joint probabilities defining the recurrence interval of the hydrograph as a whole.

RECURRENCE INTERVAL OF CAUSATIVE STORM

To relate the recurrence interval of the flood as computed by the method described above with that of the causative storm, we need to find the storm which triggered the flood! For each of the annual maximum flood peaks and their associated flood volumes (which define the hydrograph) we were able to identify the daily rainfall totals in the conjunctive period at several gauges on and around the catchment feeding the stream gauge. It remained to fit an appropriate distribution function to these data to identify them as storms of given exceedence probability.

The catchment sizes (which ranged from 65 to 901 km^2) meant that the causative storms were of reasonably large duration, i.e. between 1 and 5 days long. This meant that we were looking at storms of 1, 2, 3, 5 and 7 days duration (with a preponderance of 3 day storms) in the sample with which we were dealing. With the storms causing annual maximum floods it was unusual that there would be large variation between the rainfalls recorded at the various rain gauges spanning the catchment, however the variability was enough for us to attempt to compute the catchment input as accurately as possible, thus two techniques were used. The first was to average the daily totals of rain at the relevant rain gauges to compute the storm input to each catchment. The second method was to fit multi-quadric surfaces to the rainfall totals and integrate these data over each catchment following the treatment of Pegram & Pegram (1993). It was found that there was surprisingly good agreement between the two methods when computing the average depth of rain, even for catchments with areas an order of magnitude larger than those treated in this study. It thus appears that in applications to storms causing annual floods, it is sufficient to take the average daily totals of rainfall at gauges spanning medium sized catchments to obtain the total volume of storm input on a day-by-day basis; this work tends to validate the approach taken by the FSR (NERC 1975).

To compute the probability distributions of the storms of various lengths of days which caused the annual maximum flood peaks, it was necessary to analyse the catchment storm rainfall for all the years of record. Probability distribution functions were fitted to the 1-, 2-, ... , 7-day storms identified as the largest in each year of rainfall record available; it so happened that some of the storms causing the annual maximum floods were not in this list; in addition, the rainfall records invariably were longer than the flood records. The approach adopted was to identify the recurrence interval of the causative storm depth for a given duration from the corresponding sequence of annual maximum storms. Using the above techniques it was possible to ascribe recurrence intervals to each one of the storms causing the recorded floods. A summary of the results for the three rarest floods in each of nine catchments appears in Table 1.

RELATING RECURRENCE INTERVALS OF FLOODS & STORMS

In the FSR (Chapter 6 of Volume 1) appears the results of a simulation relating causative storms and resultant flood hydrographs in Great Britain. It was there suggested that the return period of the flood peak was somewhat less than that of the causative storm over a wide range of interest. That relationship is plotted in Figure 2 and labelled as the FSR curve. Superimposed on the curve are individual peak storm pairs (not including volume at this stage) derived in this study. It will be seen that there is a spread between the recurrence intervals - especially near the 10 year return period - that goes from 2 to 16 years, a range of a factor of 8. It will also be seen that the results for the peaks match with the FSR curve as a lower envelope, in a sense corroborating and validating their simulation methodology.

A considerable improvement in the relationship between the recurrence intervals of storm and flood is obtained if the floods are defined not by peaks alone, but by their peak-volume pairs or hydrographs. Figure 3 shows these relationships for the data analysed in this study. Two things stand out from Figure 3. The first is that the band has narrowed from that of Figure 2, indicating that the factor relating maximum and minimum recurrence intervals in the spread of data is of the order of 2 to 3, not 8 as in Figure 2. Secondly and perhaps more comfortingly, the return period of the flood resulting from a storm of given recurrence interval is larger than that of the causative storm, i.e. it is rarer, and that by only a factor of 3.

CONCLUSION

Flood hydrographs have been defined in terms of annual maximum peak and corresponding volume in this study and the distribution of flood volumes (conditional on peaks) is defined in a way that the distribution of the peaks can be determined independently of that of the volumes from the sequence of annual flood peaks after

Table 1. Data for storm-flood return period relationship plotted in Figures 2 & 3

P Peak discharge (m^3/s)
V Volume ($m^3 \times 10^6$)
T_P Return period of flood peak (years)
T_F Return period of flood hydrograph described by peak & volume (years)
R Storm depth (tenths of mm)
D Storm duration (days)
T_S Return period of storm (years)

Station	Date	P	V	T_P	T_F	R	D	T_S
A2H027	10/04/67	396	5.84	77	103	2208	3	100
($357km^2$)	04/02/66	236	2.16	22	30	1143	1	40
(24years)	22/03/87	177	3.14	12	20	1086	3	15
A2H029	21/01/67	84	1.31	60	88	1064	3	70
($129km^2$)	25/01/75	57	0.76	29	40	1750	5	40
(25years)	28/01/78	45	0.79	17	31	1035	7	10
A2H045	28/01/78	272	12.1	42	92	2015	7	70
($653km^2$)	01/02/77	210	6.15	24	40	1508	3	15
(17years)	21/03/87	127	1.28	10	10	527	5	5
A2H047	01/02/77	116	1.44	24	47	1508	3	20
($65km^2$)	09/01/87	105	0.65	20	20	765	3	10
(15years)	01/03/81	68	0.68	18	22	932	1	20
A2H049	10/12/86	73	1.13	60	100	965	1	110
($371km^2$)	28/01/78	53	2.93	18	84	1566	3	40
(16years)	05/01/85	29	0.31	8	15	605	2	20
A5H004	08/12/69	291	8.89	27	37	803	2	25
($629km^2$)	03/02/67	247	10.2	20	34	770	3	15
(26years)	12/02/76	233	17.6	18	46	850	2	30
A6H006	12/02/76	74	2.64	18	37	1326	2	40
($168km^2$)	23/01/78	61	1.14	13	14	908	3	10
(16years)	20/02/75	45	1.13	7	10	1237	5	10
B6H003	02/02/60	281	8.21	11	33	2375	2	26
($92km^2$)	16/04/67	280	4.47	11	10	1120	3	7
(29years)	08/02/85	226	9.64	8	37	2162	2	24
C1H004	05/02/75	687	76.4	18	52	835	2	30
($901km^2$)	21/01/72	632	31.1	16	27	1040	3	20
(30years)	04/05/76	467	32.1	9	17	956	3	12

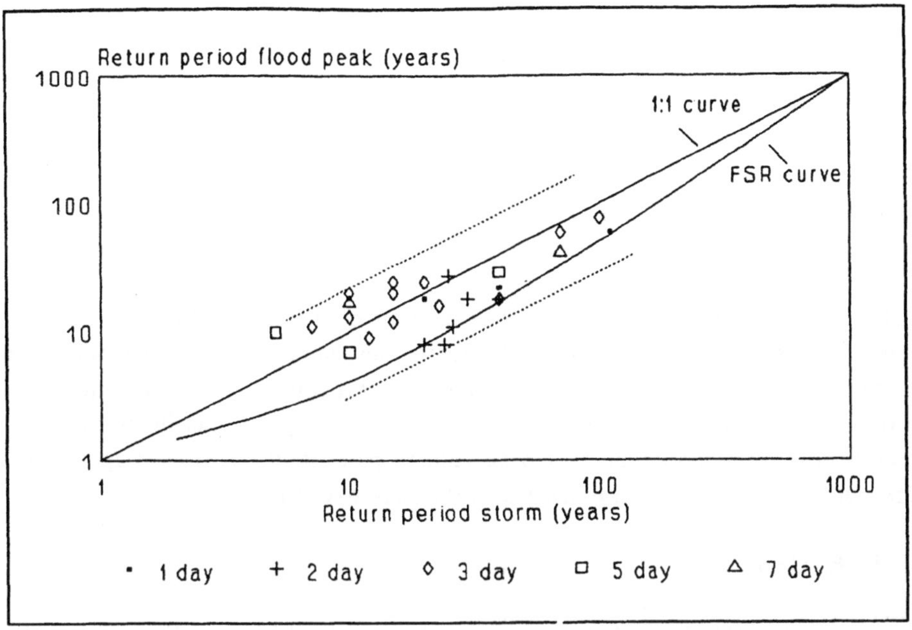

Figure 2 Return Period of Flood Peak versus Return Period of Storm

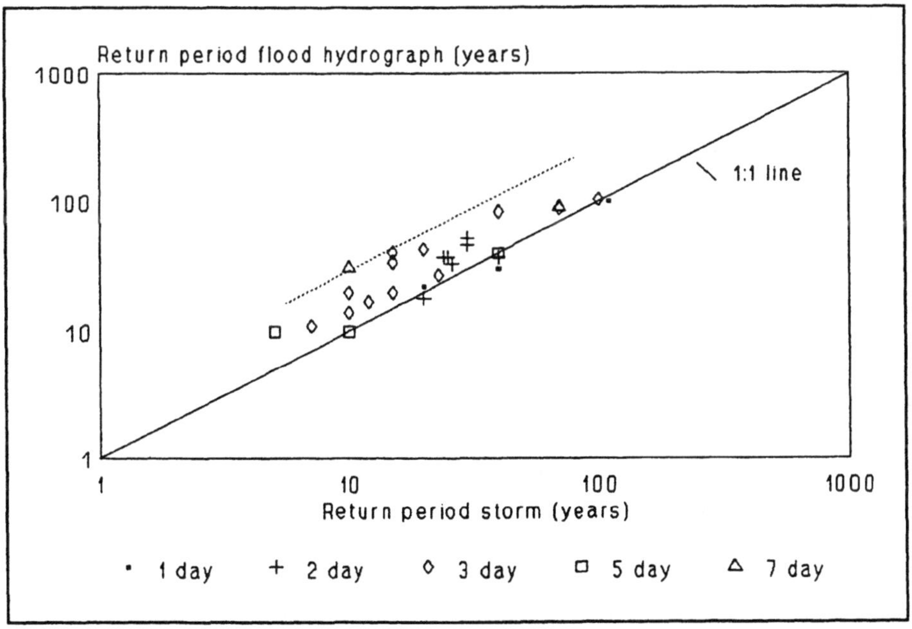

Figure 3 Return Period of Flood Hydrograph (peak & volume) versus Return Period of Storm

examination of the continuous record.

A method was described which attaches return period to the storms which caused the floods which yield annual maximum peaks. It turns out that integration of storm rainfall over a catchment using averaging or multi-quadrics yields similar total storm input. Comparison of the recurrence intervals of flood hydrographs (described by the two variables peak and volume) with those of their causative storms reveals that there is a much tighter association than between the storms and the flood peaks alone.

Finally, it was found that the recurrence interval of a flood hydrograph, defined in terms of its peak-volume pair, is longer that of the causative storm, but only up to a factor of three. From a design point of view, this is a conservative assumption because the flood caused by a storm is unlikely to occur more frequent than its progenitor, and therefore its magnitude (taken as a peak-volume pair) is likely to be bigger than that corresponding to the recurrence interval of the causative storm.

REFERENCES

Deacon, M.P. (1993). "Flood studies in South Africa - Towards a better methodology", Unpublished MScEng thesis, University of Natal. 135 pages.

Johnson, N.L. and S. Kotz (1972). Distributions in Statistics: Continuous Multivariate Distributions. Wiley.

Hiemstra, L A V and D. Francis, (1979). "The Runhydrograph - Theory and application for flood predictions", Water Research Commission Report, Pretoria.

Hiemstra, L. A. V., W.S. Zucchini, and G.G.S. Pegram (1976). "A method of finding the family of runhydrographs for given return periods", Journal of Hydrology, Vol. 30, pp. 93-103.

NERC (1975). Flood Studies Report, Vol. 1 - Hydrological Studies, Natural Environment Research Council, United Kingdom.

Pegram, Guy C. and Geoffrey G.S. Pegram (1993). "Integration of Rainfall via Multiquadric Surfaces over Polygons", Journal of Hydraulic Engineering, ASCE, Vol. 119, No.2, pp 151-163.

Tihansky, D.P. (1970). "Properties of the Bivariate Normal Cumulative Distribution", Rand Corporation Report P-4400-1, Santa Monica CA, USA.

BEST FITTING DISTRIBUTION FUNCTION FOR FLOODS IN THE UPPER INDUS BASIN (PAKISTAN)

N. IN-NA, R. LOOF, and C.S. AFZAL
Engineering (English) Program, Thammasat University
Rangsit Campus, Pathum Thani, 12121, Thailand

In Pakistan, the use of the Gumbel (EV1) distribution is a common practice. However, there was no investigation on the proper use of statistical distributions for annual maximum flood frequency analysis in Pakistan. A flood frequency analysis of the upper Indus basin was carried out for 20 unregulated stream flow stations located on various rivers with each data set having at least 20 observations. The Wald-Walfowitz test of independence and Grubbes & Beck test for detection of outliers were applied. Extreme value type 1 (EV1), General extreme value (GEV), Lognormal 2 (LN2), and Pearson Type 3 (P3) distributions were used as the parent models. The Probability Weighted Moments (PWM), and Maximum Likelihood (ML) methods were taken for parameter estimation. Commonly considered criteria to evaluate the performance of distributions such as statistical measures, goodness of fit test, and probability plots using appropriate probability papers were employed. The results of the study conclude that for flood frequency analyses of the upper Indus basin, GEV (EV2 and EV3) distribution with the PWM method of parameter estimation should be applied because it fits well on GEV probability papers with less RMSE, BIAS, and K-S statistics.

INTRODUCTION

A variety of distribution has been used in irrigation hydrology in fitting the hydrology random variables of annual stream flow [see, e.g., U.S. Water Council Committee (1967); Sangal and Biswas (1970); Natural Environment Research Council (1975); Kite (1978); Bobee (1979); In-na and Nguyen (1989); Nguyen et.al. (1989); Bobee and Ashkar (1991)]. Several investigators have different views in selecting the probability distribution functions to fit a hydrologic random variable. In Pakistan, the Gumbel distribution has been selected for flood frequency analysis. The Gringorten plotting position formula has been selected for plotting the results of frequency analysis on Gumbel paper without making any test while adopting this method. This method was selected because it is commonly used in frequency analysis. In doing so, they ignored the fact that selection of distribution in

K. W. Hipel (ed.),
Stochastic and Statistical Methods in Hydrology and Environmental Engineering, Vol. 1, 259–272.

frequency analysis is very important because results of frequency analysis can be used for many engineering purposes; i.e., 1-for the design of dams, bridges, culverts, and flood control structures, 2-to determine economic value of projects (Chow, 1988). Wrong selection of distribution may effect the design of the structure and its economic condition. Using the statistical technique, it is recommended that data still be plotted so that one can get an idea of how well the data fits the assumed statistical form and to spot potential problems.

Until now, no study was conducted in Pakistan to investigate the proper use of statistical distributions for annual maximum flood frequency analysis. Keeping in mind the importance of the use of best fit statistical distribution and plotting position formula in hydrology, a study is needed in the above mentioned area to investigate which distribution is reliable and more appropriate in the study area i.e. the upper Indus basin. Since the characteristics of the whole Indus river basin are nearly similar, the proper investigation for probability density function for the Indus basin will serve as a guideline for future and present work and a great contribution in a sense that most of the important structures of Pakistan are on the Indus river.

The main objective of the study is, therefore, to investigate best fit statistical distribution amongst Gumbel (EV1), General Extreme Value (EV2 and EV3), Pearson type 3 (P3) and Lognormal 2 (LN2) distributions. It has been found from the study that GEV (EV2 and EV3) distribution is the most reliable distribution for flood estimation in the upper Indus basin.

THEORETICAL CONSIDERATIONS

General Extreme Value Distribution (GEV)

The GEV distribution is a three parameter one which has the EV1, EV2 and EV3 distributions as a special case (Jenkinson, 1955). Its density function is:

$$F(X) = e^{-[1-K(x-u)/\alpha]^{\frac{1}{K}}} \quad ..... \quad K \neq 0 \tag{1}$$

$$F(X) = \exp[-\exp(-\frac{(x-u)}{\alpha})], \; K = 0 \tag{2}$$

Where u, α, K, are location, scale and shape parameters respectively.

If K=0, F(X) becomes Gumbel (EV1) distribution and K<0 and K>0, F(X) becomes EV2 and EV3 distributions respectively.

Extreme Value Type 1 or Gumbel Distribution

A continuous random variable x follows the Gumbel distribution if its density and probability density function are given by the following equations respectively:

$$F(x_1) = \exp[-\exp(-(x_1-u)/\alpha)] \tag{3}$$

$$f(x_1) = \frac{1}{\alpha} \exp\left[-\frac{(x_1-u)}{\alpha} - e^{(x_1-u)/\alpha}\right] \tag{4}$$

The parameter u is a location and α is scale parameter.
Where

$$u = \hat{\mu}_1 - 0.5772\alpha \tag{5}$$

$$\alpha = \hat{\sigma}\frac{\sqrt{6}}{\pi}, \ldots\ldots \tag{6}$$

$$\hat{\mu}_1 = mean$$

$$\hat{\sigma} = standard\ deviation$$

$$X_1 = u + \alpha Y_t \tag{7}$$

Where X_T is the discharge at various return periods. If T is the return period then the value of reduced variate can be obtained by the following equation:

$$Y_T = [-1n - 1n(1 - \frac{1}{T})] \tag{8}$$

Extreme Value Type 2 or Frechet Distribution

The density function is given as:

$$F(X_2) = \exp[-(1-\frac{K(X_2-u)}{\alpha})]^{1/K} \tag{9}$$

This distribution has three parameters; u and α are location and scale parameters where K is a shape parameter which must be negative.

The probability density function is given as:

$$f(x_2) = \frac{1}{\alpha}\,(1-\frac{x_2-u}{\alpha}K)^{\frac{1}{K}-1}\,.e^{-[1-\frac{K(x_2-u)}{\alpha}]^{\frac{1}{k}}} \tag{10}$$

$$X_2 = u+\frac{\alpha}{K}-(\frac{\alpha}{K})Y_2 \tag{11}$$

Where X_2 is the discharge at selected return periods and Y_2 is an EV2 reduced variate.

$$Y_2 = \exp\,[KY_1] \tag{12}$$

Extreme Value Type 3 or Weibull Distribution

The distribution function is

$$F(x_3) = \exp[-(1-\frac{K(x_3-u)}{\alpha})]^{1/K} \tag{13}$$

This distribution has three parameters; u and α are location and scale parameters where K is a shape parameter which must be positive.
Where

$$K > 0,\ \alpha > 0,\ -\infty \leq x_3 \leq u+\frac{\alpha}{K} \tag{14}$$

The probability density function is given as:

$$f(x_3) = \frac{1}{\alpha}\left(1-\frac{x_3-u}{\alpha}K\right)^{\frac{1}{K}-1} . e^{-\left[1-\frac{K(x_3-u)}{\alpha}\right]^{\frac{1}{K}}} \tag{15}$$

Where

$$X_3 = u+\frac{\alpha}{K}+\left(\frac{\alpha}{K}\right)Y_3 \tag{16}$$

and

$$Y_3 = -\exp[-KY_1] \tag{17}$$

Pearson Type 3 Distribution

The probability density function is given by the following equation:

$$f(x) = \frac{(x-x_o)^{\gamma-1}e^{-\left(\frac{x-x_o}{\beta}\right)}}{\beta^{\gamma}\Gamma(\gamma)}\,dx \tag{18}$$

and the density function is given by:

$$F(X) = \int_{x}^{x_o} \frac{(x-x_o)^{\gamma-1}e^{-\left(\frac{x-x_o}{\beta}\right)}}{\beta^{\gamma}\Gamma(\gamma)}\,d(x) \tag{19}$$

$$MEAN = x_o+\beta\gamma \tag{20}$$

$$VARIANCE = x_o+\beta^2\gamma \tag{21}$$

$$SKEWNESS = g = \frac{2}{\sqrt{\gamma}} \tag{22}$$

Bobee and Robitaille (1977) recommended an unbiased skewness coefficient especially for P3 distribution.

$$g_2 = \frac{N(N-1)^{1/2}}{N-2}(1+8.5/N)g \tag{23}$$

Where x_o, ß, and γ are location, scale and shape parameters respectively and N is sample size.

$$X_T = x_o + \beta Y_T \tag{24}$$

Where X_T is discharge at return period T and Y values having F(X) = G(Y) = 1 - 1/T.

Lognormal 2 Parameter Distribution

The lognormal with two parameters (LN2) has the following density function:

$$f(X;\mu_z,\sigma_z) = \frac{\exp\left(\frac{-(0.5)[1nX-\mu_z]}{\sigma_z}\right)^2}{[\sigma_z X\sqrt{2\pi}]} \tag{25}$$

Where μ_z, σ_z are scale and shape parameters and Y = 1nX is distributed as a normal variable with mean μ_z, and standard deviation σ_z. Let Z = $[1nX- \mu_z]/\sigma_z$ be the standard normal variable, then the F(X; μ_z, σ_z) = Φ(Z) where Φ(Z) is the standard normal distribution function.

$$Mean = \exp\left(\frac{\mu_z + \sigma_z^2}{2}\right) \tag{26}$$

$$Variance = (Mean)^2[\exp(\sigma_z)^2-1] \tag{27}$$

$$Skewness = [\exp(\sigma_z)^2+2][\exp(\sigma_z)^2-1]^{0.5} \tag{28}$$

Parameters Estimation

For Gumbel and GEV distributions, the probability weighted movement (PWM) method was found superior to all others methods of parameter estimation [Hosking et al. (1985); Phien and Emma (1989)], therefore, parameters of Gumbel and GEV (EV2 and EV3) distributions were estimated by using PWM method. It was also found that the maximum likelihood (ML) method had been used satisfactory with the P3 and LN2 distributions [Matalas and Wallis (1979); Bobee (1979); Stedinger (1980); Jain and Singh, (1987)]. Hence the ML method was used for the P3 and LN2 distributions.

Statistical Measures

Statistical measures selected are root mean square error (RMSE) and BIAS. These can be computed as:

$$RMSE = \sum_{i=1}^{N} \left(\frac{Q_o - Q_c}{Q_o}\right)^2 \tag{29}$$

$$BIAS = \sum_{i=1}^{N} \left|\frac{Q_o - Q_c}{Q_o}\right| \tag{30}$$

Where Q_o and Q_c are observed and calculated discharges and N is the sample size.

Grubbs and Beck Test for Detection of Outliers

To apply Grubbs and Beck (G-B) test, the following two quantities are calculated:

$$X_H = \exp(\hat{\mu} + Kn\,\hat{\rho}) \tag{31}$$

$$X_L = \exp(\hat{\mu} - Kn\,\hat{\rho}) \tag{32}$$

Where μ and σ are the mean and standard deviation of the natural logarithms of the sample, respectively, and Kn is the G-B statistic tabulated for various sample sizes and significance levels by Grubbs and Beck (1972). At the 10% significance level, the following polynomial approximation is used in the computer program HFA for estimating the tabulated values:

$$Kn = -3.62+6.28N^{1/4}-2.498N^{1/2}+0.491N^{3/4}-0.379N \tag{33}$$

Where N is the sample size. In applying the G-B test any sample values greater than X_H are considered to be high outliers and those less that X_L are considered to be low outliers.

Wald and Walfowitz Test for Independence

For a sample of size N (x_1 , ... , x_N) the Wald-Walfowitz (W-W) considers the statistic R such that

$$R = \sum_{i=1}^{N-1} x_i x_{i+1} + x_1 x_N \tag{34}$$

In the case where the elements of the sample are independent, R follows a normal distribution with mean and variance given by:

$$MEAN = \frac{(S_1^2 - S_2)}{(N-1)} \tag{35}$$

$$Var(R) = \frac{(S_2^4 - S_4)}{(N-1)} - (MEAN)^2 + \frac{(S_1^4 - 4S_1^2 S_2 + 4S_1 S_3 + S_2^2 - 2S_4)}{(N-1)(N-2)} \tag{36}$$

with $S_r = Nm'_r$ and m'_r is the rth moment of the sample about origin and S_2 is of order 2 about mean ($S_2 = m^1_2$).

$$\mu = \frac{(R-MEAN)}{(Var(R))^{0.5}} \tag{37}$$

This quantity follows a standardized normal distribution (Mean 0 and Variance 1) and can be used to test at the level α the hypothesis of independence, by comparing [u] with the standard normal deviate $u_{\alpha/2}$ corresponding to a probability of exceedance $\alpha/2$.

Unbiased Plotting Position Formula

This unbiased plotting position formulas used for GEV and EV1 distributions is:

$$P_m = \frac{N-m+0.05g+0.65}{N-0.08g+0.38} \quad (38)$$

Where P_m is the plotting position, m is the order number within a sample of size N and g is the coefficient of skewness [In-na and Nguyen (1989)].

The unbiased plotting position formula used for P3 and LN2 distributions can be written as follows:

$$P_m = \frac{N-m+0.3g+0.47}{N+0.3g+0.05} \quad (39)$$

Where P_m is the plotting position and m is the order number within a sample of size N and g is the skewness coefficient [Nguyen et al. (1989)].

Computer Packages and Program

Computer packages used in this study for parameter estimation of Gumbel, Generalized Extreme Value, and Lognormal distributions are respectively Gumbel, GEV, and Lognor which were developed by Phien (1990) and for the parameter estimation of Pearson type 3 distribution, computer program called "hydrological frequency analysis" (HFA) developed by Bobee and Ashkar (1991) was used. This program was also used to perform G-B test for detection of outlier and W-W test for independence.

Study Area

The upper Indus basin which covers drainage area of 640,000 km^2, was selected as a study area, and eight major rivers were selected for the study: the Indus, Gilgit, Haro, Soan, Kabul, Swat, Chitral and Hunza rivers (Table 1). The flows at these stations are natural. They have not been regulated by a reservoir upstream, and each station has at least 20 years of record. Descriptive statistics of annual maximum series are presented in Table 2.

RESULTS AND DISCUSSION

Results of the Grubbs and Beck test for the detection of outliers show that two stations, namely Gilgit and Sanjwal, have higher outliers. These two stations are, therefore, excluded from the analysis. The value of Wald and Walfowitz test statistics of 13 stations is less than the critical value at 5% significance level, and the remaining four stations accept the independence hypothesis at 1% significance level. Hence, it can be concluded that the annual flood peaks are independent of each other.

Table 3 shows values of K-S statistic, Absolute Maximum Value (D), RMSE, and Bias of EV1, GEV, LN2 and P3 distributions for 17 selected stations. It can be seen from Table 3 that based on the Kolmogorov-Smirnov test, D, RMSE and Bias of different distributions, the GEV distribution (EV2 and EV3) with unbiased plotting position formula for GEV distribution gives more reliable results than the other considered distributions.

CONCLUSIONS

Based on the results of less RMSE, Bias, and K-S statistic, the GEV (EV2 and EV3) distribution with the PWM method of parameter estimation is considered for flood frequency analysis of the upper Indus basin. It is noted also from the study that K-S test alone is not sufficient to distinguish the rank of fitness of specified distributions because almost all gauging stations are acceptable. The other test methods should be included in the further studies to improve the results.

REFERENCES

Bobee, B. (1979) Comment on "Fitting the pearson type 3 distribution in practice", Water Resources Research 15-3, 730.
Bobee, B. and Ashkar, F. (1991) "The gamma family and derived distribution applied in hydrology", Water Resources Publication.
Bobee, B. and Robitaille, R. (1977) "Use of pearson type 3 and log pearson type three distribution revisited", Water Resources Research 13-2, 427-442.
Chow, V.T. (1988) Applied Hydrology, McGraw Hill Int'l Editions, New York.
Grubbs, F.E. and Beck, G. (1972) "Extension of sample sizes and percentage points for significance test of outlying observations", Technometrics 14-4, 847-854.
Hosking, J.R.M., Wallis, J.R. and Wood, E.F. (1985) "Estimation of generalized extreme value distribution by the method of probability weighted moments", Technometric 27-3, 251-261.
In-na, N. and Nguyen, V.T.V. (1989) "An unbiased plotting position formula for the general extreme value distribution", Journal of Hydrology 106, 193-209.
Jain, D. and Singh, V.P. (1987) "Comparison of some flood frequency distributions using empirical data", Hydrologic Frequency Modelling, D. Reidel Publishing Co., USA.
Jenkinson, A.F. (1955) "The frequency distribution of the annual maximum (or minimum) values of meteorological elements", Q.J.R. Meteorol. Soc. 81, 158-171.
Kite, C.W. (1978) "Frequency and risk analysis in hydrology", Water Resources Publication, Michigan, USA.
Matalas, N.C. and Wallis, J.R. (1979) "Eureka! It fits a pearson type 3 distribution", Water Resources Research 9-2, 281-289.
Natural Environmental Research Council (1975) "Flood studies report", vol. 1,

U.K.
Nguyen, V.T.V., In-na, N., and Bobee, B. (1989) "A new plotting position formula for pearson type distribution", Journal of Hydraulics Division, ASCE 115-6, 709-730.
Phien, H.N. (1990) "A computer package for frequency analysis with the generalized extreme value distribution", Hydrosoft 13-4, 150-155.
Phien, H.N. and Emma, F.T. (1989) "Parameter estimation for the general extreme value distribution", Water SA 15-2, 147-154.
Sangal, B.P. and Biswas, A.K. (1970) "The 3 parameter lognormal distribution and its application in hydrology", Water Resources Research 16-3, 505-515.
Stedinger, J.R. (1980) "Fitting the normal distribution to hydrologic data", Water Resources Research 16-3, 481-490.
U.S. Water Resources Council (1967) "A uniform technique for determining flood flow frequencies", Bulletin 15, Washington D.C., USA.

TABLE 1. Inventory of stream gauging stations in study area

S.No.	River	Station	Period of Record	Agency
1	Indus	Kachura	1971-90	SWHP
2		Partab Bridge	1962-88	IRR
3		Darband	1960-74	DIC
4		Besham Qila	1969-88	SWHP
5	Gilgit	Gilgit	1960-72 1981-88	IRR
6		Alam Bridge	1966-88	SWHP
7	Haro	Khanpur	1960-83	IRR
8		Sanjwal	1960-79	IRR,SWHP
9		Gariala	1970-90	SWHP
10	Soan	Chirah	1960-86	SWHP
11		Rawalpindi	1960-83	SWHP
12		Dhok Pathan	1964-88	SWHP
13	Kabul	Warsak	1921-69	WDP,SWHP
14		Nowshera	1961-88	SWHP
15	Swat	Kalam	1961-88	SWHP
16		Chakdara	1961-88	SWHP
17	Chiral	Chitral	1964-88	SWHP
18	Hunza	Dainyor Bridge	1966-90	SWHP

Legend:
DIC — Dams Investigation Circle
WAPDA — Water and Power Development Authority
IRR — Department of Irrigation and Power
SWHP — Surface Water Hydrology Project (WAPDA)
WDP — Warsak Dam Project

TABLE 2. Descriptive statistic of annual maximum series

S.No.	Station	Sample Size	Mean m^3/s	SD m^3/s	Skewness (Sample)	Skewness P3
1	Warsak	49	2646.81	797.45	1.1974	1.4050
2	Alam Br.	23	3071.66	524.55	0.8316	1.1390
3	Besham	29	10254.96	1122.98	0.4382	0.5667
4	Chakdara	28	670.11	117.80	0.3637	0.4741
5	Chirah	27	226.32	143.99	1.1858	1.5590
6	Chitral	25	1036.88	139.72	0.5404	0.7242
7	Dainyor	24	1911.25	333.09	0.4343	0.5881
8	DhokPat.	25	1433.22	1072.17	2.1904	2.9352
9	Gariala	20	545.30	293.78	0.3310	0.4717
10	Gilgit	19	1244.50	191.74	0.2153	0.3120
10a	Gilgit*	20	1300.38	311.98	2.1350	3.0440
11	Kachura	20	4867.93	843.66	-0.5474	-0.7800
12	Kalam	28	368.68	61.40	-0.5610	-0.7610
13	Khanpur	24	159.88	99.77	1.8091	2.4500
14	Nowshera	28	3297.96	967.34	1.0940	1.4260
15	PartabBr	27	7671.82	1025.88	-0.0571	-0.0750
16	Rawalpindi	24	651.97	452.80	2.5353	3.4330
17	Sanjawal	19	297.37	131.00	1.0737	1.5540
17a	Sanjawal*	20	356.74	294.45	3.2400	4.6150

* = Statistics considering outlier

TABLE 3. RMSE, Bias and K-S statistic of EV1, GEV, LN2 & P3 distributions for seventeen selected stations

St.No.	DIST.	K-S STAT.	D	RMSE	BIAS
1	EV1	0.1920	0.05640	0.04700	1.17400
	GEV		0.05510	0.04700	1.17270
	P3		0.05327	0.07887	1.45900
	LN2		0.05800	0.05648	1.23000
2	EV1	0.2750	0.08430	0.00716	0.33920
	GEV		0.07512	0.00725	0.34000
	P3		0.09340	0.00810	0.35970
	LN2		0.14200	0.16750	0.51370
3	EV1	0.2400	0.08200	0.00670	0.36720
	GEV		0.07135	0.00470	0.30560
	P3		0.07930	0.01000	0.41410
	LN2		0.11200	0.00710	0.36900
4	EV1	0.2510	0.12400	0.03668	0.81140
	GEV		0.07440	0.01740	0.60900
	P3		#0.27	0.84145	2.18300
	LN2		0.08100	0.02316	0.66700
5	EV1	0.2552	0.06694	0.67880	2.22800
	GEV		0.05415	0.15830	1.33900
	P3		0.10120	0.18300	1.29340
	LN2		0.07600	0.22870	1.80460
6	EV1	0.2640	0.08400	0.01313	0.45320
	GEV		0.07455	0.00870	0.37780
	P3		0.07848	0.10840	0.74900
	LN2		0.76000	0.00910	0.38520
7	EV1	0.2700	0.09680	0.02300	0.59760
	GEV		0.08700	0.01970	0.57530
	P3		0.11100	0.04860	0.76740
	LN2		0.14300	0.02150	0.62250
8	EV1	0.2640	0.07280	1.20800	3.08300
	GEV		0.06470	0.56190	2.59700
	P3		0.11176	1.29400	3.76700
	LN2		0.08600	1.12050	3.55150
9	EV1	0.2940	0.08630	0.64650	1.95870
	GEV		0.06920	0.42640	2.06000
	P3		0.12027	0.84036	2.95370
	LN2		0.12400	0.53580	2.52600
10	EV1	0.3010	0.10171	0.01770	0.52320
	GEV		0.10730	0.01010	0.35670
	P3		0.10830	0.01937	0.52600
	LN2		0.15200	0.01200	0.40240

TABLE.3 (cont.)

10a*	EV1	0.2940	0.08630	0.04830	0.66980
	GEV		0.08300	0.02940	0.58800
	P3		0.10310	0.06010	0.86950
	LN2		0.11900	0.07200	0.87700
11	EV1	0.2940	0.21760	0.08550	1.11060
	GEV		0.08830	0.02010	0.45410
	P3		0.09550	0.07678	0.66715
	LN2		0.15700	0.05000	0.79880
12	EV1	0.2510	0.15940	0.17199	1.49000
	GEV		0.08400	0.02460	0.51770
	P3		0.11940	0.02790	0.55460
	LN2		0.09400	0.07726	1.06800
13	EV1	0.2700	0.08545	0.51500	2.56500
	GEV		0.07735	0.17390	1.62400
	P3		0.11940	0.50440	2.80660
	LN2		0.08700	0.31367	2.14700
14	EV1	0.2510	0.08116	0.03460	0.74880
	GEV		0.06970	0.03020	0.69570
	P3		0.06854	0.02323	0.54070
	LN2		0.11700	0.04920	0.90120
15	EV1	0.2552	0.14496	0.03370	0.74840
	GEV		0.10540	0.01420	0.53140
	P3		0.13510	0.04464	0.79960
	LN2		0.11400	0.01818	0.60740
16	EV1	0.2700	0.15170	0.62430	2.88060
	GEV		0.10590	0.62770	2.80600
	P3		0.14640	1.23000	3.74260
	LN2		0.13500	0.72190	3.34300
17	EV1	0.3010	0.10270	0.14060	1.38600
	GEV		0.10440	0.18965	1.37100
	P3		0.08680	0.15450	1.37200
	LN2		0.09400	0.22470	1.50450
17a*	EV1	0.2940	0.17900	0.01010	0.35680
	GEV		0.11900	0.37130	1.84500
	P3		0.18660	1.24600	4.24300
	LN2		0.14100	0.54900	2.63300

D = Absolute Max. Value # = Data is Rejected
K-S Statistic is at 5% * = Values Considering Outlier Significance Level

A PARETO-BASED FOS DISTRIBUTION FOR HYDROLOGIC FREQUENCY ANALYSIS

S. ROCKY DURRANS
Department of Civil and Environmental Engineering
The University of Alabama
260 H. M. Comer Hall
Tuscaloosa, Alabama 35487-0205
USA

Distributions of fractional order statistics (FOS), which may be viewed as pre-asymptotic extreme value distributions, have been recently defined by Durrans (1992) as plausible models for frequency analyses of hydrologic data. In this paper we examine these distributions more closely and demonstrate that they may be viewed as generalized probability models. The particular FOS distribution that arises from the Pareto distribution is used for illustrative purposes. This Pareto-based FOS distribution is shown to be more flexible than the generalized Pareto distribution discussed by Hosking (1988, 1990) and Vogel and Fennessey (1993), yet it has no more parameters.

INTRODUCTION

The problem of selection of a distribution that is appropriate for modeling of an observed data sequence is one that has plagued hydrologists for many years. The difficulty is due in large part to the typically small size of hydrologic data samples, and to the frequent presence of excessive skewness and/or kurtosis. Further compounding a statistical frequency analysis are issues related to the detection and treatment of outliers, mixed populations, utilization of exogenous information including historical and paleological data, and a host of others. Here we set aside these peripheral issues and focus on the problem of selection of a probability model. There are a large number of alternatives available, and they can often yield widely varying estimates of quantiles of interest. The choice of one alternative, often to the exclusion of others, remains both a controversial topic and one of intense interest.

Approaches to the problem of model selection are as diverse as are the models themselves. They range from subjective graphical plotting techniques to formal tests of hypotheses that are grounded in rigorous statistical theory, yet which are frequently not powerful enough to adequately accomplish their intended purpose. Theoretical arguments based on the central limit theorem (Chow, 1951) and extreme value theory (Gumbel,

K. W. Hipel (ed.),
Stochastic and Statistical Methods in Hydrology and Environmental Engineering, Vol. 1, 273–283.

1958) have been forwarded in certain instances, and in yet other cases the use of particular models has been institutionalized (WRC, 1981; NERC, 1975). However, the belief that no single model is adequate to represent all cases seems to be coming widely held (NRC, 1988, p. viii). More modern approaches to model selection are based on the concept of robustness (Kuczera, 1982) and on the use of so-called moment ratio diagrams (MRDs), which illustrate the skewness-kurtosis (and sometimes the skewness-coefficient of variation) relationships that exist among alternative models. Examples of MRDs are given by McCuen (1985) and Bobee and Ashkar (1991) for the case of conventional product moments, and by Hosking (1990) and Vogel and Fennessey (1993) for the case of *L*-moments.

Regardless of the model selection approach that one may choose to utilize, there remains a significant degree of uncertainty in the model that is selected. The concept of robustness has an intrinsic appeal, but a practical demonstration of robustness may involve a time-consuming and costly simulation study. The use of MRDs is an attractive alternative, but is hampered by one's ability to accurately estimate high-order sample moments. *L*-moments (Hosking, 1990) are attractive in this respect because they tend to be much less biased and variable than conventional product moments, but the curves indicative of alternative probability models which might be drawn on a MRD are often quite close to one another in the region of interest for most hydrologic data. That is, in spite of the relatively small variance and bias of the *L*-moment estimators, the small sample sizes typically found in hydrology often give rise to estimator variances large enough that it can remain quite difficult to discriminate between competing models. Figure 1 is an illustration of a MRD and shows the behavior of the *L*-moment ratios τ_3 and τ_4 for some selected distributions. Points representative of much hydrologic data lie in the region to the right of the point shown for the Gumbel (EV-1) distribution, and hence in a location where it can be difficult to discriminate between competing models.

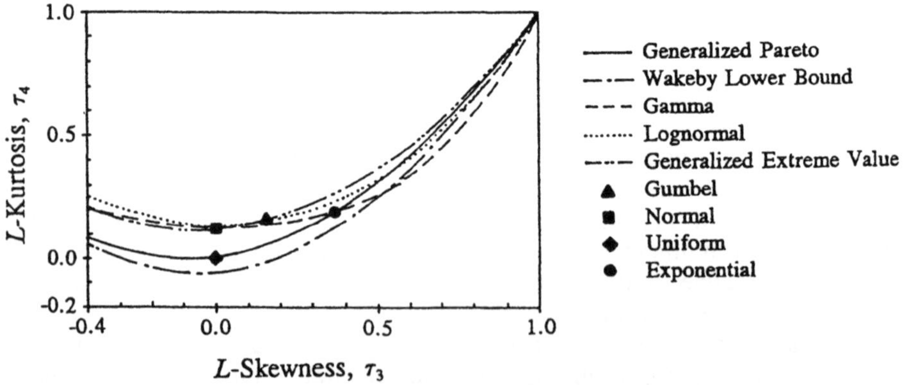

Figure 1. Moment ratio diagram.

One way to combat the uncertainty in model selection is to choose a model that is sufficiently general that it embeds other models of potential interest as special cases. In so doing one can effectively side-step the model selection problem entirely, and one can proceed with some confidence that the appropriate form of the general model will be borne out in the process of parameter estimation when it is applied to an observed data sequence. The Wakeby (Houghton, 1978) is a familiar example of a model of this general type, as are the generalized extreme value (Jenkinson, 1955) and generalized lambda (Greenwood et al., 1979) distributions. Hosking (1988, 1990) provides listings of several other generalized distributions as well. Perhaps the greatest difficulty with generalized models, however, is that they often contain an excessive number of parameters. This can lead to problems not only in estimation of the parameter values, but often leads also to increased variances of quantile estimators. Generalized models can also be difficult to regionalize because of the frequently large variations in parameter and quantile estimates. It is frequently the case as well that generalized distributions will still only represent a single curve, as opposed to a two-dimensional region, on a MRD. The Wakeby is an exception to this, as is the two-component extreme value (TCEV) distribution (Rossi et al., 1984), but these models have four or more parameters.

Distributions of fractional order statistics (FOS), which have been recently introduced by the author (Durrans, 1992), are a class of models that provide a significant degree of generality yet which are also quite parsimonious in their use of parameters. FOS distributions may be viewed as pre-asymptotic extreme value distributions, but involve a real- as opposed to integer-valued order parameter. Even with only three parameters, the skewness-kurtosis properties of FOS distributions often span two-dimensional regions on MRDs. This is true because FOS distributions often contain two shape parameters. In the sequel we discuss FOS distributions more fully, with a particular emphasis on the FOS distribution that is obtained by a generalization of the Pareto distribution. It is emphasized, however, that this is not the same as the generalized Pareto discussed by Hosking (1988, 1990) and Vogel and Fennessey (1993). The Pareto-based FOS model contains the same number of parameters but is much more general.

THE PARETO-BASED FOS MODEL

Overview

The mathematical definition of a FOS distribution is most easily stated in terms of its cumulative distribution function (cdf). For any known probability distribution with cdf F(x), the cdf Λ(x) of its FOS counterpart is defined as

$$\Lambda(x) = F^{\gamma}(x) \tag{1}$$

where γ is any non-zero positive real number. This expression may be recognized as being identical to that for the cdf of the largest order statistic in a sample of size γ drawn from the distribution $F \equiv F(x)$, but we now permit the order parameter γ to take on real as opposed to just integer values. This is the reason for the name "fractional" order statistics. Some general properties of the class of FOS distributions, and some reasons why they should be of interest to hydrologists, are described by Durrans (1992, 1993).

The Pareto-based FOS model

The Pareto distribution has a density function given by

$$f(x) = \frac{\alpha \beta^{\alpha}}{x^{\alpha+1}} \tag{2}$$

where $\alpha > 0$ is a scale/shape parameter, $\beta > 0$ is a location parameter that defines the lower bound of the distribution, and $\beta \leq x < \infty$. The cdf of the Pareto distribution is

$$F(x) = 1 - \left(\frac{\beta}{x}\right)^{\alpha} \tag{3}$$

and the quantile function (inverse cdf) can be stated explicitly as

$$x = \frac{\beta}{(1 - F)^{1/\alpha}} \tag{4}$$

The conventional non-central moments μ_r' of the Pareto distribution exist provided that $\alpha > r$ and may be stated as

$$\mu_r' = \frac{\alpha \beta^{r}}{\alpha - r} \tag{5}$$

The FOS counterpart of the Pareto distribution is obtained following (1) and has a cdf given by

$$\Lambda(x) = \left[1 - \left(\frac{\beta}{x}\right)^{\alpha}\right]^{\gamma} \tag{6}$$

where $\gamma > 0$ is an additional shape parameter. The other parameters and the range of X remain as previously given. It may be shown (see Appendix A) that the conventional moments of this FOS distribution also exist when $\alpha > r$, and that they may be stated as

$$\mu'_r = \alpha\gamma\beta^r \sum_{i=0}^{\gamma-1} \binom{\gamma-1}{i} \frac{(-1)^i}{\alpha(i+1) - r} \tag{7}$$

when γ is an integer. Either numerical integration or interpolating functions can be used when γ is non-integral and greater than unity. Interpolating functions are probably better as numerical integration of the defining equation tends to be quite prone to round-off errors. Interpolation cannot be done when γ is in the range (0,1), however, as γ can never be equal to zero. This is an area in which additional work is needed.

PWMs and *L*-moments of the Pareto-based FOS model

Greenwood et al. (1979) have introduced probability weighted moments (PWMs) as an alternative to conventional moments. They have noted that the use of PWMs is often attractive in cases where the quantile function (inverse cdf) of a probability distribution may be written explicitly. Hosking (1990) has introduced *L*-moments, which are linear combinations of PWMs, and which can also be used for parameter estimation. Because sample estimators of PWMs and *L*-moments are formed as linear combinations of sample values, they are much less biased and variable than are conventional product moments. They are also much less sensitive to outliers which may be present in a data sequence.

Using the notation of Greenwood et al. (1979), we consider here the $M_{1,j,0}$ PWMs for $j = 0,1,2,\ldots$ These are defined as

$$M_{1,j,0} = E[X\Lambda^j(x)] = \int_0^1 x(\Lambda)\Lambda^j d\Lambda \tag{8}$$

Note that when $j = 0$ this represents the mean of the FOS distribution. For the case of the Pareto-based FOS distribution, the PWMs may be shown to be

$$M_{1,j,0} = \alpha\beta\gamma \sum_{i=0}^{b} \binom{b}{i} \frac{(-1)^i}{\alpha(i+1) - 1} \tag{9}$$

provided that $\alpha > 1$ and that b is a positive integer, where b is defined as

$$b = \gamma(j+1) - 1 \tag{10}$$

(see Appendix B). When b is not integral either numerical integration of the defining equation (B.4) must be accomplished, or interpolating functions must be devised. As in the case for conventional moments, interpolating functions are probably preferable, but again, additional work is needed for the case where γ is in the range (0,1).

An interesting point that may be noted here is that the PWMs exist whenever the value of the parameter α is greater than unity. Recall that the conventional moments exist only in cases where $\alpha > r$, where r is the order of the moment. An additional point of interest relates to a well-known result for conventional moments which states that if μ_r' exists, then so does μ_s' for $s < r$. Also, if μ_r' does not exist, then neither does μ_s' for $s > r$ (Stuart and Ord, 1987, p 72). In the case of PWMs, Greenwood et al. (1979) state that if $M_{1,0,0}$ exists, then so does $M_{1,j,0}$ for $j > 0$. That is, unlike conventional moments, higher-order PWMs exist as long as the zero-order PWM exists. It is not clear, however, from the discussion by Greenwood et al. whether a higher-order PWM can exist when the zero-order PWM does not exist. To address this issue, consider the case where $\gamma = 1/2$ and $\alpha > 1$. In order for the quantity b to be an integer, it is necessary that the moment order j take on one of the values $j = 1,3,5,\ldots$ (in fact, for any rational value of γ, a set of j values may found that will yield integral values of b). Hence, when $\gamma = 1/2$ and $\alpha > 1$ all positive and odd-ordered PWMs exist. This is true whether the zero-order PWM exists or not, and thus demonstrates that the existence of higher-order PWMs is not contingent upon the existence of the zero-order PWM. It can be shown that the zero-order PWM in this case does in fact exist, but the existence of the higher-order PWMs has been shown independently of this.

The first four *L*-moments of the Pareto-based FOS model are defined in terms of a linear combination of PWMs and may be stated as

$$\lambda_1 = M_{1,0,0} \tag{11a}$$

$$\lambda_2 = 2M_{1,1,0} - M_{1,0,0} \tag{11b}$$

$$\lambda_3 = 6M_{1,2,0} - 6M_{1,1,0} + M_{1,0,0} \tag{11c}$$

$$\lambda_4 = 20M_{1,3,0} - 30M_{1,2,0} + 12M_{1,1,0} - M_{1,0,0} \tag{11d}$$

L-moment ratios, which are analogous to the conventional coefficients of variation, skewness, and kurtosis, are defined as

$$\tau_2 = \frac{\lambda_2}{\lambda_1} = \text{L-coefficient of variation} \tag{12a}$$

$$\tau_3 = \frac{\lambda_3}{\lambda_2} = \text{L-skewness} \tag{12b}$$

$$\tau_4 = \frac{\lambda_4}{\lambda_2} = \text{L-kurtosis} \tag{12c}$$

Fully expanded expressions for the *L*-moments and *L*-moment ratios for the Pareto-based FOS model are lengthy and so are not given here. Since they are formed as finite summations, however, they may be easily computed.

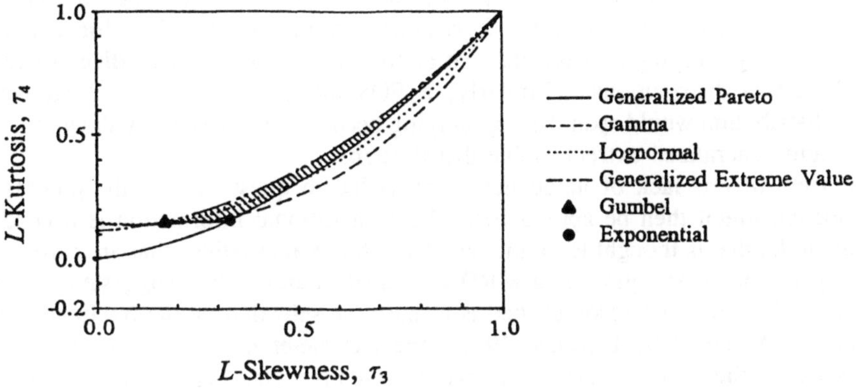

Figure 2. Moment ratio diagram. The shaded region can be represented by the Pareto-based FOS model when $\gamma \geq 1$.

DISCUSSION

Figure 2 is an illustration of a MRD that shows the two-dimensional region in *L*-skewness and *L*-kurtosis space that can be represented by the Pareto-based FOS model when $\gamma \geq 1$. It is clear from this figure that the FOS model, which may be viewed as a generalized distribution containing three parameters, embeds as special cases several other probability models that have been widely used in hydrology. Among these are the shifted exponential, Gumbel (EV-1), and Frechet (EV-2) models. The Pareto-based FOS model can also emulate the 3-parameter lognormal model in certain cases.

It may be observed that the curve representative of the generalized Pareto distribution, which forms one boundary of the region shown for the Pareto-based FOS model, corresponds to the case where the fractional order parameter γ is equal to unity. For any other value of γ that is greater than unity one also obtains a curve (not shown) lying between and behaving in the same manner as the curves for the generalized Pareto and generalized extreme value distributions. When $\gamma = 10$ the curve for the Pareto-based FOS model is for all practical purposes the same as the one for the generalized extreme value model. This implies that convergence of the distribution of the largest order statistic to the limiting asymptotic form is quite fast for the case of a Pareto initial distribution. This fast convergence is not true, however, for many other distributions that have been frequently used in hydrology. Fisher and Tippett (1928), for example, have stressed the extremely slow convergence for the case of a normal initial distribution, and J.R. Stedinger (personal communication, 1992) has found that the rate of convergence is even slower when the initial distribution is lognormal.

It should be noted that the Pareto-based FOS model is not unique in the sense that it is the only model with the ability to represent the two-dimensional region depicted in Figure 2. A FOS model based on the 2-parameter gamma distribution, for example, would span the region lying between the two curves shown for the generalized extreme value and gamma distributions. Similarly, a FOS model based on the 2-parameter lognormal distribution would span the region lying between the two curves shown for the lognormal and generalized extreme value distributions.

In view of this lack of uniqueness, one is led to question how the process of model selection might then be approached. Some additional information is needed to select one model that is thought to be the best from those alternatives that are capable of representing the point or region on a MRD that is of interest. It is suggested that one way in which this might be accomplished is to base the selection on the fractional order parameter γ. As noted by Durrans (1993), the parameter γ can be interpreted in a physical sense as the average number of severe events per year that occur at the gaging location of interest. Since the rate of convergence to the limiting generalized extreme value curve depends on the initial distribution F that is chosen for the FOS model, and since for any point lying in the two-dimensional region on a MRD that can be represented by a particular FOS model there is a corresponding value of the order parameter γ, it should be possible in principle to make the model selection based on this physical interpretation of γ.

SUMMARY

It has been shown in this paper, using the Pareto-based FOS distribution as an example, that FOS distributions can provide a significant degree of flexibility for frequency modeling without the burden imposed by an excessive number of unknown parameters. Even with only three parameters, FOS models are often capable of spanning two-dimensional regions on moment ratio diagrams. As a consequence of this, they embed as special cases many of the models that have been commonly applied in hydrology.

In order to illustrate the generality of the Pareto-based FOS model in this paper, the focus has centered on the definition of several moment-based measures (conventional, PWM, and *L*-moments) in terms of the parameters of the model. In practice, when fitting a model to an observed data sequence, what one needs are the inverse relationships. That is, one needs relationships that can be applied to estimate the model parameters based on knowledge of sample estimates for the moments. For the case of the Pareto-based FOS model, it can be seen by inspection of the expressions developed here that a system of nonlinear equations must be solved regardless of the type of moments employed. Because of this, there is probably no significant computational benefit to be gained by use of any particular type of moments. It is recommended, however, that either PWMs or L-moments be employed because the sample moment estimators for those methods are much more well behaved than are those for conventional moments.

An additional point that may be noted with respect to the issue of parameter estimation concerns the parameter γ. If this parameter can be estimated based on exogenous information, such as the average number of severe events per year, then the problem of estimating the other two parameters α and β is greatly simplified. This is particularly true in the case of PWMs when the value of γ is non-integral, because the orders j of the moments for which a closed form expression can be found are dependent on the value of γ.

APPENDIX A

By definition, the conventional non-central moments of the Pareto-based FOS distribution are

$$\mu_r' = \int_0^1 x^r d\Lambda = \beta^r \int_0^1 \frac{d\Lambda}{(1 - \Lambda^{1/\gamma})^{r/\alpha}} \tag{A.1}$$

Letting $a = r/\alpha$ and $u = 1 - \Lambda^{1/\gamma}$ so that $d\Lambda = -\gamma(1 - u)^{\gamma-1}du$, this integral can be transformed into

$$\mu_r' = \gamma\beta^r \int_0^1 \frac{(1 - u)^{\gamma-1}}{u^a} du \tag{A.2}$$

This integral exists for all $a \in (0,1)$ and for all $\gamma > 0$ despite the fact that there is always a singularity at $u = 0$, and that there is one at $u = 1$ as well when $\gamma < 1$. For integral values of γ the numerator of the integrand in (A.2) can be expanded binomially and the integral evaluated term by term. This yields

$$\mu_r' = \gamma\beta^r \sum_{i=0}^{\gamma-1} \binom{\gamma-1}{i} (-1)^i \int_0^1 u^{i-a} du \tag{A.3}$$

in which the integral converges for all values of $a = r/\alpha$ less than unity, or for all values of α greater than r. When this condition is met we obtain the result

$$\mu_r' = \alpha\gamma\beta^r \sum_{i=0}^{\gamma-1} \binom{\gamma-1}{i} \frac{(-1)^i}{\alpha(i+1) - r} \tag{A.4}$$

APPENDIX B

The probability weighted moments (PWMs) of the Pareto-based FOS distribution are defined as

$$M_{1,j,0} = E[X\Lambda^j(x)] = \int_0^1 x\Lambda^j d\Lambda \tag{B.1}$$

where $\Lambda(x)$ is the cdf of the FOS distribution. Since, by definition, $\Lambda(x) = F^\gamma(x)$ where $\gamma > 0$ and F(x) is the cdf of the Pareto distribution, this can be restated as

$$M_{1,j,0} = \int_0^1 xF^{j\gamma}d\Lambda \tag{B.2}$$

Now, the differential $d\Lambda = \gamma F^{\gamma-1}dF$ and the quantile function (inverse cdf) of the Pareto distribution is given by (4) so

$$M_{1,j,0} = \beta\gamma\int_0^1 \frac{F^{\gamma(j+1)-1}}{(1-F)^{1/\alpha}}dF \tag{B.3}$$

To evaluate this integral, let $a = 1/\alpha$ and $b = \gamma(j + 1) - 1$. Also let $u = 1 - F$ so $F = 1 - u$ and $dF = -du$. Transformation of the integral therefore yields

$$M_{1,j,0} = \beta\gamma\int_0^1 \frac{(1-u)^b}{u^a}du \tag{B.4}$$

The integral in this expression is of the same form as the integral in (A.2). Despite the singularities, it exists for all $a \in (0,1)$ and $b > -1$. For integral values of b the numerator of the integrand in (B.4) can be expanded binomially and the integral evaluated term by term. This yields

$$M_{1,j,0} = \beta\gamma\sum_{i=0}^{b}\binom{b}{i}(-1)^i\int_0^1 u^{i-a}du \tag{B.5}$$

in which the integral converges for all values of $a = 1/\alpha$ less than unity, or for all values of α greater than unity. When this condition is met we obtain the result

$$M_{1,j,0} = \alpha\beta\gamma\sum_{i=0}^{b}\binom{b}{i}\frac{(-1)^i}{\alpha(i+1)-1} \tag{B.6}$$

CITED REFERENCES

Bobee, B., and Ashkar, F. (1991) The Gamma Family and Derived Distributions Applied in Hydrology, Water Resources Publications, Littleton, Colorado.
Chow, V.T. (1951) "The log-probability law and its engineering applications", Proceedings, American Society of Civil Engineers 80, 536-1 to 536-25.
Durrans, S.R. (1992) "Distributions of fractional order statistics in hydrology", Water Resources Research 28(6), 1649-1655.
Durrans, S.R. (1993) "Flood frequency analysis with distributions of fractional order statistics", Proceedings, 13th Annual AGU Hydrology Days, Colorado State University, March 30 - April 2, pp. 287-298.
Fisher, R.A., and Tippett, L.H.C. (1928) "Limiting forms of the frequency distribution of the largest or smallest member of a sample", Proceedings, Cambridge Philosophical Society 24, 180-190.
Greenwood, J.A., Landwehr, J.M., Matalas, N.C., and Wallis, J.R. (1979) "Probability weighted moments: Definition and relation to parameters of several distributions expressable in inverse form", Water Resources Research 15(5), 1049-1054.
Gumbel, E.J. (1958) Statistics of Extremes, Columbia University Press, New York.
Hosking, J.R.M. (1988) Fortran Routines for Use with the Method of *L*-Moments, IBM Research Report RC 13844, Yorktown Heights, New York.
Hosking, J.R.M. (1990) "*L*-moments: Analysis and estimation of distributions using linear combinations of order statistics", Journal of the Royal Statistical Society B, 52, 105-124.
Houghton, J.C. (1978) "Birth of a parent: The wakeby distribution for modeling flood flows", Water Resources Research 14(6), 1105-1109.
Jenkinson, A.F. (1955) "The frequency distribution of the annual maximum (or minimum) values of meteorological elements", Quarterly Journal of the Royal Meteorological Society 81, 158-171.
Kuczera, G. (1982) "Robust flood frequency models", Water Resources Research 18(2), 315-324.
McCuen, R.H. (1985) Statistical Methods for Engineers, Prentice-Hall, Englewood Cliffs, New Jersey.
NERC. (1975) Flood Studies Report, Natural Environment Research Council, London.
NRC. (1988) Estimating Probabilities of Extreme Floods: Methods and Recommended Research, National Academy Press, Washington, D.C.
Rossi, F., Fiorentino, M., and Versace, P. (1984) "Two-component extreme value distribution for flood frequency analysis", Water Resources Research 20(7), 847-856.
Stuart, A., and Ord, J.K. (1987) Kendall's Advanced Theory of Statistics, Vol. 1, Oxford University Press, New York.
Vogel, R.M., and Fennessey, N.M. (1993) "*L*-moment diagrams should replace product moment diagrams", Water Resources Research 29(6), 1745-1752.
WRC. (1981) Guidelines for Determining Flood Flow Frequency, Bulletin 17B of the Hydrology Committee, U.S. Water Resources Council, Washington, D.C.

THE GRADEX METHOD: FOR A STATISTICALLY ROBUST EVALUATION OF EXTREME-VALUE FLOODS BASED ON LOCAL CHARACTERISTICS

R. GARCON
Electricité de France - Generation and Transmission Division
General Technical Division - Water Resources Department
37 Rue Diderot - BP 41 - 38040 Grenoble - France

After rapidly summarizing the various problems involved in design flood assessment, we present the broad lines of the Gradex method. This method estimates the probability of extreme floods by extrapolating rare events on the basis of the empirical distribution of precipitation for a drainage basin. The basis for this extrapolation is examined from a statistical point of view.

GENERAL REMARKS WITH REGARD TO DETERMINING DESIGN FLOOD DISCHARGES

When designing a structure which may be exposed to flooding, engineers are obliged to take a certain risk, in part a result of the incomplete knowledge of the hydrological events which may affect the structure during its lifetime.

We must therefore come up with some higher value, referred to as the project design flood, knowing that there is always some risk that it may be exceeded: the higher the value, the smaller the risk of a disaster but the greater the initial investment. Attempts to resolve this dilemma mathematically encounter two obstacles.

First, the extent of the damage caused for example by a damburst is often such that, particularly when human life is at risk, one is tempted to look for a solution offering "absolute" safety. Unfortunately, if there is a physical limit to a river's discharge level, it is extremely high and, were it possible to evaluate it correctly, it would result in most schemes never actually being built. As the design flood value must therefore be lower than the physical limit, then the most serious danger lies in forgetting that this choice has been made, which does nothing to eliminate the associated risk. Consequently it does seem preferable to face facts and attempt to assess flood frequency as objectively as possible.

This leads us to consider the flow of a river as a random variable. We shall attempt to determine its distribution and then retain a few values located in the outlying portion. A flood with a 1 000 or 10 000 year return period deduced from this type of analysis has no real physical consistency and must be treated as an artifice, the aim of which is to

K. W. Hipel (ed.),
Stochastic and Statistical Methods in Hydrology and Environmental Engineering, Vol. 1, 285–298.

compensate for our ignorance with regard to the largest event that will actually occur within a given period. This in no way reduces the practical value of such a probalistic view of natural phenomena and we will hereafter use this approach without any further reference to its physical shortcomings.

Having said that, the problems encountered estimating probabilities of this sort present an additional obstacle. The extreme complexity of the mechanisms involved in atmospheric circulation and drainage processes is such that the relevant values cannot be calculated using an entirely deterministic approach. We must therefore attempt to make the best possible use of discharge data from the past. Starting with a minimalist hypothesis of unvarying climate and run-off conditions for the catchment area under study, these observations represent a statistical sample offering information about the population from which it was taken. Unfortunately the return periods which concern hydraulic engineers are generally much too large to allow the use of limited hydrological records without forming additional assumptions.

Even observations over a 3000 year period only provide the basis for a very rough assessment of events with a 1000 year return period: there is only a 90% probability that the flood with a 1000 year return period lies within the interval marked by the largest and seventh largest observed floods. The enormous flood discharge variations observed in a large number of examples suggests that this interval is often very wide. Our aim is not to deny the value of historical hydrology data. Even when it is imprecise, whether based on surveys conducted at the time of the event or on subsequent analysis of sediment, this type of data is always precious and should be taken into account. But without additional assumptions, it could only provide an extremely approximate assessment of floods with a high return period.

It would be tempting, at this point, to compensate the relative poverty of observations, in temporal terms, by using their spatial richness. In other words, when studying a particular drainage basin, one could use measurements from other localities. This approach, which is intended to extend precipitation data artificially, requires considerable care. If one restricts attention to neighbouring areas, the increase in the volume of data is illusory, on account of the high spatial correlation of hydrological events. If, on the other hand, data is taken from drainage basins at some considerable distance, it is difficult to claim that one is dealing with identical climatological characteristics, in particular with regard to infrequent events.

If one abandons any attempt to use spatial data, the estimation will be extremely dependent on the sample covering the observed period. Extrapolation to rare frequencies in an empirical distribution of flood discharge values is extremely risky. This is clear when one considers fifty-year records of rainfall, in which the maximum value is almost ten times higher than all the others. The presence or absence of this sort of outlying value, in the sample used to determine the design flood value, is critically important.

It is nevertheless possible, using several simple, reasonable assumptions to improve the robustness of low frequency flood estimations, basing them on exclusively local data, which is entirely typical of the area under study. The GRADEX method, which is described below, provides a valuable compromise solution for a certain type of drainage basin.

THE GRADEX METHOD

This method was developed by engineers at *Electricité de France* in 1966 and has been used frequently ever since. It deliberately ignores the complexity of the mechanisms which initiate floods. Instead it offers a statistical model of the process taken as a whole. Its underlying principle is to use the statistical distribution of precipitation to deduce the distribution of extreme-value discharges, setting aside all but the most essential features of the rain-discharge transformation process. This method has already been explained in detail by Guillot, P., Duband, D. (1967), and Duband, D. et al (1988), so we only recall here its broad lines.

Selecting a time step value

With this method, one must start by estimating the time base, **H**, which represents the average duration of the hydrograph representing direct run-off from the basin. **H**, which will be used as the time base for the study of rainfall and discharge, can be estimated approximately. Experience shows that it can be doubled, without greatly affecting the results. Although, for practical reasons a twenty-four hour time base is often used when studying average-sized drainage basins, it generally ranges from 4 to 72 hours, except for certain particular cases.

Determining rare precipitation values

The basic assumption concerns the shape of the precipitation distribution: the frequency of occurrence of rainfall **P**, falling in **H** hours, at a given place and in a given season, can be expressed by a simple exponential decay function. When **P** is large, the probability of exceeding **P** is given by:

$$1 - F(P) = \alpha \exp(-P/a) \qquad (1)$$

This approximation is assumed to be valid for an average precipitation falling on a uniform drainage basin. The **a** parameter is called the GRADEX (GRADient of EXtreme values) hence the name given to this method. It corresponds to the slope of the asymptote of the rainfall distribution plotted on Gumbel paper. It can be calculated analytically, as is the case for α, on the basis of rainfall records. It constitutes a climatological characteristic of the site or the basin under study.

Only the single, highest, seasonal Gradex value will be used, out of all the precipitation data, for it can easily be demonstrated that the maximum annual precipitation distribution, which is the product of the seasonal distributions, decays asymptotically in line with the highest seasonal Gradex value.

Extreme rainfall values, which can easily be calculated at this stage, will not play an explicit role in determining extreme flood values.

Calculating flood volume for a given return period

A second assumption is now required before the above data can be used to calculate runoff. We shall assume that for events above a certain return period $\mathbf{T_0}$ the basin tends towards a saturation limit and that any increase, **dP**, in rainfall results in an increase, **dQ**, in runoff, where **dP** = **dQ**. This assumption, combined with the previous one, implies that from an asymptotic point of view, the frequency distribution functions **F** and **G**, corresponding to rainfall and average runoff for time base **H** can be expressed as :

$$\mathbf{G}(\mathbf{Q}) = \mathbf{F}(\mathbf{Q} + \mathbf{D}) \qquad (2)$$

where **D** is the expected runoff deficit value or the quantity of water temporarily retained by the basin during floods for the period **H**.

A common mistake, stemming from the apparent simplicity of this result, is to consider the 1000-year flood value as the 1000-year precipitation value minus **D**. In fact, the flood may equally be caused by a 10 000-year precipitation falling on dry soil or a 100-year precipitation falling on a basin which is already saturated. One must bear in mind that the assumption that rainfall distribution is subject to exponential decay is essential for the previous equation to be accurate.

The end result of the two assumptions is that, when plotted on Gumbel paper, the asymptotes for the distributions of **P** and **Q** form two parallel, straight lines, separated vertically by **D**. It is thus possible to extrapolate the flood distribution starting from the return period $\mathbf{T_0}$ corresponding to saturated conditions. The larger the drainage basin and the greater its permeability, the longer this return period may be. This is equally true for gentler slopes and infrequent heavy rainfall. For those cases best suited to this approach, it may range from ten to fifty years.

As is the case when choosing the time base, the starting point for extrapolation has, relatively speaking, little effect on the estimation of extreme floods.

Calculating instantaneous discharge

Using the largest possible number of hydrographs, corresponding to significant floods, the mean ratio **k** between peak instantaneous discharge and average flow over a period **H** can be calculated. The ratio can on the average be assumed to be unrelated to the discharge. This is the third and last assumption. If the ratio dispersion is not too great, peak discharge can be deduced from the average flood simply by its affinity with the **k** ratio.

High **k** dispersion or an average value of over 2 generally mean that the time base has been overestimated. This affects the robustness of extreme flood assessment. On the other hand, if **k** is lower than 1.3 the time base has probably been underestimated and the design flood is likely to be unnecessarily high.

Limits to the validity of the GRADEX method

The relevance of the method is limited by its assumptions. It is not suitable for large drainage basins, even in cases where major floods are to a large extent formed by direct rainfall runoff. In locations where the surface area exceeds a certain size, the length of the time base and the heterogenous nature of rainfall events throw doubt on the exponential approximation. As far as we know, the method has not so far been used for drainage basins of more than 20 000 km^2.

In the case of highly permeable basins, with very gentle slopes or very low rainfall, one encounters another problem. The return period used to extrapolate the empirical discharge distribution, using the rainfall Gradex, is often too high in relation to the size of the available rainfall records. 10-year or even 100-year events fall too short of the theoretical saturation level of the basin.

Strongpoints and shortcomings of the GRADEX method

The simplicity of the method, combined with a certain intellectual elegance, makes it extremely attractive. Thanks to its fundamental simplicity, which also contributes to its robustness, it is easy to assess the validity of the results which depend only on a small number of essential points.

Its explicitly probabilistic approach is also an advantage.

As it relies exclusively on local data, it reflects the significant disparities, with regard to meteorological and hydrological risks, between one site and the next.

Finally, discharge data is used in a rational way and hydrometric records covering periods of between 10 and 30 years are generally sufficient. Large floods, for which measurements are sometimes inaccurate and certainly subject to high sample dispersion, do not have a decisive effect on extrapolation. On the contrary, extrapolation is based on rainfall data which is plentiful for a given area, generally reliable and statistically more regular.

If local data, in particular rainfall data, is not available the interest of the method is clearly smaller.

Apart from the restrictions concerning the areas in which the method can be employed and its limited use, the main shortcoming of the Gradex method is that it is based on three assumptions, and that one of them never fails to provoke the distrust of possible users. We refer, of course to the precipitation distribution assumption.

Reactions to this last point have led us to concentrate research work over the last few years on the processing of rainfall data, and more specifically to answer two fundamental questions : How good is the exponential approximation and how statistically robust are the estimations of extreme precipitations obtained using this method?

EXPONENTIAL DECAY ASSUMPTION

A method which bases the estimation of extreme floods on the equivalent extreme rainfall values clearly cannot avoid the pitfalls discussed page 2 :

- Process too complex to be apprehended by an entirely deterministic approach;
- Samples too short to identify rainfall events with a relevant return period;
- By using "regional" data, one runs the risk of "smoothing" over local particularities (differences of up to 100% have been obtained for the Gradex of sites only 20 kilometers apart).

In view of this, an assumption which enables us to determine the probability of extreme rainfall by extrapolating from empirical distributions is clearly extremely valuable. But it must have a solid basis. Can one demonstrate that the rainfall distribution, for a given place and time, gives a linear asymptote, when plotted on Gumbel paper?

Scientists are entitled, indeed obliged, to treat this sort of claim with a certain scepticism. As we are not in a position to provide a physical or statistical demonstration in support of our claim, it would be pretentious to require a natural phenomenon to comply with our law.

One may be much more modest and accept that the suggested probabilistic model provides a reasonable approximation to reality, for return periods of up to 10 000 years. On the other hand it is necessary to return to the asymptote rapidly - for rainfalls with a two year return period - so that the slope of **P** in relation to **Log (1-F(P))** can be estimated on small samples and extrapolated to sufficiently long return periods for it to be possible to calculate the design flood.

The least that one can do before recommending the using of this type of assumption is to check its coherence with available rainfall records, for frequency distributions which can be observed. Which is exactly what we have done for a certain number of daily precipitation series from stations located, for the most part, in France.

After eliminating any records which either appeared unreliable or were too short, 192 series, with an average length of 41 years, were processed. A single season was selected, covering four months, from August to November. In a majority of cases this corresponds to the period during which the rainfall risk is highest. For each measurement site, we selected the empirical distribution of the highest rainfall values over a 24 hour period during the season. In order to be able to compare the series, we reduced the average for each sample to 0 and its standard deviation to 1.

Figure 1. Average empirical distribution for 192 samples expressed in standard measures.

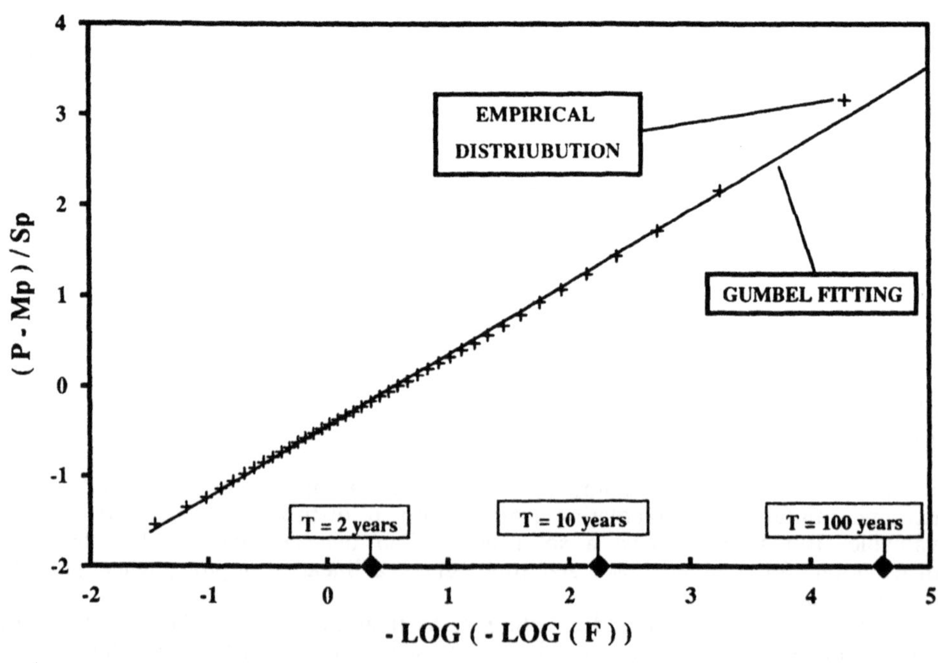

It was thus possible to calculate the "average" distribution for the highest autumn rainfall and plot it on Gumbel paper (Figure 1), associating each point with the empirical frequency distribution found using Gringorten's formula, thus:

$$\mathbf{F_{i,n} = (i - 0.44)/(n + 0.12)} \quad (3)$$

The average distribution has been calculated as the average of all points with the same empirical frequency in the different samples.

It is immediately apparent from this graph why the method based on exponential approximation is so appealing to the authors. A slight difference is noted for the last point, but it is difficult to assess its significance. A few erroneous values could result in a phenomenon of this sort and one can never be sure, despite careful selection criteria, that this is not the case. What is more, the 192 series are not unrelated and a few famous, widespread rainfall events are responsible for several maximum values (October 1960, September 1980, etc.).

The problem of sample dispersion

Unfortunately the empirical distributions which most engineers have to deal with only rarely resemble the almost ideal shape of this average distribution. When the former are plotted on Gumbel paper they frequently diverge from a theoretical straight line, in particular in the upper part of the distribution. This problem is accentuated as one moves closer to the extremity of the empirical distribution reaching a climax at the last point. As it is this part which is used to discern the slope or - worse still - the curve which is supposed to indicate extreme rainfall values, one often has the impression that the exponential decay assumption is not suitable for the current sample.

However, the theory does allow for, at least in part, deviations of this type. It is unusual for a sample to possess the same characteristics as the population from which it was drawn.When one generates samples, using an exponential decay law, it is easy to calculate for the last point in the samples the probable deviations from its average value . Figure 2 shows that the distribution for the last data point is very wide, showing significant positive asymmetry. Note, in passing, that the size of the sample does not influence the result.

Figure 2. Distribution of sample end points.

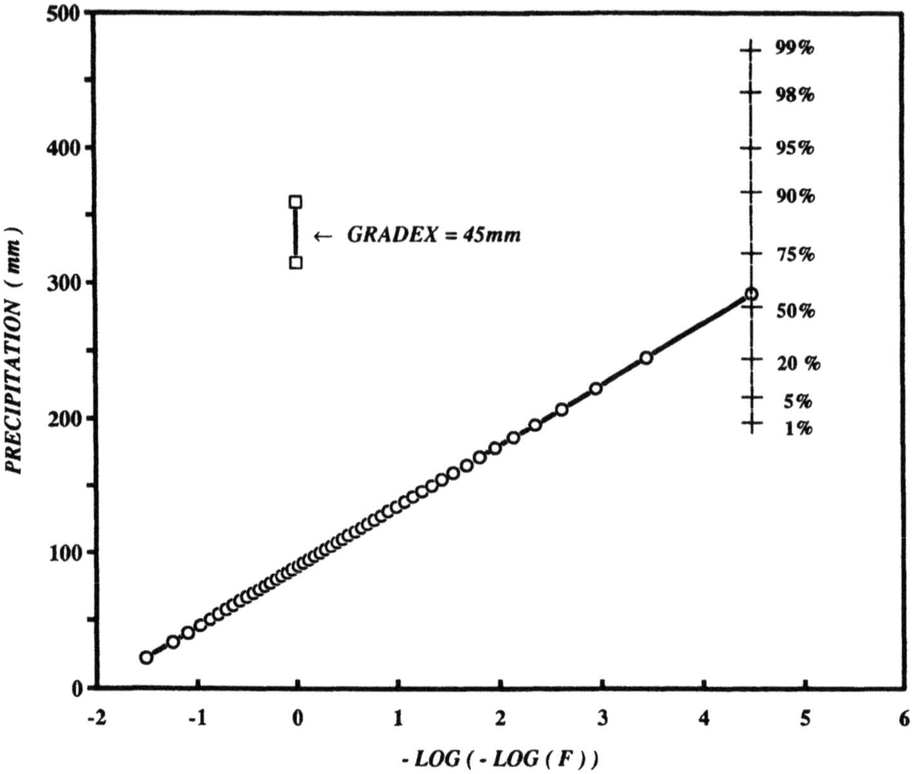

In order to obtain additional theoretical confirmation, a large number of samples were generated using Gumbel's law and sorted according to the difference between the first and sixth data points. In Figure 3 we have plotted the average distribution of the 20% which showed the greatest difference, and in Figure 4, the 20% with the smallest difference. The two classical shapes with the end points respectively above and below the line should therefore not be systematically interpreted as the symptoms, respectively, of "another law for heavy rainfall" or "a physical limit to rainfall events". The same result would be obtained by replacing Gumbel's law with another simple exponential decay law and it does not depend on the size of the sample either.

Figure 3. Generation of 1000 samples according to a Gumbel function - ***Average distribution of 20% of the samples.***

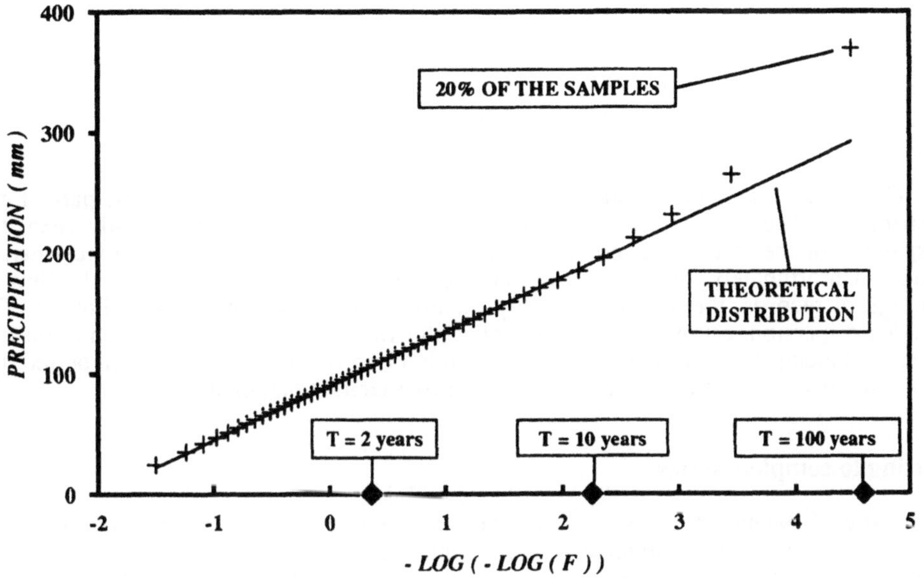

Figure 4. Generation of 1000 samples according to a Gumbel function - ***Average distribution of 20% of the samples.***

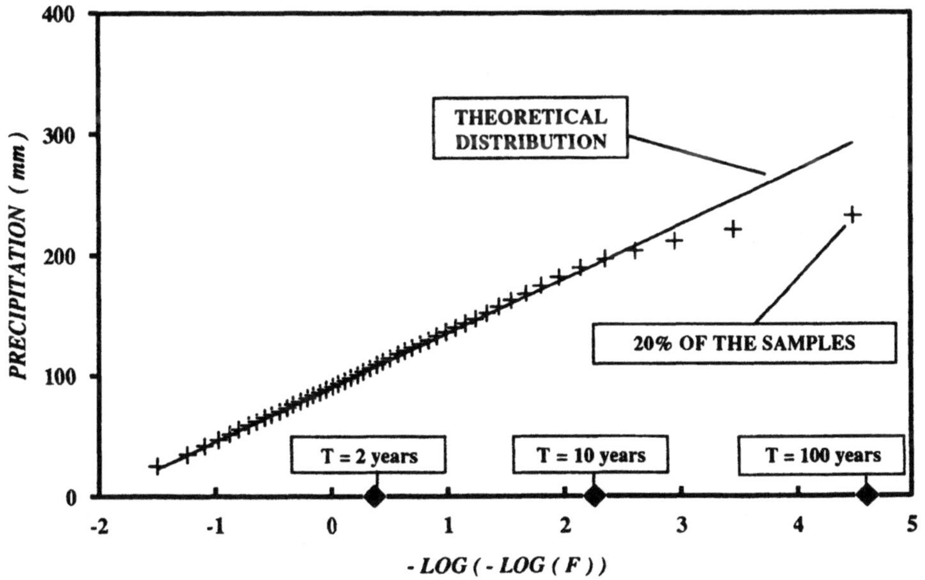

LOOKING FOR STATISTICAL ROBUSTNESS

The experiment discussed above illustrates the extent to which samples may diverge (40% of all cases) from theoretical predictions. It should encourage us to be extremely careful. Even after subjecting data to criticism, one must resist a natural temptation to attribute too much weight to the upper end of the empirical distribution. There is little statistical robustness to be gained by referring exclusively to the most serious event on record, but intuitive extrapolation of the slope over the last few points is equally unsatisfactory. More sophisticated methods, where a curve is calculated - using the exponent of a Weibull function - on a single sample, are not better.

On the contrary, one should make the best possible use of the robust part of the distribution. The longest series are obviously the best. As we have already seen, the dispersion of the last data points is just as great for long records, but the robust part is longer. Unfortunately, one cannot, in practice, prolong samples indefinitely. In view of the age of records, the statistical quality of our estimations is necessarily limited. The Gradex dispersion, estimated from a Gumbel plot fitted to a series of maximum seasonal or annual precipitation values, is extremely high. For a thirty year observation period, the standard error of estimation attains 17% of the true Gradex parameter.

Fitting to complete series

One way of extending the size of a series involves replacing the series of maximum seasonal values by the complete series. In the case of a series of records for 24 hour periods, the number of observations actually used can be multiplied by about 100. However this approach is extremely optimistic as distributions of this sort contain a large number of observations which are completely useless for calculating extreme values. When plotted on Gumbel paper the result curves steeply and the linear part only concerns heavy rainfall, or in other words only between 2 and 5% of the sample. Gumbel's function cannot provide a satisfactory fit for these distributions.

Which raises the question of whether the additional data is of any practical use.

We have attempted to find a way of estimating the Gradex, based on fitting just the last, linear part of the distribution. In practice, the estimation can be obtained by calculating the difference between a selected threshold and the average of all observations which exceed the threshold. Needless to say, choosing the threshold involves a delicate compromise.

The higher the threshold, the fewer the number of observations and the less robust the estimation will be.

If it is too low one runs the risk of using the exponential model in a context where it is not applicable and underestimating the Gradex.

Consequently, this approach tends to ignore part of the data corresponding to intermediate rainfall values. In order to remedy this problem, we developed a group of functions for fitting complete series of 24-hour rainfall Figures. The best results were obtained by modelling the probability of exceeding P using:

$$1 - F(P) = \alpha \exp(-P/a) + \beta \exp(-P/c) \qquad (4)$$
(**c** is lower than **a**, thus setting the limiting decay rate)

This model fits empirical distributions very well for the parts which are not sensitive to sampling dispersion and can be used for quite different climatic regions. It has been used a great deal in France, in areas where an oceanic (daily Gradex ranging from 5 to 20 mm) or Mediterranean climate prevails (daily Gradex ranging from 20 to 60 mm). It has also been used, although less frequently, in equatorial or tropical areas with record dailyGradex values of 200 mm on Reunion Island.

The 4 parameters α, β, **a** and **c** are calculated using the 3 empirical distribution characteristics: zero value frequency distribution $\mathbf{F_0}$, the mean **m** and the standard deviation **s**. The fourth equation we require can be obtained by adding a constraint affecting the value of α which is set to $(\mathbf{m/s})^2$. This condition was suggested by the convergent results obtained for a large number of fits.

This approach makes Gradex estimation more accurate. For a total of 4000 24-hour rainfall records and current values for α, β, **a** and **c**, the standard error of the Gradex estimation varies between 4 and 9%. The least favourable cases (relatively infrequent) are those where the values of **a** and **c** are close. By replacing the traditional Gumbel fit for maximum seasonal values with this method we reduced the Gradex confidence interval by an order of between 2 and 4.

Identifying the 3rd moment can replace the $\alpha = \mathbf{m^2/s^2}$ condition. The latter was only validated for a 24-hour time step and may result in a certain bias when estimating the Gradex for other time steps or in less suitable regions, but doing without this assumption will reduce robustness.

Testing robustness on real cases

Before concluding, it seems appropriate to review the robustness of various ways of estimating rainfall values with a given return period.

Unlike the theoretical approach discussed above, we have conducted tests under field conditions. We took the same 192 series of rainfall records and concentrated on the autumn data. Each sample was then split into two successive sub-periods with a length of between 20 and 22 years. Then each of the two sub-periods was used to estimate 24-hour rainfall with a 40-year return period, $\mathbf{P_{40}}$, adopting 3 different methods :

- using the highest rainfall in the sample as the estimate;
- by fitting a Gumbel function to the empirical distribution for the 20 to 22 maximum seasonal rainfalls;
- by fitting the distribution of all the autumn 24-hour rainfall values using the sum of two exponentials.

The comparison of the estimates obtained by applying the same method to two distinct samples, from a single population, provides an objective idea of the statistical robustness of the method. The results are summarized on the 3 correlation graphs shown in Figures 5, 6 and 7. Each of the 192 series analysed is marked by a point; the abscissa corresponds to the estimation of $\mathbf{P_{40}}$ taken from the first sub-period and the ordinate corresponds to the equivalent value from the second part of the sample.

Figure 5. 40 year precipitation estimates - Correlation between estimates using 2 different periods - *1st Method : Maximum Precipitation.*

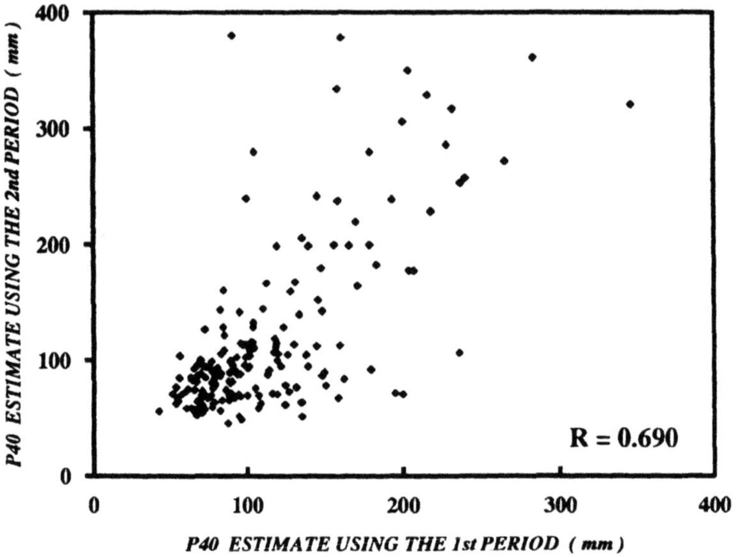

Figure 6. 40 Year Precipitation Estimates - Correlation between estimates using 2 different periods - *2nd method : Gumbel fitting of seasonal maximums.*

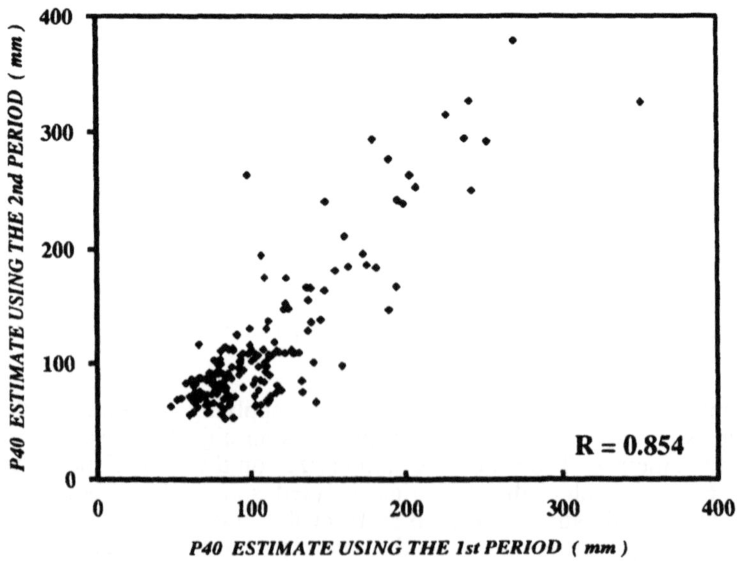

Figure 5 further demonstrates the massive dispersion caused by the maximum value in a rainfall sample. The quality of the estimation obtained using the Gumbel fit is far superior, as one can see in Figure 6. However a comparison of Figures 6 and 7 shows how much progress can be made using all the 24-hour rainfall records. The standard error of estimation, which is over 30% of the true value in the first method, drops to 18% for the second and 12% for the last method.

This degree of inaccuracy for a rainfall value with relatively short return period shows how important it is that these methods be robust when used to assess extreme rainfall events.

Figure 7. 40 Year Precipitation Estimates - Correlation between estimates using 2 different periods - *3rd method : Fitting of the distribution of all rainfalls.*

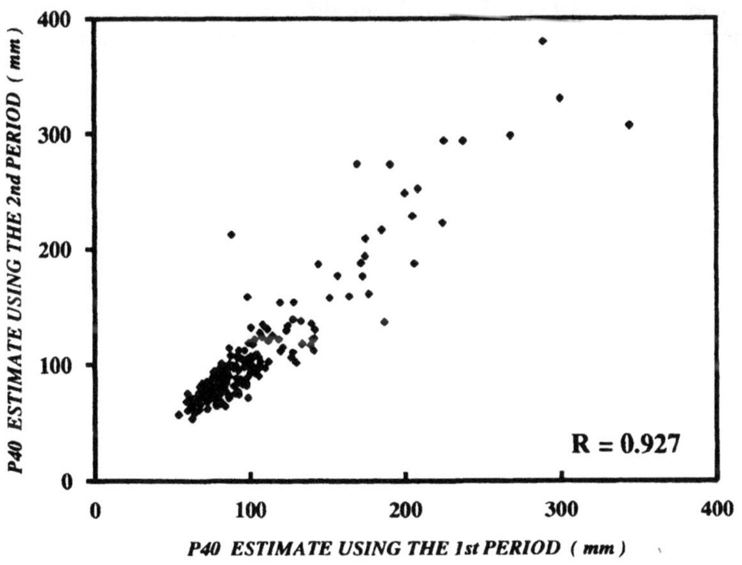

CONCLUSION

The Gradex method is particularly suitable for calculating design flood values for average-sized drainage basins (less than 10 000 km^2). Based on a few, simple assumptions it enables us to deduce the probability of extreme floods from rainfall probabilities.

As it only takes into account local data, it allows for the major topological disparities affecting hydrological risks.

Finally, it combines this quality with an intrinsic robustness, based on effective statistical processing of available data and considerable ease of use.

No single method can be guaranteed to provide an accurate or even unbiased result for a flood with a 10 000-year return period. But it is essential, for the general control of safety and the corresponding investments, that the ratio between the calculated risk and the actual risk vary as little as possible from one basin to the next. For these reasons, the Gradex method and the compromise it represents, is an instrument which deserves attention.

REFERENCES

Bernier, J., and Veron, (1964) "Sur quelques difficultés rencontrées dans l'estimation d'un débit de crue de probabilité donnée", Revue de statistique appliquée.

Duband, D., Michel, C., Garros, H., and Astier, J. (1988) "Evaluation des crues extrêmes et de la crue de projet par la méthode du Gradex" / "Estimating extreme value floods and the design flood by the Gradex method", CIGB, 16th Congress, Vol IV, 1009-1047, Q.63, R.60, San Francisco.

Guillot, P., and Duband, D. (1967) "La méthode du gradex pour le calcul de la probabilité des crues à partir des pluies", IASH, Publication n° 84.

Commission Internationale des Grands Barrages (1992) "Selection of the design flood - Current Methods" / " Choix de la crue de projet - Méthodes actuelles", CIGB, PARIS.

A KERNEL QUANTILE FUNCTION ESTIMATOR FOR FLOOD FREQUENCY ANALYSIS

YOUNG-IL MOON and UPMANU LALL
Utah Water Research Laboratory
Utah State University, Logan, UT 84322-8200
U.S.A.

Nonparametric Kernel Estimation Methods have recently been introduced as viable and flexible alternatives to parametric methods for flood frequency estimation. A kernel estimator (KQ) of the quantile function is presented here. Boundary kernels are used for extrapolation of tail probabilities. The bandwidth of the estimator is chosen using an automatic, "plug-in" method. Comparisons of the estimator with selected tail probability estimators are offered. The KQ estimator presented here is shown to be competitive with other estimators.

INTRODUCTION

The objective of flood frequency analysis is to obtain an estimator of flood quantile magnitude (Q_T) for one or more locations on a river system. Correspondingly, a flood magnitude may be specified and an estimate of its return period (T) desired. In this paper, we focus on determining the flood quantile relationship using data from a gaged site. The flood quantile analysis procedures are classified by number and type of data used. Traditionally an annual maximum frequency model is proposed and calibrated from the N-year record of annual maximum flood peaks at a site. The quantile Q_T is then estimated as

$$\hat{Q}_T = \hat{\theta}_1 + \hat{\theta}_2\, y_T(\hat{\theta}_3) \tag{1}$$

where $\hat{\theta}_1$, $\hat{\theta}_2$, and $\hat{\theta}_3$ are estimates of location, scale, and shape parameters of a selected distributional form $f(q;\theta)$, and $y_T(\theta_3)$ is a standardized variate value of return period T from the f(x) distribution.

Past and current research into methods of flood quantile estimation at a gaged site has concentrated mostly on the statistical aspects of the problem, based on the assumption that the sample of flood observations comes from a population form with a known probability density function (pdf). However, no unique pdf or procedure will work best in all circumstances. Classical parametric estimation procedures are also most heavily weighted towards fitting the main body of the assumed probability density, and accord a

K. W. Hipel (ed.),
Stochastic and Statistical Methods in Hydrology and Environmental Engineering, Vol. 1, 299–311.

negligible weight to the estimation of the tail of the distribution. Even if the parametric pdf fits well, considerable uncertainties for the magnitude of the floods at some frequencies exist. Furthermore, if there are mixed populations of floods, then the flood frequency analysis becomes harder. Annual maximum flows in a number of locations may come from different causes (e.g., snowmelt, rainfall runoff, cyclonic activity). This leads to statistically heterogeneous populations or mixture distributions. The identification of finite mixtures of arbitrary or unknown populations from short records (typically n= 30-70) is not an attractive proposition and is not usually pursued. Webb and Betancourt (1992) tried to develop storm type classifications, separate events on that basis, fit a parametric pdf to each storm, and combine estimates. While a good demonstration of mixed population of floods, it is not usually easily implemented by a field engineer. Often, discriminating between different parametric probability models for the sample sizes typically available using standard tests such as the Chi-square and the Kolmogrov-Smirnov (e.g., Kite 1977) is difficult. Such tests are rather insensitive to the behavior of the distribution in the tails. This is a mismatch in objectives that is particularly onerous since such parametric fits are not robust to outliers in either tail of the sample distribution. The tail behavior of a mixture is often dictated by the tail behavior of the mixture component that has the heaviest tail. We must look for methods that are robust to such situations (mixtures) and can give reasonable answers for a limited extrapolation of the data (e.g., 100-year flood). One way of circumventing the problem is to consider techniques that are based on tail estimators.

Two types of such methods are considered in this paper. The first method focuses on the tail of the distribution as defined by the k upper order statistics of the sample, and seeks to develop estimators of right tail quantiles under the assumption of tail behavior (e.g., an exponential tail). Thus a parametric form is assumed for the tail, but not for the entire probability model. Such models typically encompass a number of probability distributions with the same limiting behavior in the tails. We are not aware of past hydrological investigations into such methods for flood frequency analysis. The second estimation method considered is the use of kernel quantile estimators based on the full sample data set. These estimators are fully nonparametric in that no assumption as to the underlying probability density is made and these estimators are based on a kernel smoothing of the empirical quantile function of the data. Kernel estimators in the flood frequency context have been considered by Adamowski (1985, 1989), Adamowski and Feluch (1990), Adamowski and Labatiuk (1987), Schuster and Yakowitz (1985), Bardsley (1988, 1989), Guo (1991), Lall and Niu (1989), Lall et al. (1993), and Moon et al. (1993) to seek alternatives to parametric methods for flood frequency estimations. Briefly, this is a desire to deal with multimodal flood distributions, model robustness, and to sidestep the dilemma of model choice from a finite set of ill-fitting alternatives. Lall et al. (1993) showed that flood frequency estimators based on kernel density estimation may not perform well as distribution function estimators. Moon et al. (1993) showed that variable kernel distribution function estimators (VK-C-AC) can be superior to the tail probability estimators proposed in the literature. A quantile estimator can be defined through the inverse of the distribution function or through kernel averaging of the sample quantile function. Related recent work on kernel quantile function estimation within the range of the data is presented by Sheather and Marron (1990), Zelterman (1990) and Takeuchi and Yokoyama (1991). The kernel quantile function approach in this paper was suggested by Hans-Georg Müller. Parzen (1979) provided the rudiments of such an estimator and argued for quantile function estimation rather than density function estimation. We compared this estimator with selected tail estimators. We present the kernel quantile estimator, discuss its application with a data set purported to have mixtures of mechanisms, and present results from some Monte Carlo analysis.

The tail estimators proposed by Hill (1975), Hosking and Wallis (1987), and Breiman and Stone (1985), an extreme value (EV1), and a kernel quantile method (KQ) were tested with data from Gaussian, Pearson III, and an asymmetric mixture of Gaussian distributions. This paper investigated the properties of kernel quantile estimator relative to tail estimators.

Since only the upper part of the data is used for estimating upper tail probabilities or upper tail quantiles, tail estimators do not care whether or not the lower data values follow the distribution. A disadvantage is that tail estimators still need to specify parametric family behavior and the place at which tail starts. Pickands (1975), Hill (1975), Hall (1982), and Hall and Welsh (1985) examined the problem of estimating the number of extreme values or the cutoff point for the tail to achieve optimal performance, and showed that, in general, this number depends on unknown properties of the tail. Therefore, the size of the extreme subsample used to construct the estimators must also be estimated from the sample. These methods are based on asymptotics and one must consider whether or not asymptotics for the small sample sizes available in practice can be invoked. In our simulations, we used samples of size 20 and 100 since these sample sizes were chosen on the basis that it is not reasonable to consider less than 20 values for flood frequency analysis, and samples of size greater than 100 are rare. We found a simple strategy of specifying 5 and 10 upper order statistics, respectively, outperformed the sophisticated asymptotic strategies presented by these authors. We suspect that this is due to the high variance associated with these methods for such small samples. The proposed KQ estimator depends smoothly on the whole data set with local adaptation to the quantile function. The choice of the bandwidth determines the number of extreme values that enter into a particular estimate. An objective method for choosing the bandwidth is presented.

Many of the proposed estimators of tail probabilities (Hill 1975, and Breiman et al. 1981) assume that a distribution function F(x) is in the domain of attraction of a known distribution function G(x) for all values greater than some predetermined value x_{p_0}. Hill (1975) and Hosking and Wallis (1987) developed tail probability estimators (PT1 and PT2) by forming an estimate of an extreme right tail quantile under the assumption that the behavior in the upper tail follows the Pareto distribution. The second method proposed by Breiman and Stone (1985) assumes that the tail of the distribution is exponential or approximately linear in log(p). To obtain a more accurate approximation, they also considered the quadratic tail method in which x_p is assumed to be a quadratic function of log(p).

The reader is referred to Moon et al. (1993) for algorithms of selected tail probability estimators. Hill's method (PT1) is presented for historical reasons, for comparison with a recent Pareto model due to Hosking and Wallis (1987) (PT2). The Exponential and Quadratic tail methods (ET, QT) due to Breiman and Stone are also presented. The Type I Extreme Value distribution (EV1) is also discussed in this paper because EV1 is often used for maximum flood events in practice, and can be considered a tail estimation method.

THE KERNEL QUANTILE ESTIMATOR

The Kernel Quantile estimator (KQ) is based on a kernel smoothing of the empirical quantile function of the data. The empirical quantile function is prescribed through a standard "plotting position formula". Let y_i, i=1...n, be the observed sequence of n annual maximum flows, arranged in ascending order. Let p_i, i=1...n, be the corresponding plotting positions estimated using a standard formula (e.g., the Weibull, Beard or Adamowski formula). Here we use Adamowski's (1981) formula :

$$p_i = \frac{i - 0.25}{n + 0.5} \tag{2}$$

The empirical quantile function $x(p_i)$ is defined by the sample values y_i corresponding to each p_i. The quantile function x(p) to be estimated is defined as the event magnitude corresponding to the pth quantile.

For flood frequency analysis, we are interested in the upper quantiles, i.e., p between 0.5 and 1, and in particular, for p>0.9. Typical sample sizes for flood frequency analysis range from 20 to 100. An extrapolation of the data to $p>p_n$ is usually needed.

The KQ estimator is based on the Gasser-Müller (1984) kernel regression estimator. It considers a convolution of the empirical quantile function with a kernel or weight function.

$$x(p) = \sum_{i=1}^{n} \frac{1}{h_p} \int_{s_{i-1}}^{s_i} y_i \, K\left(\frac{p-u}{h_p}\right) du \tag{3}$$

where s_i is an interpolating sequence of the p_i, given as $s_i=(p_i+p_{i+1})/2$, i=1...n-1, $s_0=0$, $s_n=1$; h_p is a bandwidth associated with the point p; and K(.) is a kernel or weight function, and $p \in [0,1]$. The kernel function K(.) is usually taken to satisfy the requirements, $\int K(t)dt = 1$, $K(t) = - K(t)$, and $\int K^2(t)dt < a$, i.e., it is a symmetric probability density with finite variance.

It is preferred that K have compact support to minimize the effect of the asymmetric distribution of data near the boundaries (p=0 and p=1) since the curve to be estimated has a discontinuity at an endpoint. A specialized boundary kernel that corresponds to the kernel used in the interior is needed within a bandwidth of the boundary to take care of the bias in the weighted convolution in the boundary region. Müller (1991) develops boundary kernels corresponding to specific interior kernels. Here we use the Epanechnikov kernel in the interior and the corresponding Müller boundary kernel. These kernels for $-1<t<1$ are:

Epanechnikov kernel: $K(t) = 0.75(1 - t^2)$ (4)

Boundary kernel:

$$K(t) = 6(1-t)(q+t)\frac{1}{(1+q)^3}\left\{1 + 5\left(\frac{1-q}{1+q}\right)^2 - 10\frac{1-q}{(1+q)^2}t\right\} \qquad (5)$$

where $q=(1-t)/h_p$ for the right boundary if $t \in [1-h_p,1]$; $K(-t)$ and $q=t/h_p$ for the left boundary if $t \in [0,h_p]$; $0<q<1$ and $t=(p-p_i)/h_p$.

The boundary effect is illustrated through the following example. Let x(1)...x(n) be the data and the corresponding p values are 1/(n+1)....n/(n+1). To estimate a quantile at p = 0.8 with the Epanechnikov kernel function K(t) with the smoothing parameter h_p=0.1, then the x(i) values for 0.7<p<0.9 will contribute to the estimate. The kernel estimate for $x_{0.8}$ is not in the boundary area in Figure 1. The curve in Figure 1 is the Epanechnikov kernel function. If the data from x(k) to x(m) are in the range 0.7<p<0.9, then the estimate of $x_{0.8}$ is :

$$x_{0.8} = w_1\, x(k) + w_2\, x(k+1) + \ldots\ldots + w_{m-1}\, x(m-1) + w_m\, x(m) \qquad (6)$$

where the $w_1, \ldots\ldots, w_m$ are weights obtained from solving the integration in equation (3).

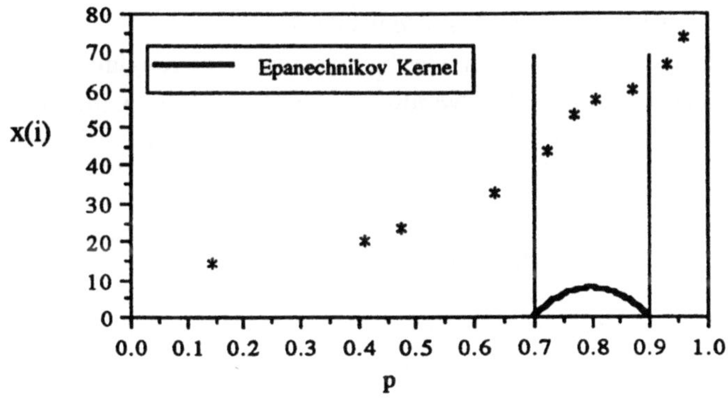

Figure 1. Kernel estimate for $x_{0.8}$.

However, if we are interested in the $x_{0.99}$ then we have a boundary problem. From Figure 2, the x(i) value for 0.89<p<1 produces a large bias. In this case, we used Müller's boundary kernel, that is, the equation (5). These interior and boundary kernels are shown for q=0.1 and 0.9 in Figure 2.

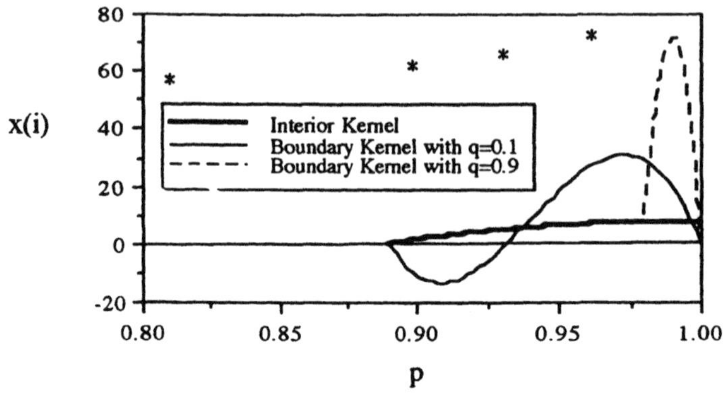

Figure 2. Kernel estimate for $x_{0.99}$.

The quantile $x_{0.99}$ by KQ for Santa Cruz River data with 71 data points (Webb and Betancourt, 1992) without considering the boundary effect is: $x_{0.99}$=0.0001x(66) + 0.037x(67) + 0.092x(68) + 0.130x(69) + 0.154x(70) + 0.202x(71) = 524 m3/s. This is a biased estimate since the kernel center at p=0.99 extends past 1.0 where there are no observations. However, if we use the boundary kernel (equation (5)) then $x_{0.99}$ = -0.001x(66) - 0.177x(67) - 0.105x(68) + 0.245x(69) + 0.545x(70) + 0.494x(71) = 1094 m3/s. The bandwidth (h_p) of KQ for this case is 0.064.

The bandwidth or smoothing parameter h_p determines the roughness or smoothness of the estimated function. Smaller bandwidths result in fewer data points contributing to the estimate at any point, and hence a rougher on more bumpy estimator. Larger bandwidths however allow averaging over a larger data space resulting in a smoother estimator. As bandwidth increases, bias increases and variance decreases. Some methods to find an optimal bandwidth that balances bias and variance are reported by included the Generalised Cross Validation (GCV) method proposed by Craven and Wahba (1979) and the Plug-In method by Gasser et al. (1991) as well as local least absolute deviation and least squares cross validation aimed at minimizing the mean square error of x(p). We found the Gasser et al.'s Plug-In method with an Epanechnikov kernel worked better than the others in our tests.

APPLICATIONS

A comparison between a kernel quantile estimator (KQ) and tail estimators (PT1, PT2, ET, QT, EV1) for the Santa Cruz's River data from Webb and Betancourt (1992) is presented in Figure 3. Classical parametric estimates of the 100-year flood for the Santa Cruz River at Tucson, Arizona range from 572 to 2,780 m^3/s. Webb and Betancourt

(1992) separated floods above base discharge (48 m^3/s) by storm types into three categories: monsoonal storms (56 data points), frontal systems (18 data points), and dissipating tropical cyclones (19 data points). The censored data for each type of flood were fit to three-parameter log-Pearson type III distribution using maximum-likelihood analysis and then they determined the 100-year flood by combining populations of floods caused by Webb and Betancourt's three categories. The 100-year flood estimated from a mixed-population analysis was 1,050 m^3/s. In Figure 3, the KQ is close to the empirical c.d.f., and none of the tail estimators except QT appear reasonable. The bandwidth (h_p) of KQ for this case is 0.064. KQ and QT perform similarly. QT is close to the empirical distribution function in $0.92<p<0.98$ and KQ is close in $0<p<0.92$ and $p>0.98$. The KQ estimate of the 100-year flood is 1094 m^3/s and that for QT is 1102 m^3/s. We worked with a single population of annual maximum data of 1914 to 1986 (71 data points) and did not break it down into different storm events. The KQ results in this case is comparable with that from an analysis of the mixture.

A comparison between kernel quantile estimator (KQ) and variable kernel distribution function estimator (VK-C-AC) (See Lall et al. (1993) or Moon et al. (1993)) is shown in Figure 3. KQ and VK-C-AC have similar performance up to P = 0.9. The quantile value of VK-C-AC is lower than that of KQ from P=0.9 to P=0.985 and higher after P=0.985. The 100-year flood of VK-C-AC for Santa Cruz data is about 1,250 m3/s. This value is larger than that of KQ (1094 m3/s). VK-C-AC appears to be sensitive to outliers and interpolates the empirical quantile function within the range of the data. KQ smooths the quantile function and appears more resistant to outliers.

A Monte Carlo approach was used to compare the performance of selected methods. The experiments tested the estimators described in the previous section and compared their performance with parametric estimators. We considered three parent populations - Normal (*N*) (mean 0, variance 1), 3 parameter Gamma (*G*) (mean 0, variance 1, skew 1) and a Mixture (*M*) of Normals (0.5N(0,1) + 0.5 N(3,1)). One hundred replicates (ns) of samples of sizes 20 and 100 were generated from each population. This is not a large sample size for Monte Carlo studies. We were restricted by available computational resources. Consequently our results may be viewed as indicative rather than conclusive. Estimation in the tail of each distribution sampled was considered at values of x_p corresponding to these values of p: 0.9, 0.95, 0.98, 0.99, 0.995 (i.e., recurrence intervals (T) are 10, 20 ,50, 100, and 200 years). We recorded the following statistics relevant to the estimation of x_p in each case:

$$\text{bias} = \sum_{m=1}^{ns} \frac{(\hat{x}_p - x_p)_m}{ns} \tag{7}$$

$$\text{rmse} = \left(\sum_{m=1}^{ns} \frac{(\hat{x}_p - x_p)_m^2}{ns} \right)^{1/2} \tag{8}$$

The average bias and RMSE for PT1, PT2, ET, QT, EV1, and KQ was computed. Results for the mixture data, size 100, are shown in this paper. Details for other cases are available from the authors. The kernel quantile estimator (KQ) gives relatively consistent and competitive results in terms of bias and RMSE for each of the estimation situations considered. The performance of the tail estimation depends on the underlying model. The bias for QT and KQ is small and stable for a range of p values while that of PT1, PT2, ET, and EV1 increases with p in Figure 4. In Figure 5, KQ and QT have very similar behavior but KQ is better in terms of bias at all points and in terms of RMSE up to T=50 years. PT1 has good results for mixture data but poor results for normal and gamma data since PT1 has a thick tail as shown in Figure 3 and the mixture data has the heaviest tail. The results for sample size 20 are similar to those of sample size 100.

CONCLUSIONS

KQ and QT performed similarly. They appeared relatively robust with respect to the other methods for different situations tested. This is not surprising, since both these methods consider a local quadratic extrapolation of the empirical quantile function, and are described through linear weighted combinations of order statistics. The weight sequence and the number of order statistics used by the two methods differ. The analysis of the Santa Cruz River data revealed that these methods can give reasonable results with data from mixed populations. However, they also illustrated the futility inherent in flood frequency estimation - it is easy to design innumerable schemes that are equally plausible within the range of the data and quite different under extrapolation.

Robustness in performance across different situations is a desirable attribute. We feel kernel quantile estimators can be developed that are superior to KQ. Such developments may require some theoretical analysis of tail properties to determine the bandwidth and the appropriate kernel functions. Sensitivity of KQ to the choice of a plotting formula also needs to be formally examined.

ACKNOWLEDGEMENT

The work reported here was supported in part by the U.S. Geological Survey through Grant No. 14-08-0001-G1738 and in part through their funding of the second author's 1992-3 sabbatical leave, when he worked with the Branch of Systems Analysis, WRD, USGS, National Center, Reston VA.

REFERENCES

Adamowski, K. (1981) “Plotting formula for flood frequency”, Water Resources Bulletin 17(2), 197-202.

Adamowski, K. (1985) “Nonparametric kernel estimation of flood frequencies”, Water Resources Research 21(11), 1585-1590.

Adamowski, K. and Labatiuk, C. (1987) “Estimation of flood frequencies by a nonparametric density procedure”, Hydrologic Frequency Modeling, 97-106.

Adamowski, K. (1989) “A Monte Carlo comparison of parametric and nonparametric estimation of flood frequencies”, Journal of Hydrology 108, 295-308.

Adamowski, K. and Feluch, W. (1990) “Nonparametric flood-frequency analysis with

historical information", Journal of Hydraulic Engineering 116(8), 1035-1047.
Bardsley, W. E. (1988) "Toward a general procedure for analysis of extreme random events in the earth sciences", Mathematical Geology 20(5), 513-528.
Bardsley, W. E. (1989) "Using historical data in nonparametric flood estimation", Journal of Hydrology 108, 249-255.
Breiman, L., Stone, C. J., and Ginns, J. (1981) "Further developments of new methods for estimating tail probabilities and extreme value distributions", Technical Report #TSD-PD-A243-1, Tech. Service Corp., Santa Monica, California 90405.
Breiman, L., and Stone, C. J. (1985) "Broad spectrum estimates and confidence intervals for tail quantiles", Technical Report 46, Univ. of California, Berkeley, California.
Craven, P. and Wahba, G. (1979) "Smoothing noise data with spline functions", Numerische Mathematik 31, 377-403.
Gasser, T. and Müller, H. G. (1984) "Estimating regression functions and their derivatives by the kernel method", Scandinavian Journal of Statistics 11, 171-185.
Gasser, T., Kneip, A., and Kohler,W. (1991) "A flexible and fast method for automatic smoothing", Journal of the American Statistical Association 86(514), 643-652.
Guo, S. L. (1991) "Nonparametric variable kernel estimation with historical floods and paleoflood information", Water Resources Research 27(1), 91-98.
Hall, P. (1982) "On some simple estimates of an exponent of regular variation", J. R. Statist. Soc. B 44(1), 37-42.
Hall, P. and Welsh, A. H. (1985) "Adaptive estimates of parameters of regular variation", Annals of Statistics 13, 331-341.
Hill, B. M. (1975) "A simple general approach to inference about the tail of a distribution", Ann. Math. Statist. 3, 1163-1174.
Hosking, J. R. M. and Wallis, J. R. (1987) "Parameter and quantile estimation for the generalized pareto distribution", Technometrics 29(3), 339-349.
Kite, G. W. (1977) Frequency and Risk Analyses in Hydrology, Water Resources Publications, Fort Collins, Colorado.
Lall, U. and Niu, J. (1989) "Variable bandwidth kernel density estimation", AGU Spring Meeting, Baltimore, Maryland, May 7-12, 1989.
Lall, U., Moon, Y.-I., and Bosworth, K. (1993) "Kernel flood frequency estimators: Bandwidth selection and kernel choice", Water Resources Research 29 (4), 1003-1015.
Moon, Y.-I., Lall, U., and Bosworth, K. (1993) "A comparison of tail probability estimators", Journal of Hydrology, in press.
Müller, H.-G. (1991) "Smooth optimum kernel estimators near endpoints", Biometrika 78(3), 521-530.
Parzen, E. (1979) "Density quantile estimation approach to statistical data modelling" Smoothing Techniques for Curve Estimation 757, 155-180, Springer-Verlag, New York.
Pickands, J. (1975) "Statistical inference using extreme order statistics", Ann. Stat. 3(1), 119-131.
Schuster, E. and Yakowitz, S. (1985) "Parametric/nonparametric mixture density estimation with application to flood-frequency analysis", Water Resources Bulletin 21(5), 797-803.
Sheather, S. J. and Marron, J. S. (1990) "Kernel quantile estimators", Journal of the American Statistical Association 85(410), 410-416.
Takeuchi, H. and Yokoyama, S. (1991) "A note on quantile estimation by the kernel method", Communications in Statistics - Simulation 20(1), 149-156.

Webb, R. H.and Betancourt, J. L. (1992) "Climatic variability and flood frequency of the Santa Cruz River, Pima county, Arizona", USGS Water-Supply Paper 2379, 1-40.
Zelterman, D. (1990) "Smooth nonparametric estimation of the quantile function" Journal of Statistical Planning and Inference 26, 339-352.

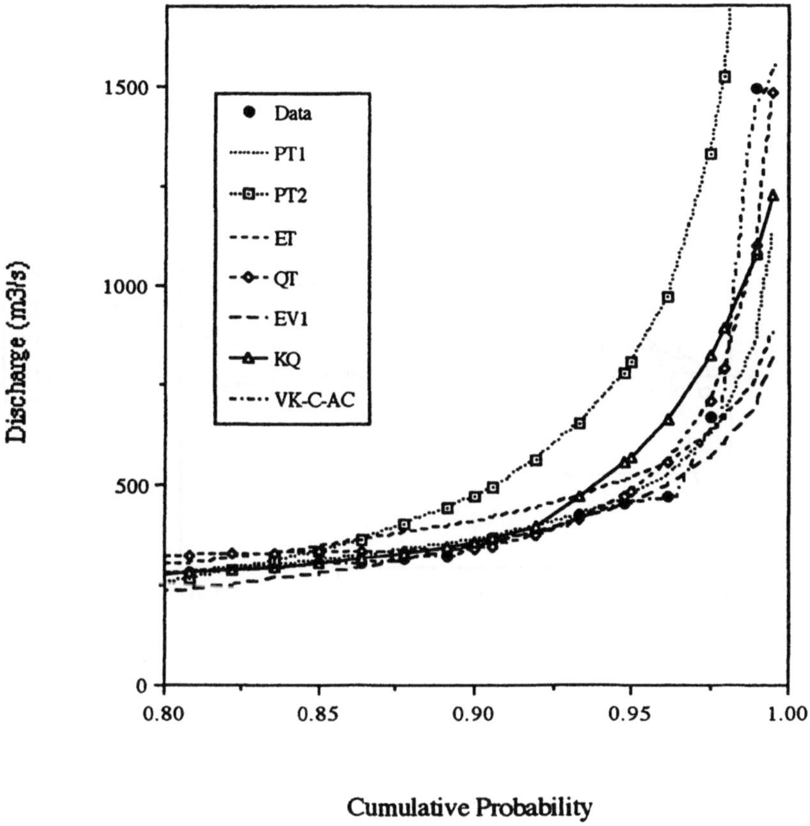

Figure 3.
Comparison of frequency curves between tail estimators, KQ, and VK-C-AC for Santa Cruz River at Tucson, Arizona (1914-1986).

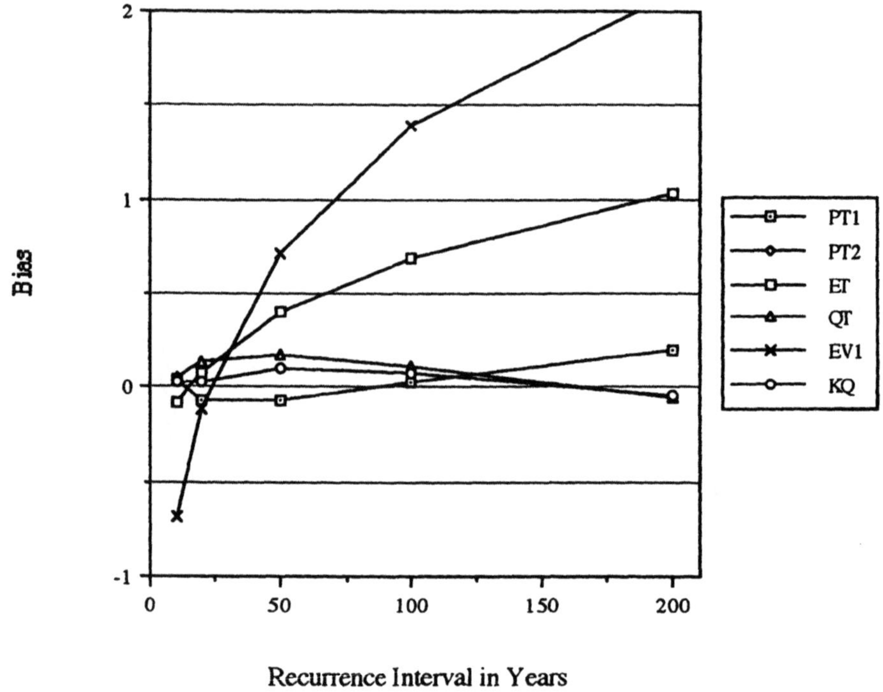

Figure 4.
Bias for Mixture data {0.5N(0,1) + 0.5N(3,1)}, n = 100.

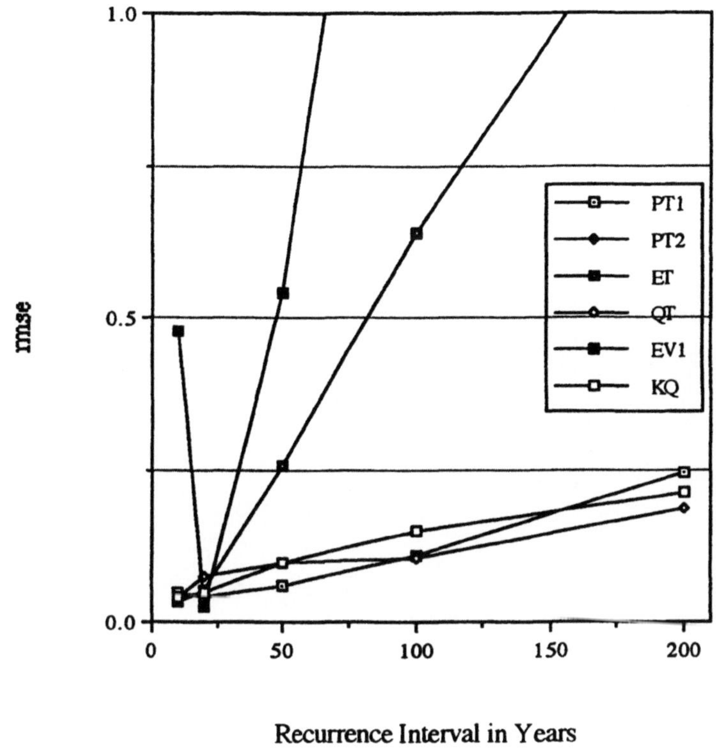

Figure 5.
RMSE for Mixture data {0.5N(0,1) + 0.5N(3,1)}, n = 100.

ESTIMATION OF OCCURRENCES OF HIGH AND STEEP WAVES OFF THE CALIFORNIA COAST

DAVID W. WANG[1] and DAVID B. GILHOUSEN[2]
[1]Computer Sciences Corporation, Bldg 3203,
Stennis Space Center, Mississippi 39529-6000
U.S.A.
[2]National Data Buoy Center, Bldg 1100,
Stennis Space Center, Mississippi 39529-6000
U.S.A.

This study presents a theoretically derived joint probability distribution function of wave height and wave steepness. Applying this function to long-term wave measurements from a buoy moored off California, the probability of wave height and steepness exceeding given thresholds was estimated. This study demonstrates a practical approach to derive vital information from existing long-term wave data for the operational safety of small vessels and offshore structures.

INTRODUCTION

Past experiences among civil engineers and naval architects show that capsizing of smaller vessels and damage to marine structures are due to forces resulting from waves with high height and high steepness (Myrhaug and Kjeldsen, 1987). An encounter with a very high, steep wave can suddenly capsize small vessels. This is the most probable reason for the loss of 26 Norwegian vessels and 72 lives in the period 1970-1979 (Myrhaug and Kjeldsen, 1984, 1986, 1987). Hamilton (1980) attributed the capsizing of four instrumented discus buoys to high and steep waves. Because the wave profile changes with time in a random fashion, the individual wave magnitudes and the times of their occurrence are usually unpredictable. Information on the occurrence of such waves can only be estimated through the probabilistic approach, which requires a probability function that describes the random phenomenon (Ochi, 1980).

Only a few studies have been carried out on the joint probability distribution function of wave height and wave steepness. Myrhaug and Kjeldsen (1984, 1987) developed a joint probability distribution function of wave height and steepness based on field data from the Norwegian continental shelf. Their development focused on the joint distribution of wave crest front steepness and wave height. The parameters in the joint probability distribution function were empirically determined from the field data. Dahle et al. (1988) applied the results of Myrhaug and Kjeldsen (1987) to estimate the probability of capsizing in steep and high waves.

A joint probability distribution function of wave height and steepness using the Rayleigh distribution function for wave height (Longuet-Higgins, 1952), and the steepness probability distribution function developed by Overvik and Houmb (1977) is presented in this paper. This joint probability distribution function has two parameters, significant wave height and average wave period, that are relevant to the sea state. Using this joint probability distribution function of wave

K. W. Hipel (ed.),
Stochastic and Statistical Methods in Hydrology and Environmental Engineering, Vol. 1, 313–325.

height and steepness along with the joint probability distribution of significant wave height and average wave period, the probability of waves exceeding a certain height and steepness was estimated. The joint probability distribution of significant wave height and average wave period was derived from 5 years of wave measurements at a National Data Buoy Center (NDBC) buoy station off the California coast.

JOINT DISTRIBUTION OF WAVE HEIGHT AND WAVE STEEPNESS

Figure 1 is a sketch of the ocean surface profile. The individual wave height, h, is defined as the distance from crest to trough between two zero-upcrossings of mean sea level. The associated wave period, T, is defined as the time interval between these two upcrossings of the mean sea level. The associated steepness, s, is defined as the ratio of wave height, h, to wave length, L. Using the dispersion relationship of linear wave theory, the wave length is computed by

$$L = \frac{gT^2}{2\pi} \tag{1}$$

Hence, the steepness, s, is defined as

$$s = \frac{h}{L} = \frac{2\pi h}{gT^2} \tag{2}$$

Based on the low correlation coefficient between wave height and wave steepness given in the study by Myrhaug and Kjeldsen (1984) as 0.05, we can assume that wave height and steepness are statistically independent. The joint probability distribution function of wave height and steepness, $p(h, s)$, can be approximated as

$$p(h, s) = p(h) \cdot p(s) \tag{3}$$

where $p(h)$ is the probability density function of wave height and $p(s)$ is the probability density function of wave steepness.

Overvik and Houmb (1977) derived the probability density function of wave steepness based on the theoretical work by Rice (1944). Assuming a two-dimensional Gaussian surface consisting of an infinite number of sine waves with random amplitude, frequency, and phase, Overvik and Houmb (1977) derived the probability density function of steepness as

$$p(s)=(\pi g)^2\left(\frac{s}{m_4}\right)\exp\left(-\frac{(\pi g)^2 s^2}{2m_4}\right) \tag{4}$$

where m_4 is the fourth moment of the wave spectrum, $S(\omega)$, which can be obtained from

$$m_n=\int_0^\infty \omega^n S(\omega)d\omega; \quad n=0,1,2,3,4,... \tag{5}$$

where ω is radian frequency. For general engineering applications, the significant wave height, H_s, and the average zero-crossing wave period, T_z, can be expressed as

$$H_s=4\sqrt{m_0}, \qquad T_z=2\pi\sqrt{\frac{m_0}{m_2}} \tag{6}$$

Using the above relations, the probability density function of steepness can be expressed as

$$p(s)=4(1-\varepsilon^2)\frac{s}{S_m^2}\exp[-2(1-\varepsilon^2)(\frac{s}{S_m})^2] \tag{7}$$

where S_m is the significant steepness (Tucker, 1991, p. 97), which is expressed as

$$S_m = \frac{2\pi H_s}{gT_z^2} \tag{8}$$

and ε is the spectral width parameter expressed as

$$\varepsilon^2=1-\frac{m_2^2}{m_0 m_4} \tag{9}$$

Because the ocean wave can be generally considered as a narrow-banded ($\varepsilon = 0$) Gaussian random process (Ochi, 1980), the $p(s)$ then can be further expressed as

$$p(s)=4\frac{s}{S_m^2}\exp[-2(\frac{s}{S_m})^2] \tag{10}$$

Overvik and Houmb (1977) showed that the probability density function of wave steepness shown in Equation (10) fitted well to field data.

Considering the ocean wave as a narrow-band Gaussian random process, the probability distribution of wave height can be fitted by the Rayleigh probability function (Longuet-Higgins, 1952), which is

$$p(h)=2(\frac{h}{\sigma^2})\exp(-\frac{h^2}{\sigma^2}) \tag{11}$$

where σ is the root-mean-square of wave height, which can be approximated by

$$\sigma=\frac{H_s}{\sqrt{2}} \tag{12}$$

Hence, the wave height probability function can be written as

$$p(h)=4(\frac{h}{H_s^2})\exp(-2\frac{h^2}{H_s^2}) \tag{13}$$

Using the relations shown in Equations (3), (10) and (13), the joint probability density function of wave height and steepness then can be expressed as

$$p(h,\ s)=16\left(\frac{h}{H_s^2}\right)\left(\frac{s}{S_m^2}\right)\exp\left(-2\frac{h^2}{H_s^2}\right)\exp\left(-2\frac{s^2}{S_m^2}\right) \tag{14}$$

Figure 2 shows the contour plot of $p(h,\ s)$ as a function of wave height normalized by H_s and steepness normalized by S_m.

The probability of the wave height exceeding an arbitrary height, h_c, and an arbitrary steepness, s_c, can be obtained from

$$P_1=\int_{h_c}^{\infty}\int_{s_c}^{\infty}p(h,\ s)dsdh\ =\ \exp\left(-2\frac{h_c^2}{H_s^2}\right)\exp\left(-2\frac{s_c^2}{S_m^2}\right) \tag{15}$$

Notice that S_m is related to H_s and T_z, as shown in Equation (8). The above equation indicates that P_1 is a conditional probability distribution function for a given sea state represented by H_s and T_z, which can be expressed as

$$P_1=P[(h\geq h_c)\cup(s\geq s_c)|H_s,\ T_z] \tag{16}$$

Figure 3 shows the contour plot of P_1 as a function of H_s and T_z for $h_c = H_s$, and $s_c = 0.1$. Figure 4 shows the contour plot of P_1 for $h_c = 5$ m and $s_c = 0.1$, which were the design criteria for smaller vessels used by Myrhaug and Kjeldsen (1987). Figures 5a and 5b show the contour plots of P_1 for $h_c = 4$ m and $s_c = 0.1$ and $h_c = 4$ m and $s_c = 0.12$, respectively. As shown in the figures, by changing h_c and s_c, the area circled by the contour lines changes but the general pattern of contour lines remains unchanged. It is noted that 0.12 is the limit of wave steepness for breaking waves (Ochi and Tsai, 1983).

APPLICATIONS

The probability distribution function, P_1, is related to the sea state, expressed as H_s and T_z. However, for a specific location, the estimation of the probability of occurrence of high and steep waves cannot be made without having the probability of occurrence of such a sea state for that location. Hence, the probability distribution of occurrence of waves exceeding a given height and steepness in a given coastal area is expressed as

$$P_{h_c,\ s_c}=P_1\cdot P_2 \tag{17}$$

where P_1 is the conditional probability distribution shown in Equation (16), and P_2 is the joint probability distribution of H_s and T_z. P_2 can be obtained from long-term wave measurements for a specified area.

In the present study, 55 months (December 1987 through June 1992) of wave climate data collected from an NDBC buoy station were used. The wave data were collected by a 3-meter pitch-roll buoy moored 60 km off the coast of Monterey, California (station 46042), in a water depth of 2100 m. Figure 6 shows the location of station 46042 as well as other NDBC buoy stations. Details of this data set can be found in Wang (1992). The wave climate data consist of significant wave height, H_s, and average wave period, T_z, which were computed from the measured wave spectrum using Equation (6). A total of 37,683 hourly significant wave height and average wave period recordings were taken in this study, which covered approximately 96% of the 55-month period.

Figure 7 shows the scatter plot of significant wave height versus average wave period. The lines representing the upper and lower boundary of the data scatter have the following relations

$$H_s = 0.1T_z^2 \text{ (upper boundary)} \tag{18}$$

and

$$H_s = 0.01T_z^2 \text{ (lower boundary)} \tag{19}$$

Figure 8 shows the contour plot of the joint distribution of H_s and T_z. Compared to the probability distribution of P_I , as shown in Figures 4 and 5, most of data were outside the region where the value of P_I is high. Figure 9 shows the contour plot of the probability distribution, $P_{hc,\, sc}$. The total probability of occurrence of high and steep waves for a given h_c and s_c at the specified location can then be computed from

$$P[\, h \geq h_c,\ s \geq s_c\,] = \int_0^\infty \int_0^\infty P_{h_c,\, s_c}\, dH_s dT_z \tag{20}$$

Because of the seasonal variation of sea state shown in the data from this buoy station (Wang, 1992), the probability of occurrence of high and steep waves was also computed for the summer seasons (June, July, and August) and winter seasons (December, January, and February). Table 1 shows the computed probability of occurrence from Equation (20) for two selected thresholds of wave height and steepness.

TABLE 1. Summary of computed probability of occurrence

	average H_s (m)	average T_z (sec)	$P[h \geq h_c, s \geq s_c]$	
			$h_c = 4$ m, $s_c = 0.1$	$h_c = 5$ m, $s_c = 0.1$
Summer Seasons	1.65	6.67	$0.15 * 10^{-5}$	$0.04 * 10^{-5}$
Winter Seasons	2.62	8.31	$2.36 * 10^{-5}$	$1.33 * 10^{-5}$
Total Data	2.17	7.35	$1.09 * 10^{-5}$	$0.52 * 10^{-5}$

Note that the average significant wave height during the winter seasons was 2.62 m, which was about 60% higher than that of the summer seasons. The probability of occurrence of high and steep waves ($h_c = 4$ m, and $s_c = 0.1$) in the winter seasons was 15 times greater than the summer seasons. The difference was even higher for $h_c = 5$ m. Figures 10, 11, and 12 show the estimated probability of occurrence of high and steep waves for various h_c and s_c for the 55-month period, for the summer seasons, and for the winter seasons, respectively. For evaluating the operational safety of smaller vessels in the area around the buoy station, the results displayed in Figure 10 can be used to define the risk level based on the design h_c and s_c of smaller vessels. To consider seasonal factors, the results shown in Figures 11 and 12 should be used to obtain a proper evaluation.

CONCLUDING REMARKS

Information concerning the probability of occurrence of high and steep waves is crucial to the safety of smaller marine vessels and offshore operations. In this study, a joint probability distribution of wave height and steepness was theoretically derived. Applying the derived function with the joint probability distribution of significant wave height and average wave period derived from long-term wave measurements from a single moored buoy station off the California coast, the probability of occurrences of high and steep waves was estimated. By applying this method to long-term wave measurements from other NDBC buoy stations along the coasts of the United States, a quantitative representation of sea severity in terms of the probability of encountering a very steep and high wave can be provided. This study presents a practical approach to derive vital information from existing long-term wave data.

REFERENCES

Dahle, E.A., Myrhaug, D., and Kjeldsen, S.P. (1988) "Probability of capsizing in steep and high waves from the side in open sea and coastal waters", Ocean Engineering, Vol. 15, No 2, 139-151.

Hamilton, G. (1980) "Environmental conditions associated with the capsizing of discus buoy", NDBC Report F-820-1.

Longuet-Higgins, M.S. (1952) "On the statistical distribution of the heights of sea waves", J. Mar. Res., Vol., 11, No. 3, 245-266.

Myrhaug, D. and Kjeldsen, S.P. (1984) "Parametric modelling of joint probability density distribution for steepness and asymmetry of deep water waves", Applied Ocean Research, Vol. 6, No. 4, 207-220.

Myrhaug, D. and Kjeldsen, S.P. (1986) "Steepness and asymmetry of extreme waves and the highest waves in deep water", Ocean Engineering, Vol. 13, No. 6, 549-568.

Myrhaug, D. and Kjeldsen, S.P. (1987) "Prediction of occurrences of steep and high waves in deep water", J. Waterway, Port. Coastal Ocean Engng., ASCE, Vol 113, No 2, 122-138.

Ochi, M.K. (1980) "Stochastic analysis and probabilistic perdition of random seas", UFL/COEL-TR/042, Department of Coastal and Oceanographic Engineering, University of Florida.

Ochi, M.K. and Tsai, C.H. (1983) " Prediction of occurrence of breaking waves in deep water", J. Physical Oceanography, Vol. 13, November, 2008-2019.

Overvik, T., and Houmb, O.G. (1977) "Distribution of wave steepness", in Proceedings of POAC '77, 984-994.

Rice, S.O. (1944) "Mathematical analysis of random noise", Bell. System Tech. Journ., Vol. 23, 282-332.

Tucker, M.J. (1991) Waves in Ocean Engineering: Measurement, Analysis, Interpretation, Ellis Horwood Limited, England.

Wang, D.W. (1992) "Long-term wave statistics from an NDBC buoy station", in Proceeding of MTS '92, 477-485.

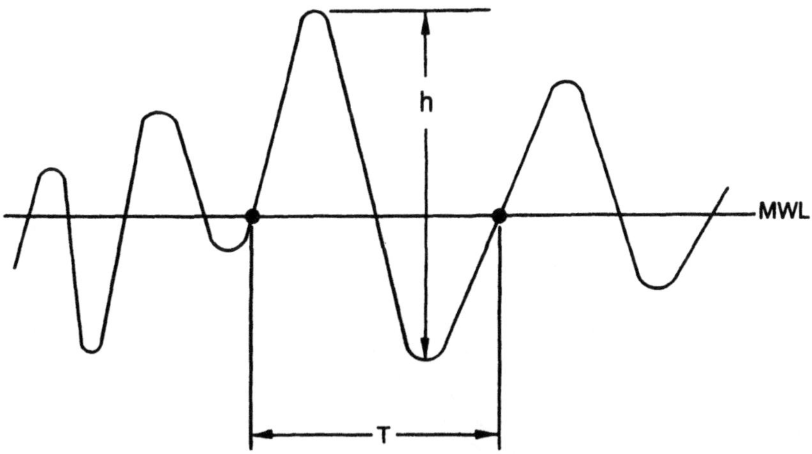

Figure 1. Explanatory sketch of the definition of wave height and wave period.

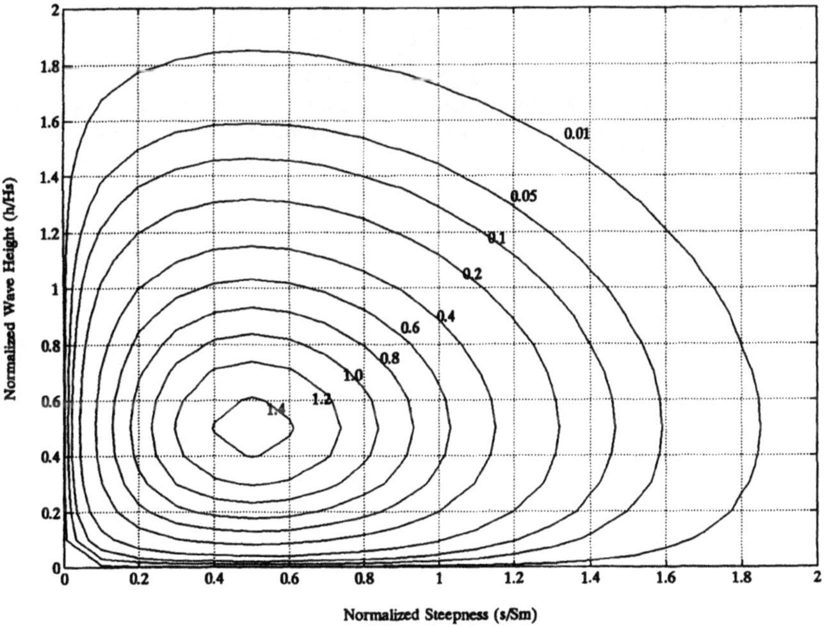

Figure 2. Contour plot of joint probability density function of normalized wave height, h/H_s, and steepness, s/S_m.

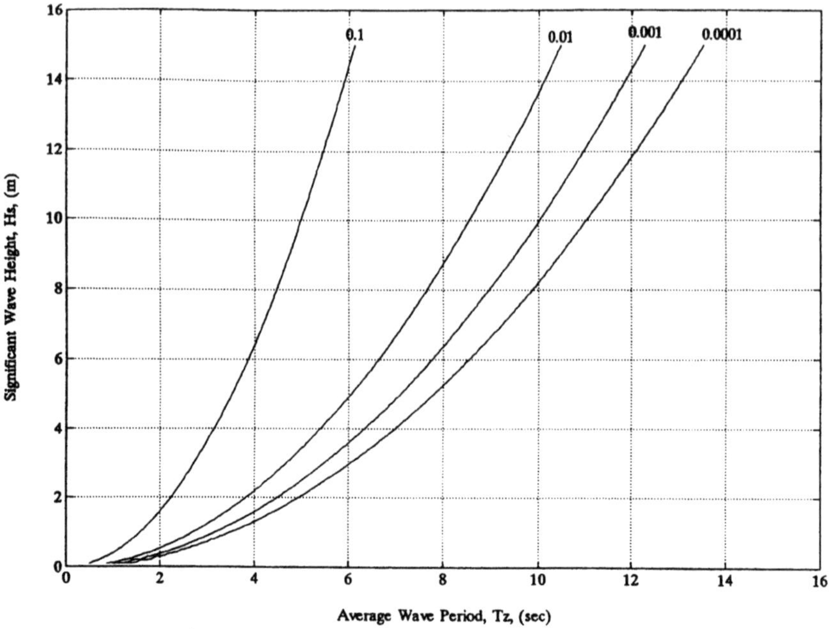

Figure 3. Contour plot of probability distribution, as function of H_s and T_z, of occurrence of wave for $h_c = H_s$, and $s_c = 0.10$. The value on the contour line indicates the probability density.

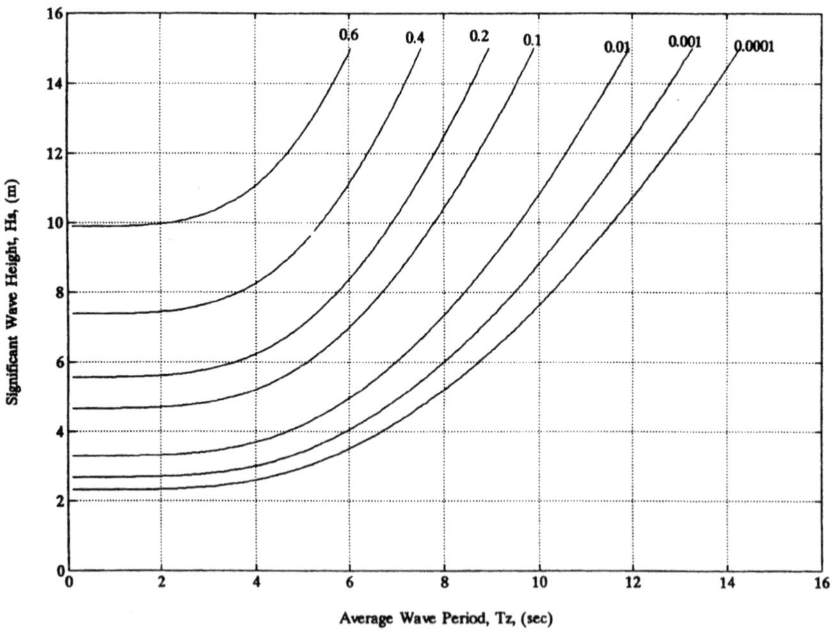

Figure 4. Contour plot of probability distribution, as function of H_s and T_z, of occurrence of wave for $h_c = 5$ m, and $s_c = 0.10$. The value on the contour line indicates the probability density.

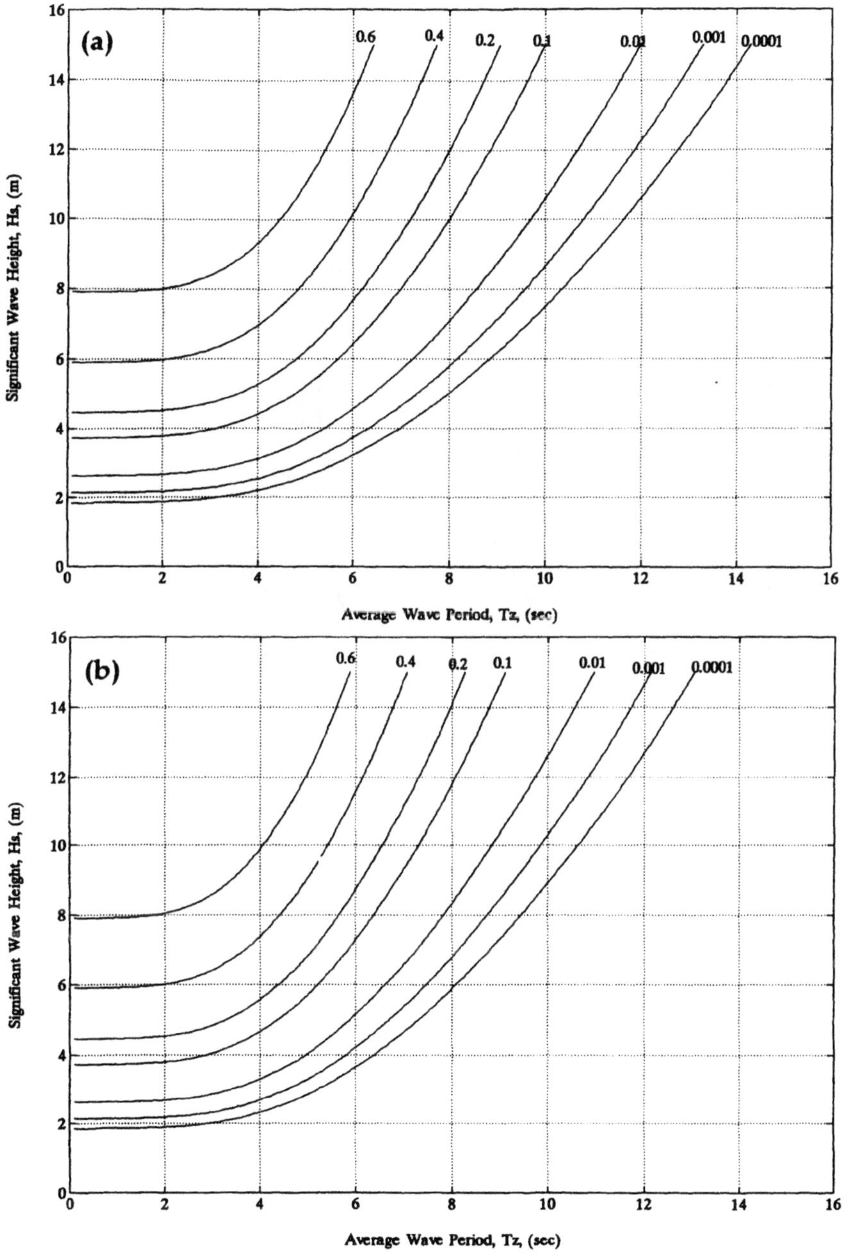

Figure 5. Contour plot of probability distribution, as function of H_s and T_z, of occurrence of wave for (a) $h_c = 4$ m, and $s_c = 0.10$, and (b) $h_c = 4$ m, and $s_c = 0.12$. The value on the contour line indicates the probability density.

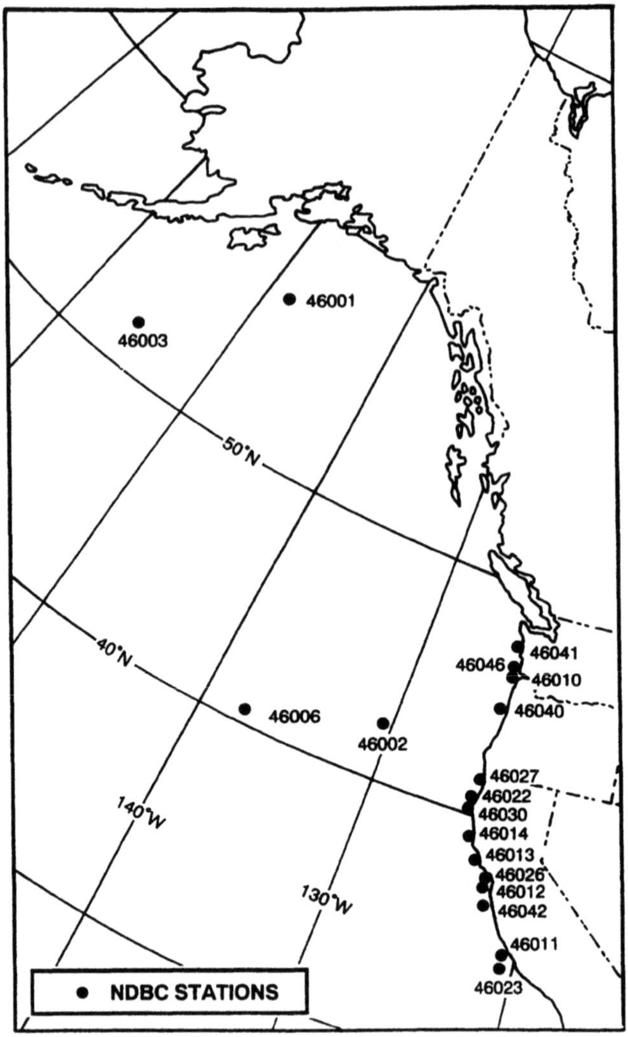

Figure 6. Location map of NDBC buoy stations along the west coast of the United States.

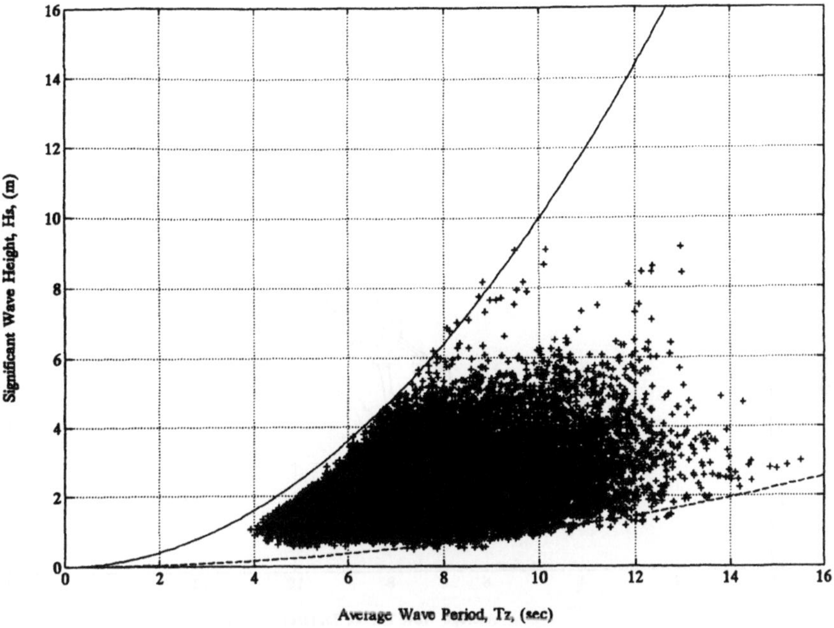

Figure 7. Scatter plot of average wave period versus significant wave height for the 55-month wave data.

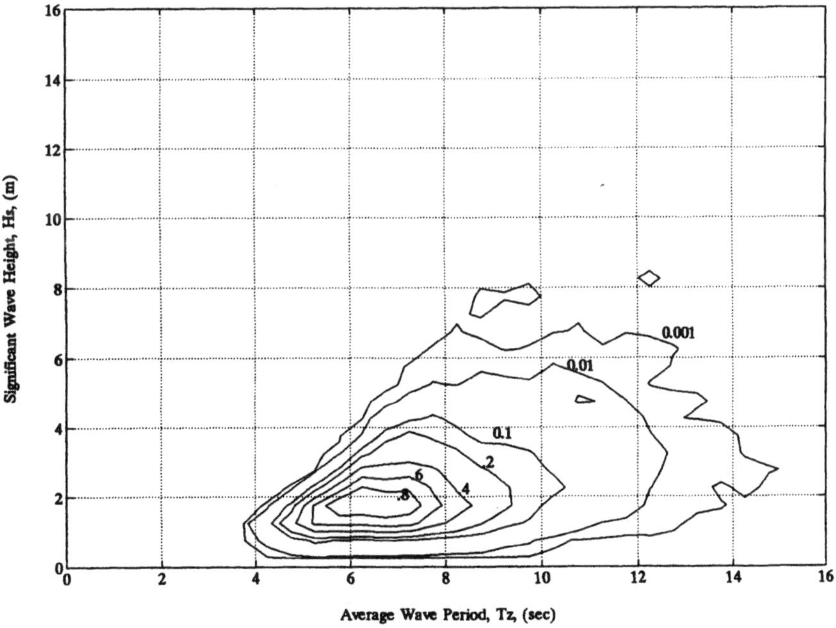

Figure 8. Contour plot of the joint probability distribution of H_s and T_z from the 55-month wave data. Contour lines indicated the 0.001, 0.01, 0.1, 0.2, 0.4, 0.6, and 0.8 of the peak value of the probability distribution.

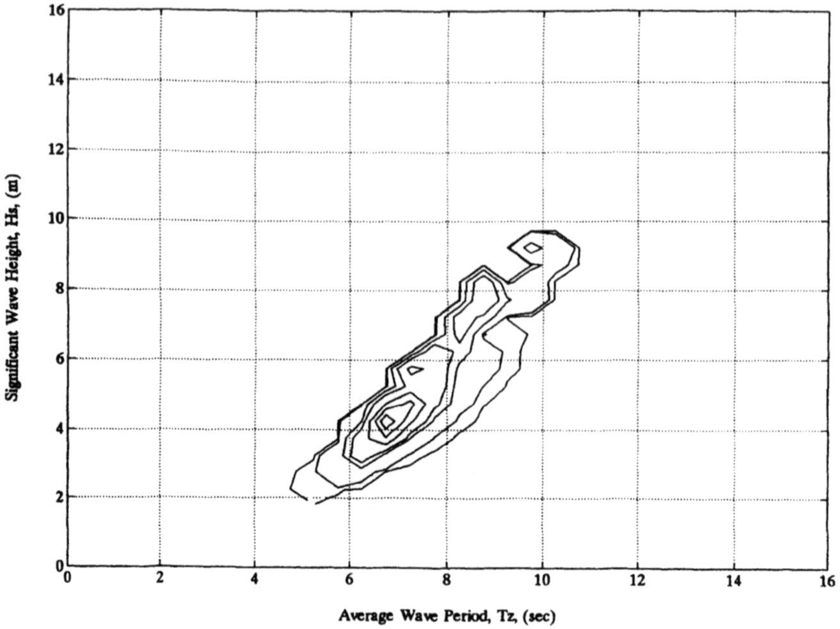

Figure 9. Contour plot of the probability distribution of occurrence of wave for h_c = 4m and s_c = 0.10. Contour lines indicated the 0.001, 0.01, 0.1, 0.2, 0.4, 0.6, and 0.8 of the peak value of the probability distribution.

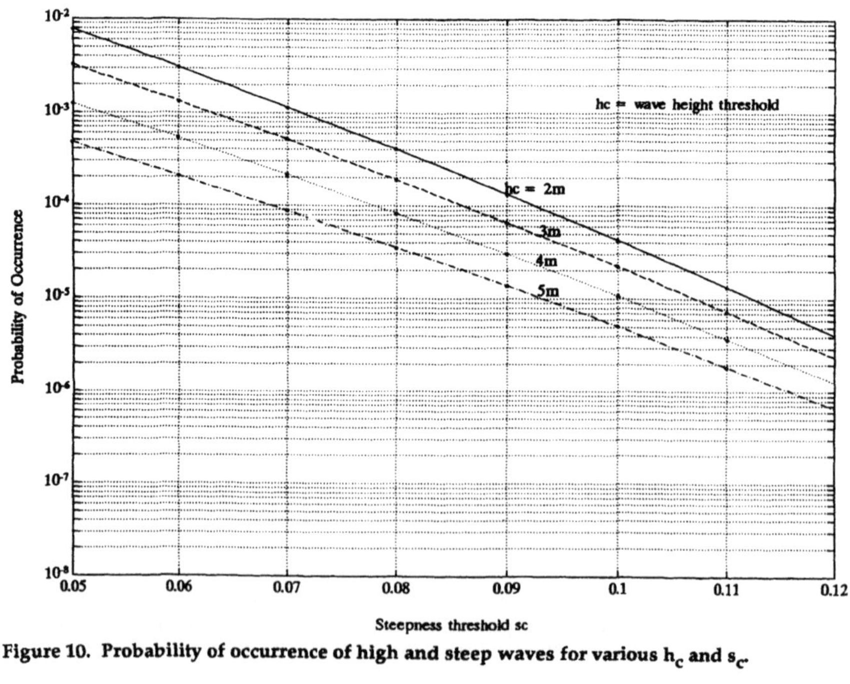

Figure 10. Probability of occurrence of high and steep waves for various h_c and s_c.

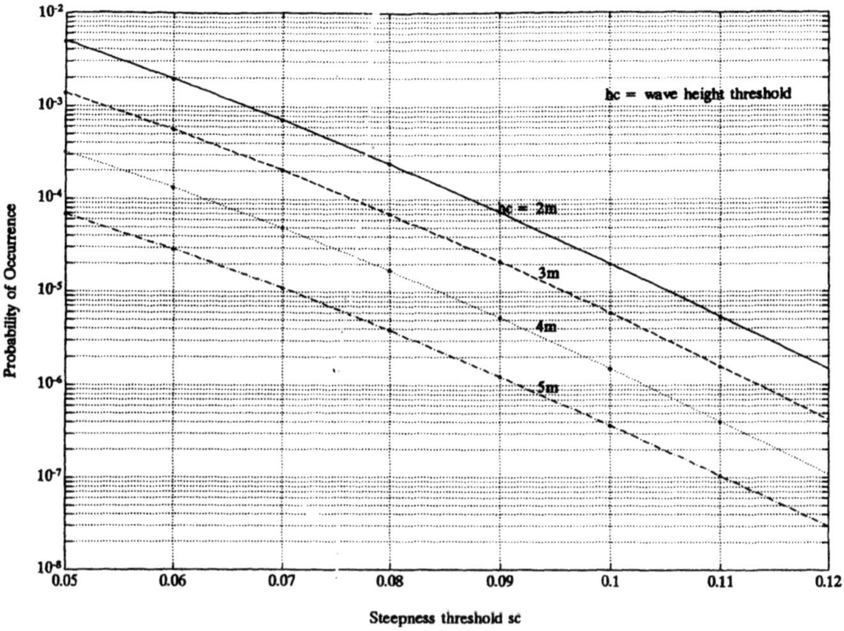

Figure 11. Probability of occurrence of high and steep waves for various h_c and s_c during summer seasons.

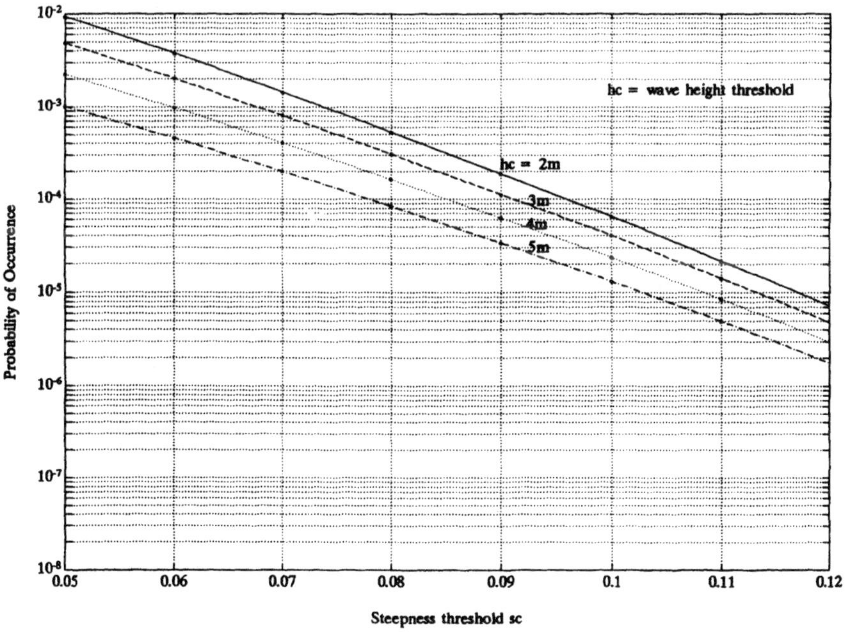

Figure 12. Probability of occurrence of high and steep waves for various h_c and s_c during winter seasons.

AJUSTE II, SOFTWARE FOR HYDROLOGICAL DATA ANALYSIS

PERRON, H.[1] LEGENDRE, P.[2], BOBÉE, B.[1], PERREAULT, L.[1]
[1] INRS-Eau, University of Québec, Institut national de la recherche scientifique, Carrefour Molson, 2800, rue Einstein, Québec (Québec), Canada, G1X 4N8
[2] Hydro-Québec, Division Hydrologie, Place Dupuis (17e étage), 855, Ste-Catherine Montréal (Québec), Canada, H2L 4P5

INTRODUCTION

In order to carry out the design and sizing of hydraulic installations, the management and rational operation of reservoirs as well as flood predictions, it is important to correctly estimate extreme flows in general or flows for a given return period T in particular, such that:

$$p = P(X \geq X_T) = 1 / T$$

where p is the exceedance probability associated to the hydrological risk (Kite, 1978). The fitting of statistical distributions to annual maximal flows is a common tool for determining X_T and its confidence intervals. Many distributions may be examined to fit the flows such as:

- Gamma(G)
- Log Gamma(LG)
- Pearson (P)
- Log Pearson(LP)
- Generalized Gamma (GG)

Bobée and Ashkar (1991) present a review of methods developed at INRS-Eau and elsewhere for fitting the distributions G, LG, P, LP and GG, as well as an automatic distribution fitting software. This software, HFA (Hydrological Frequency Analysis), of which a french version also exists (AJUSTE), was developed mainly as a tool to be used for research purposes in various studies. This software is presently being used at Hydro-Québec for :

- the design of future hydraulic works;
- the verification of the adequacy of existing installations;
- the reservoir management;
- the evaluation of floodplain regions as well as the risk of flooding.

Following the practical use of HFA, which was initially developed for research purposes, we have produced a new version of the software AJUSTE, through a partnership project with Hydro-Québec. This version, AJUSTE II, is more adapted to

K. W. Hipel (ed.),
Stochastic and Statistical Methods in Hydrology and Environmental Engineering, Vol. 1, 327–338.

the needs of Hydro-Québec and includes new theoretical elements concerning the estimation of the flows X_T associated with a given return period T.

Since AJUSTE contains only the distributions of the Gamma family at the present time, we have introduced in AJUSTE II the following distributions to make it more general and more widely applicable :

- The GEV (General Extreme Value), initially proposed by Jenkinson (1955), as well as its particular cases, the Gumbel and Weibull, which has been systematically recommended in Great Britain (NERC, 1975).
- The log-normal (2 and 3 parameter) , which is frequently used in North America (Chow, 1954 and Stedinger, 1983)
- The exponential (2 parameter) recommended by Kelman (1987)
- The 3 types of Halphen distributions (Halphen, 1941), recommanded by Bobée *et al.* (1993).

Halphen (1941) proposed the distribution named after him and Morlat (1956) extended it to the full system of Halphen distributions (types A, B, B^{-1}). These 3 parameter distributions have interesting properties (see below) but are not very widely used in practice and very little theoretical development has been pursued because of the complexity of the analytical form of the probability distribution function.

Besides allowing the estimation of flows X_T, AJUSTE II includes a complete array of testing tools for verification of null hypotheses. The new software has been developed using recent computer technologies, and may be used under the *Windows*[1] environment, which makes it more flexible and easier to use. The first part of this paper shows the concept of risk and return period T as well as general considerations concerning the estimation of X_T for a return period T. In the second part, we introduce the principal characteristics of the Halphen distributions. Finally, in the last part, we present the software AJUSTE II, with an emphasis on its new and improved features in comparison with its predecessor HFA.

SOME BASIC DEFINITIONS

Definition of risk

It is important to evaluate the risk of flooding when designing hydraulic works. Therefore, we must be able to estimate the exceedance probability of a critical value x_C. Considering the maximum annual flood as a random variable X with p.d.f $f(x;\theta_1,\theta_2,\theta_3)$, (general case with three parameters), the distribution function F may be defined as follows:

$$F(x_C;\theta_1,\theta_2,\theta_3)=\text{Prob}\{x\le x_C\}=\int_{-\infty}^{x_c} f(t;\theta_1,\theta_2,\theta_3)dt=1-p \qquad (1.)$$

In general, the hydrological risk is defined as the probability that an annual flood equals or exceeds x_C, i.e. the probability that the event $A=\{x\ge x_c\}$ (failure) occurs:

$$\text{Prob}(x\ge x_c)=1-F(x_c;\theta_1,\theta_2,\theta_3)=p \qquad (2.)$$

[1]*Microsoft Windows*, copyright © 1985-1992 Microsoft corp.

The return period T

If we consider the two following complementary and mutually exclusive events:

$$A = \{x \geq x_c\} \text{ and } \overline{A} = \{x < x_c\} \tag{3.}$$

where A represents a failure event (an overflow for example), and $\overline{A}$ the no-failure event. These two events describe a Bernouilli experiment such as equation (2) :

$$\text{Prob}\{A\} = p \text{ and } \text{Prob}\{\overline{A}\} = 1 - p \tag{4.}$$

If Z is the time (in years) between two events $A = \{x \geq x_c\}$, we can show that Z is a random variable distributed according to a geometric distribution with p.d.f. $\text{Prob}(Z = z) = (1-p)^{z-1} p$. We deduce that:

$$E[Z] = T = \frac{1}{p}, \tag{5.}$$

where $E[Z] = T$ is the mean time (computed over a long period) between two events $A = \{x \geq x_c\}$. T is the return period of the event $A = \{x \geq x_c\}$. If the critical value is defined according to the return period T, i.e. if we have

$$x_c = x_T \tag{6.}$$

Then from (2) and (5) :

$$\text{Prob}\{x \geq x_T\} = 1 - F(x_T; \theta_1, \theta_2, \theta_3) = p = \frac{1}{T} \tag{7.}$$

For maximum annual floods, x_T is the flow of the return period T (in years), i.e. whose exceedance probability is $p = 1 / T$.

Estimation of x_T

In practice, x_T is estimated by fitting a statistical distribution to a series of maximal annual floods constructed from maximum daily floods over n years. This is possible when the observations :

a) are independent (no autocorrelation), which may be verified by the Wald-Wolfowitz test (1943);

b) come from the same statistical population $F(x; \theta_1, \theta_2, \theta_3)$, which implies :

- Stationarity (Kendall and Stuart test, 1987);
- Homogeneity (Mann and Withney test, 1947); and
- Absence of outliers (discordancy test adapted to each distribution, for example the Grubbs and Beck test (1972) for the normal distribution).

If the observations are independent and identically distirubted (iid) according to F, then from (7) we get:

$$x_T = F^{-1}\left(1-\frac{1}{T};\theta_1,\theta_2,\theta_3\right) \tag{8.}$$

where F^{-1} is the inverse function of the distribution function F.

To estimate x_T, it suffices to estimate the distribution parameters by $\tilde{\theta}_1, \tilde{\theta}_2, \tilde{\theta}_3$ and then use equation (8) in order to obtain the estimator $\tilde{x}_T$ of x_T. Many parameter estimation methods exist for the distributions often used in hydrology, in the case where none of the methods is optimal regardless of sample size. Many studies have been conducted to compare the estimation methods for a given distribution. We note in particular, the study related the gamma family (G, LG, P, LP, GG) by Messaoudi (1993).

HALPHEN DISTRIBUTIONS

Despite the number of distributions already in use in hydrology since the middle of the century, the French hydrologist E. Halphen felt the need to develop new density functions (Halphen, 1941). Past experience had convinced him that none of the traditional distributions used at that time had shapes that were globally appropriate for flood series in France. Consequently, Halphen developed three new types of distribution: type A, B and B^{-1}. Although they are largely based on empirical justifications, the Halphen distributions, which have three parameters, possess interesting theoretical properties. In particular, they each form an exponential family. Therefore, they possess joint sufficient and complete statististics for the parameters. Among the various 3-parameter distributions used in hydrology, the Halphen distributions are the only ones which have this property.

The reason why only little attention has been paid to Halphen's distributions since their introduction fifty years ago has to do with practical problems of estimating the parameters. The systems of ML-equations are nonlinear and must be solved using iterative procedures. They involve functions of Bessel and Hermite functions which are difficult to manipulate, and the numerical solutions turn out to be rather tedious. To illustrate the problems, we discuss in the following the estimation of parameters and quantiles for the type A distribution.

The Halphen type A distribution introduced by Halphen (1941) has a pdf given by :

$$h_A(x;\alpha,m,\upsilon) = \frac{1}{2m^{\upsilon}K_{\upsilon}(2\alpha)}x^{\upsilon-1}\exp\left[-\alpha\left(\frac{x}{m}+\frac{m}{x}\right)\right], \qquad x>0 \tag{9.}$$

where υ and $\alpha > 0$ are shape parameters, and $m > 0$ a scale parameter. The constant $K_{\upsilon}(2\alpha)$ is a modified Bessel function of the second kind (Watson, 1966), defined as:

$$K_{\upsilon}(2\alpha) = \frac{1}{2m^{\upsilon}}\int_0^{+\infty} x^{\upsilon-1}\exp\left[-\alpha\left(\frac{x}{m}+\frac{m}{x}\right)\right]dx \tag{10.}$$

$h_A(x)$ is unimodal with positive skewness, and its decrease at the lower and upper tails is exponential. It can be shown that the gamma (G) and the inverted gamma (IG) distributions constitute limiting forms of $h_A(m,\alpha,\upsilon)$ for specific values of the parameters: if $\alpha \to 0$, $h_A \to G$ for $\upsilon > 0$, and $h_A \to IG$ for $\upsilon < 0$. Between these limiting cases, the type A distribution can take a variety of forms which are all of interest for hydrological modeling. In particular, the harmonic distribution is a special case of Halphen type A corresponding to $\upsilon = 0$. An interesting observation is that the generalized inverse gaussian distribution (GIG), proposed independently by Good (1953) for modeling biological data, is in fact a Halphen type A distribution. The theoretical properties of the GIG were studied recently by Jorgensen (1982). We can deduce from his work that the inverse gaussian (Wald, 1946), the hyperbola, and the generalized hyperbola distributions (Barndorff-Nielsen, 1977) are special cases of the Halphen type A.

Now, assume that a record of n floods is available, and that these are iid Halphen type A distributed. Their joint density is obtained from equation (9) and can be written as:

$$h_A(\underline{x};m,\upsilon,\alpha) = \prod_{j=1}^{n} h_A(x_j;m,\upsilon,\alpha) = \prod_{j=1}^{n} \frac{1}{2m^{\upsilon}K_{\upsilon}(2\alpha)} x_j^{\upsilon-1} \exp\left(-\alpha\left[\frac{x_j}{m}+\frac{m}{x_j}\right]\right)$$

$$= \exp\left\{\sum_{i=1}^{n} c_i(m,\upsilon,\alpha)T_i(\underline{x}) + d(m,\upsilon,\alpha)\right\} \tag{11.}$$

where :

$$c_1(m,\upsilon,\alpha) = n(\upsilon-1) \qquad T_1(\underline{x}) = \frac{1}{n}\sum_{j=1}^{n} \log(x_j) = \log(G) \tag{12.}$$

$$c_2(m,\upsilon,\alpha) = -n\alpha/m \qquad T_2(\underline{x}) = \frac{1}{n}\sum_{j=1}^{n} x_j = A \tag{13.}$$

$$c_3(m,\upsilon,\alpha) = -n\alpha m \qquad T_3(\underline{x}) = \frac{1}{n}\sum_{j=1}^{n} \frac{1}{x_j} = H^{-1} \tag{14.}$$

$$d(m,\upsilon,\alpha) = -\log\left[2^n m^{n\upsilon}\left(K_{\upsilon}(2\alpha)\right)^n\right] \tag{15.}$$

It is seen that the Halphen Type A distribution form an exponential family with sufficient statistics T_1, T_2 and T_3 which are respectively the geometric, arithmetic and harmonic means. Therefore we can use the result that if $E\lfloor T_i(\underline{X})\rfloor = T_i(\underline{x})$ has a solution for parameter vector $\underline{\theta} = (\theta_1, \theta_2, \theta_3)$, then this solution is the unique ML-estimators of $\underline{\theta}$ (Bickel and Doksum,1977). For the Halphen type A, it can be shown (Perreault *et al*, 1993) that :

$$E\left[\frac{1}{n}\sum_{j=1}^{n}X_j\right]=m\frac{K_{\upsilon+1}(2\alpha)}{K_{\upsilon}(2\alpha)} \tag{16.}$$

$$E\left[\frac{1}{n}\sum_{i=1}^{n}\frac{1}{X_j}\right]=\frac{1}{m}\frac{K_{\upsilon-1}(2\alpha)}{K_{\upsilon}(2\alpha)} \tag{17.}$$

$$E\left[\frac{1}{n}\sum_{j=1}^{N}\log X_j\right]=\log m+\frac{1}{K_{\upsilon}(2\alpha)}\frac{\partial K_{\upsilon}(2\alpha)}{\partial \upsilon} \tag{18.}$$

The system of ML-equations is then obtained by equating the three expressions (16), (17) and (18) respectively to A, H^{-1} and $\log G$:

$$m\frac{K_{\upsilon+1}(2\alpha)}{K_{\upsilon}(2\alpha)}=A \tag{19.}$$

$$\frac{1}{m}\left(\frac{K_{\upsilon-1}(2\alpha)}{K_{\upsilon}(2\alpha)}\right)=\frac{1}{H} \tag{20.}$$

$$\log m+\left(\frac{1}{K_{\upsilon}(2\alpha)}\right)\left(\frac{\partial K_{\upsilon}(2\alpha)}{\partial \upsilon}\right)=\log G \tag{21.}$$

Note that the ML-estimators of the Halphen type A parameters are equivalent to those obtained by applying the Sundry Average Method (Bobée and Ashkar, 1991, p. 102). The scale parameter, m, can be eliminated from the system, leading to the following two equations involving only the shape parameters υ and α:

$$\frac{K_{\upsilon+1}(2\alpha)\,K_{\upsilon-1}(2\alpha)}{K_{\upsilon}^{2}(2\alpha)}-\frac{A}{H}\quad=\quad 0 \tag{22.}$$

$$\frac{K_{\upsilon}(2\alpha)}{K_{\upsilon+1}(2\alpha)}\exp\left\{\frac{\frac{\partial K_{\upsilon}(2\alpha)}{\partial \upsilon}}{K_{\upsilon}(2\alpha)}\right\}-\frac{G}{A}\quad=\quad 0 \tag{23.}$$

Solving simultaneously equations (22) and (23) to determine the ML-estimators $\hat{\alpha}$ and $\hat{\upsilon}$, we can deduced $\hat{m}$ from eq. (19). Although standard routines are available for calculating Bessel functions and their derivatives, the above system (equation 22 and 23) is difficult to solve. We therefore considered a two step approach (Perreault *et al.*, 1993). Initially, the parameters α and m are estimated numerically when υ is fixed using equations (22) and (19) (equation (22) is well suited for Newton-Raphson iteration). Then, the partially maximised log-likelihood funtion $\log L(\upsilon; \hat{\alpha}, \hat{m})$ obtained by inserting the estimate $\hat{\alpha}$ and $\hat{m}$ in the log-likelihood function is used to determine $\hat{\upsilon}$. In practice, having implemented the estimation of α and m for fixed values of υ (step 1) on the computer, it is straight forward to determine $\hat{\upsilon}$ numerically (step 2), e.g. by a simple tabulation $\log L(\upsilon; \hat{\alpha}, \hat{m})$.

From standard asymptotic theory for exponential families (Bickel and Doksum, 1977) $\sqrt{n}\,\lfloor(\hat{m}, \hat{\alpha}, \hat{\upsilon}) - (m, \alpha, \upsilon)\rfloor$ is asymptotically distributed as a multivariate normal distribution with mean vector (0, 0, 0) and covariance matrix $\mathbf{I}_f^{-1}(m, \alpha, \upsilon)$, where $\mathbf{I}_f(m, \alpha, \upsilon)$ is the information matrix given by (Perreault *et al.*, 1993):

$$\mathbf{I}_f(\alpha, m, \upsilon) = \begin{bmatrix} \frac{n}{m^2}\left\{2\alpha\frac{K_{\upsilon+1}}{K_\upsilon} - \upsilon\right\} & \frac{n}{mK_\upsilon}\left\{K_{\upsilon-1} - K_{\upsilon+1}\right\} & \frac{n}{m} \\ & \frac{n}{K_\upsilon^2}\left\{\frac{\partial^2 K_\upsilon}{\partial\alpha^2}K_\upsilon - \left(\frac{\partial K_\upsilon}{\partial\alpha}\right)^2\right\} & \frac{n}{K_\upsilon^2}\left\{\frac{\partial^2 K_\upsilon}{\partial\alpha\partial\upsilon}K_\upsilon - \left(\frac{\partial K_\upsilon}{\partial\alpha}\right)\left(\frac{\partial K_\upsilon}{\partial\upsilon}\right)\right\} \\ & & \frac{n}{K_\upsilon^2}\left\{\frac{\partial^2 K_\upsilon}{\partial\upsilon^2}K_\upsilon - \left(\frac{\partial K_\upsilon}{\partial\upsilon}\right)^2\right\} \end{bmatrix} \quad (24.)$$

where K_υ is used for $K_\upsilon(2\alpha)$ to simplify the notation. The ij-th element of $\mathbf{I}_f(\theta_1, \theta_2, \theta_3)$ is obtained by calculating :

$$-E\left[\frac{\partial^2 \log L}{\partial\theta_i\,\partial\theta_j}\right] \quad (25.)$$

where L is the likelihood function. Note that the asymptotic variances and covariances of the ML-estimators are functions of Bessel functions as well as of the first- and second-order derivatives of these functions.

It is obvious from equation (9) that the cumulative distribution function for Halphen type A does not exist in closed form. Therefore, it is not possible to develop an explicit expression for quantiles, or T-year events. An estimate of x_T must be determined by solving numerically the equation :

$$\Pr(X \le \hat{x}_T) = \int_0^{\hat{x}_T} h_A(x; \hat{m}, \hat{\alpha}, \hat{\upsilon})\,dx = 1 - \frac{1}{T} \quad (26.)$$

For engineering practice, evaluation of the variance of $\hat{x}_T$ is very important. If we write $x_T = g(m, \alpha, \upsilon)$, then the asymptotic variance of $\hat{x}_T$ can be calculated from the general formula (Bobée and Ashkar, 1991):

$$Var\left[\hat{X}_T\right] = \sum_{j=1}^{3}\left(\frac{\partial g}{\partial\theta_j}\right)^2 Var\left[\hat{\theta}_j\right]+\sum_{j=1}^{3}\sum_{\substack{i=1\\ i\neq j}}^{3}\frac{\partial g}{\partial\theta_i}\frac{\partial g}{\partial\theta_j}Cov\left[\hat{\theta}_i,\hat{\theta}_j\right] \quad (27.)$$

In addition to the asymptotic variances and covariances of the ML-estimators, the derivatives of x_T with respect to the parameters must be determined. We note that these derivatives are complicated functions involving integrals that must be computed numerically (Perreault *et al.*, 1993).

In summary, the Halphen Type A distribution is complex, and a lot of numerical routines are needed to apply it in practice. Similar complexities are encountered when using the Halphen Type B and B^{-1} distributions. Current research by the authors aims at developing efficient routines for estimating parameters, quantiles, and asymptotic variances.

THE AJUSTE II SOFTWARE PACKAGE

The AJUSTE II software package is an improved version of HFA, the latter containing many parameter estimation methods for each of the distributions from the Gamma family. For example, ten methods are included for the Log-Pearson Type 3. Even though it is interesting from a research point of view to have a software which contains all the available methods, it is not necessary in practice. Following the use of HFA and new theoretical developments, we have added some distributions often used in hydrology as well as the three types of Halphen distribution. Furthermore, only the estimation parameter methods which yielded the best results were included for the Gamma family (Messaoudi, 1993). We have also added some tests to verify the basic assumptions, while improving the user friendliness of the software (*Windows* environment). Table 1 (next page) presents the main differences (regarding the distributions, the parameter estimation methods and the statistical tests) between HFA and AJUSTE II.

Features of AJUSTE II

The user friendliness of the software has been greatly improved. As a matter of fact, AJUSTE II was written using the C++ language (whereas HFA was written in Fortran) and the library XVT. This library allowed the development of AJUSTE II under the graphics environments *Windows* (DOS) and *X-Windows* (UNIX), making the package easy to use and portable.

Furtermore, AJUSTE II presents the following improvements over HFA :

- the sample size is now limited only by the available memory whereas the sample size limit in HFA was 200. Furthermore, the user may work with many samples in the same work session;
- all the operations are executed from a simple menu. Quick commands have been defined for the most common operations;
- an on-line help is available at all times. This is technical help (use of the software) as well as theoretical (to guide the user in his or her decisions);
- the user may print a summary table of all the fittings conducted during a work session;

- the user may modify graphical features and zoom on certain parts of the graphs;
- It is planned that an expert system will be implemented .

TABLE 1. Comparison of the software packages HFA and AJUSTE II

	HFA	AJUSTE II
Distributions	Gamma Pearson Type 3 Log-Pearson Type 3 Generalized Gamma Log-Gamma	Gamma (ML[2], MM[1]) Pearson Type 3 (ML, MM) Log-Pearson Type 3 (BOB[1], SAM[1], WRC[1]) Generalized Gamma (MM, MM1[1]) GEV (ML, MM) Gumbel (ML, MM) Weibull (ML, MM) Halphen (type A, B and B^{-1}) (ML) Normal (ML) Log-normal 2 param. (ML, MM) Log-normal 3 param. (ML, MM) Exponential 2 param. (ML)
Tests	Wald-Wolfowitz Mann-Withney Grubbs and Beck	Wald-Wolfowitz Mann-Withney Kendall Discordancy test (for each distribution) Goodness of fit test(for each distribution)

Data Capturing

Data may be captured in two different ways :

- by reading an existing file in various formats;
- data may be input directly in the software.

The user may rank data in increasing order (for flood studies) or decreasing order (for drought studies).

Furthermore, when capturing data, the user may discard certain data points believed to be erroneous or simply ones which he she wishes to ignore for various reasons. This

[2]ML=Maximum Likelihood, MM=Method of Moments, WRC=Method of moments on the log of the observations, BOB=Direct method of moments , SAM=Sundry Averages Method, MM1=Method that uses the two first moments of the sample and the first moment of sample of the logarithms.

option allows the user to study the behaviours of various fittings and tests without taking into account the discarded data records .

Available output

Within AJUSTE II, the user may print at any time the contents of the active window, whether these are results or graphs. These outputs include:

- Sample data;
- results of tests conducted on the data;
- the fittings graphs;
- the results (parameters, $\hat{x}_T$, var $\hat{x}_T$, etc...) found during the fittings;
- a summary table of the fittings.

Basic statistics and tests

As in HFA, the user may obtain the sample and population basic statistics (mean, standard deviation, skewness coefficient, coefficient of variation and coefficient of kurtosis), and view the empirical probabilities of the sample elements (cumulative probability of each element). The empirical probabilities may be determined using the Cunnane, Weibull, Chegodayev or Hazen formula , depending on the user.

The software provides many tests :

- Mann-Withney test (to verify the homogeneity of the observations);
- Wald-Wolfowitz test (to verify the independence of the observations);
- Kendall test (to verify stationarity);
- a discordancy test for each distribution (to check for outliers);
- a goodness of fit test for each distribution (khi-square test).

Fittings

When conducting the fittings, the user has a choice of many distributions, and various parameter estimation methods depending on the distribution used. Table 1 presents the available distributions and their corresponding estimation methods. The user may also carry out many fittings on the same sample for comparison purposes.
For each fitting, the user may view a graph of the results. This graph includes:

- The data sample;
- the theoretical curve as given by the fitting;
- the confidence intervals at a level which may be fixed a priori by the user.

The above elements may be added or removed from the graph depending on the user preferences.

Furthermore, the user may easily change the general presentation of the graph by modifying:

- the graph title;
- the axis titles;
- the curves colors;
- the confidence level of the confidence intervals;
- the graph probability paper (Gamma, Gumbel and Normal).

The software also provides the estimated value $\hat{x}_T$ of x_T, var $\hat{x}_T$ and the confidence intervals of the unknown theoretical value x_T at a fixed confidence level, for 21 exceedance probabilities in the case of floods and non-exceedance probabilities in the case of droughts.

CONCLUSION

The software package AJUSTE II is an efficient and user-friendly tool, which takes into account the most recent theoretical developments. It is very useful for the planning and design of hydraulic works, the management and rational operations of reservoirs as well as flood predictions, since it provides estimates of flows corresponding to a return period T, both in the cases of floods and droughts.

Furthermore, AJUSTE II contains an important number of probability distributions. While being useful for the analysis of hydrological data, it may be successfully and easily applied to data in other areas.

REFERENCES :

Barndorf-Neilsen (1977). Exponentially decreasing log-size distributions. Proc. R. Soc. London, A, 353: 410-419.

Bickel, P.J. and Doksum K.A. (1977). Mathematical statistics : Basic Ideas and Selected Topics. Holden-Day, 492p.

Bobée, B. and Askhar, F. (1991). The gamma family and derived distributions applied in hydrology. Water Resources Publicationss, 203p.

Bobée, B., Rasmussen, P.F., Perreault, L., and Ashkar, F. (1993). Risk analysis of hydrologic data : Review and new developments. To appear in the proceedings of the Nato Advanced Study Institute on "Engineering Risk and Reliability in a Changing Physical Environment", Deauville, France, May 23 - June 5, 1993.

Chow, V.T. (1954). The log-probability law and its engineering applications. Proc. Am Soc. Civ. Eng., 80(5):1-25.

Good, I. J., (1953). The population frequencies of species and the estimation of population parameters. Biometrika, 40: 237-260.

Grubbs, F. and Beck G. (1972). Extension of sample sizes and percentage points for significance tests of outlying observations. Technometrics, 14(4): 847-854.

Halphen, E. (1941). Sur un nouveau type de courbe de fréquence. Comptes rendus de l'Académie des Sciences, tome 213:634-635. (Published under the name of D. Dugué due to the war)

Jenkinson, A.F. (1955). The frequency distribution of the annual maximum (or minimum) values of meteorological elements. Quart. J. Roy. Meteor. Soc., 81: 158-171.

Jorgensen, B., (1982). Statistical properties of the generalized inverse gaussian distribution. Lecture notes in statistics, Springer Verlag.

Kelman, J. (1987). Statistical approach to floods. Macneil and G.J Unmphrey (Eds), Stochastics Hydroloy, D. Reidel Publishing Company: 193-225.

Kendall, M.G. and Stuart, A. (1987). The Advanced Theory of Statistics, vol. 1. Charles Griffin, London, 5th ed. 604 p.Kite, G.W. (1978). Frequency and risk analysis in hydrology, 2nd printing, Water Resources Publications, 224p.

Messaoudi, H. (1993). Comparaison des méthodes d'estimation des distributions de la famille Gamma. Thèse de doctorat à l'INRS-Eau (in preparation).

Morlat, G. (1956). Les lois de probabilités de Halphen. Revue de statistique appliquée, 4(3):21-43.
Mann, H.B. and Whitney, D.R. (1947). On the test of weather one of two random variables is stochastically larger than the other. Ann. Math. Statist., 18: 50-60.
NERC (Natural Environment Research Council) (1975). Flood studies report, Vols. 1-5, NERC, London.
Perreault. L., Bobée, B. and Perron, H. (1993). La loi de Halphen de Type A. Scientific report, INRS-Eau (in preparation).
Stedinger, R. (1983). Confidence intervals for design events. Journal of Hydraulic Engineering, 109 (1): 13-27.
Wald , A. and Wolfowitz, J. (1943). An exact test for randomness in the non-parametric case based on serial correlation. Ann. Math. Statist., 14: 378-388.
Wald, A., (1946). Sequential analysis. John Wiley & Son, Inc, New York.
Watson, G.N., (1966). A treatise on the theory of Bessel functions. Cambridge University Press.

PART VI

REGIONALIZATION TECHNIQUES AND DISTRIBUTED MODELS

THE REGIONALIZATION OF RUNOFF COEFFICIENTS FOR THE PRAIRIES OF ALBERTA

M. A. SABUR and D. D. ANDRES
Surface Water Engineering
Environmental Research and Engineering Department
Alberta Research Council
PO Box 8330
Edmonton, Alberta
Canada T6H 5X2

The design of bridge and culvert crossings of rivers and streams requires an estimate of the design discharge. A strategy has been developed to estimate the magnitude of the design floods for catchments which do not lend themselves to the regional analysis of flood peaks. The approach is founded on the premise that all flood producing rain storms tend to some upper limit and that the variability of the flood peaks is affected more by the variation of the antecedent moisture condition than by the variation in the precipitation intensity. When the design antecedent moisture condition (runoff coefficient) is identified, the peak flow is determined by applying a predetermined amount of percipitation to a runoff model that requires only physiographic information to calculate the runoff hydrograph. At present, the design runoff coefficient is the least well quantified of the runoff parameters.

The analysis suggests that the normal maximum 24 hour precipitation is between 150 and 200 mm and varies on a regional basis. Furthermore, the examination of some 450 precipitation/runoff events over 22 small catchments indicates that the upper limits of the runoff coefficient vary between 1% and 53% and are a function of the ratio of the normal annual precipitation to the potential evapotranspiration of the basins. This allows for a regionalization of the design runoff coefficients and the ultimate calculation of regionally-based design floods.

INTRODUCTION

Hydraulic structures such as culverts and bridges must be designed to withstand a certain severity of flooding. In designing these works, the magnitude of the flood which is to be withstood during the projected life of the structure must be quantified in a systematic way. Hudson (1989) has shown that statistical relationships between peak flows and a variety of basin characteristics for the basins in the mountains and foothills can be made within a regional perspective. However, the statistical analysis of annual flood peaks in basins in the prairies is difficult. There appears to be too much variability in the characteristics of these basins to produce significant regional correlations.

K. W. Hipel (ed.),
Stochastic and Statistical Methods in Hydrology and Environmental Engineering, Vol. 1, 341–353.

A change in strategy is required to get closure. Rather than considering the stochastic nature of the flood peaks, the problem is better addressed from a process point of view. The fundamental philosophy of this approach is to assume that most floods in small basins are derived from severe precipitation that has a consistent upper limit which is somewhat less than the maximum probable precipitation. Furthermore, the variability in runoff is caused by the variation in the runoff coefficient, as reflected by the antecedent moisture conditions, and that severe floods (design floods) occur when both high rainfall and high runoff coefficients are apparent. Thus by defining a nominal 'design' precipitation event and a 'design' runoff coefficient the 'design' peak flood can be determined if the response of the catchment is known.

This paper summarizes the work that has been undertaken to develop an understanding of the appropriate basin storage (runoff coefficient) for use in the application of rainfall-runoff models for small prairie watersheds. This paper will (1) provide a brief description of some of the methods currently being used to determine design floods, (2) summarize the significant characteristics of more than 400 runoff events from over 20 small prairie basins, and (3) define a relationship between design runoff coefficients and climatic characteristics for these prairie basins.

TECHNOLOGICAL CONTEXT

Currently, the fundamental strategy used for design flood estimation in ungauged basins is regional flood frequency analyses. This approach has been developed on the basis of accepted statistical analysis of the recorded data. Once developed, the procedure is quick and easy to use and the estimates are reliable for watersheds with hydrological characteristics similar to those used in the derivation. Some disadvantages are that the methods require a knowledge of both hydrology and statistics in derivation and utilization, and the method cannot evaluate the effects resulting from modifications in the system. Furthermore, the analysis can be misleading if an extrapolation is made to a basin whose characterics are not within the data set on which the regional criteria are based. Another disadvantage is that this method only provides estimates of peak flow and contains no information on the shape of the hydrograph.

It is therefore more appropriate to use a simulation approach. This may include single event methods, continuous simulation methods, and other hybrid methods which include elements of both. The continuous simulation models attempt to develop a continuous streamflow record based on a continuous rainfall record. Although in concept this category of models appears plausible (Sabur, 1990), their success has not been established. Besides, these are expensive, time consuming, and impractical for agencies with limited resources because large amounts of data are required. For practical applications, where hydrologic information (volume, hydrograph shape, time to peak, etc.) is required in addition to the peak discharges, single event simulation techniques are preferred. These have been used for developing runoff hydrographs from which flood frequency curves are developed.

An important advantage of the simulation methods is that the shape of the flood hydrograph is determined, not merely the peak rate of flow. This is of importance in routing studies, in drainage design and in other situations where it is necessary to know the length of time the water level is above the particular stage. The main limitation of the method is the subjectivity in determining the volume and distribution of the rainfall excess.

The approach that is described herein is very much a hybrid. It incorporates the more or less deterministic approaches of evaluating the response characteristics of individual basins with a stochastic description of the runoff coefficients to estimate the design peak floods and volumes for a narrow range of major flood-producing rain events. A large amount of work has been directed to the characterization and regionalization of rainfall events in Alberta. Numerous studies have been directed towards identifying relationships between the response function of a drainage basin and its physical characteristics. On the other hand, there is very limited information on runoff coefficients for basins located in the prairie region. The remainder of this paper will focus on this issue.

STUDY AREA

Considering the availability of unregulated streamflow data within the prairie region of Alberta, 22 basins of sizes ranging from 30 km^2 to 580 km^2 were selected from the southern half of the Province, excluding the mountain and foothills regions. It should be noted that only the smallest basins were chosen. It was intended that these basins be representative of typical prairie catchments and that the basins are small enough to represent homogeneous physiographic units which can be defined by some simple physiographic descriptors. In addition; (1) the basin had to be in close proximity to a number of precipitation gauges, (2) the periods of record of the hydrometric stations and the meteorological stations had to overlap to provide a reasonable number of rainfall events, and (3) the rating curves at the hydrometric stations had to be sufficiently stable and well defined to allow for a confident estimate of low flows.

The locations of the gauges associated with each of these catchments are shown in Figure 1. The area that encompasses these catchments is classified as an interior plain. The surficial material consists mainly of undulating hummocky moraines, glaciofluvial, glaciolacustrine, or eolian landforms. The average landscape exhibits very small slopes, with extensive relatively flat areas, ill defined drainage, and extensive areas of hummocky terrain with internal drainage. Three main different groups of soil are found in the region: brown, dark brown and black. Many of the physiographic and climatic characteristics in the region seem to correlate with the distribution of these soils. Extensive agricultural development has replaced much of the Aspen parkland. The western margin of the plains remains mostly natural, but grazing and cropland occur throughout the remainder of the study area.

The mean annual precipitation tends to increase dramatically from east to west as the elevation increases from the plains to the mountains. In winter (October to April) the plains receive less than 150 mm of precipitation. Summer precipitation totals are about twice the winter precipitation. Peak precipitation usually occurs in May and June (Laycock, 1957) when single event rainfalls approach 150 mm (Hudson, 1989). The actual evaporation is apparently greatest in the western margins of the plains where it is in excess of 500 mm per year. It is less than this in the warmer plains to the east because the seasonal precipitation limits the amount of evaporation which might occur.

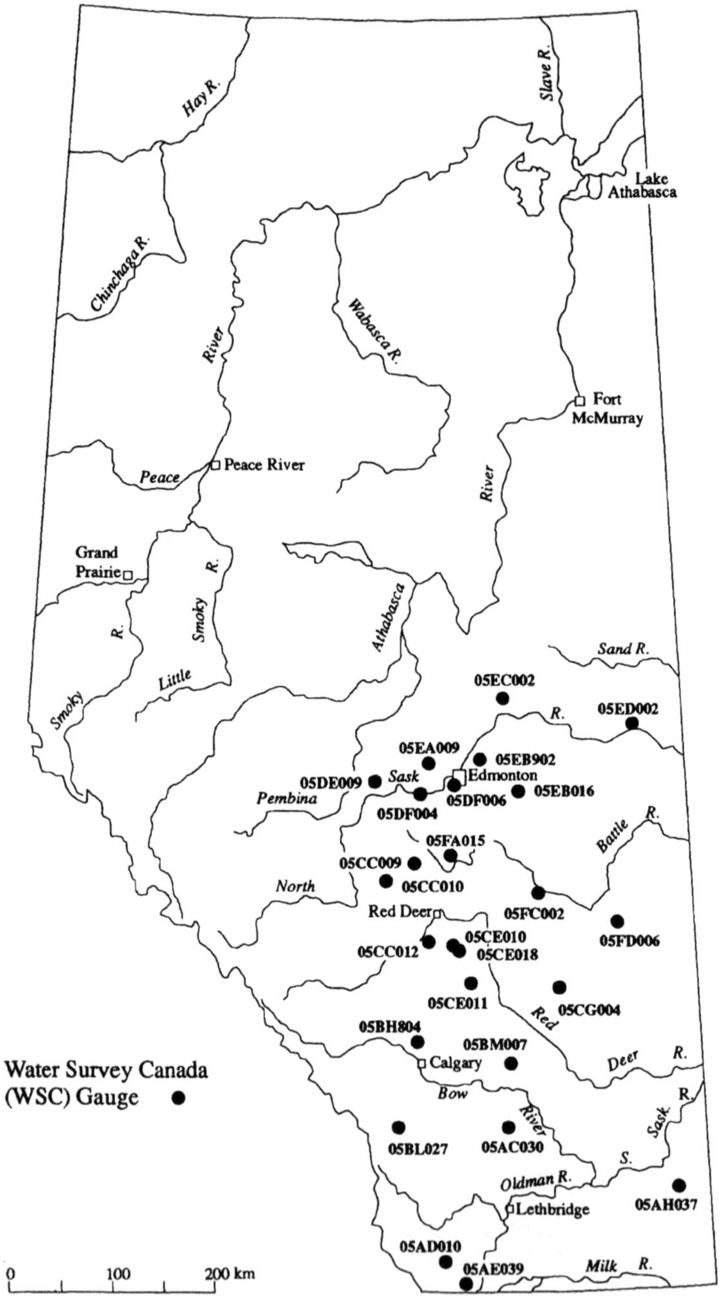

Figure 1. Study area and location of the WSC gauges representative of the basins.

Streamflow varies greatly within these regions. Water yields are greatest in the western and northern plains where precipitation is high and evaporation is low, and low in the eastern plains where precipitation is almost balanced by evaporation. Water yield ranges from over 300 mm per year in the west to less than 5 mm per year in the east.

In short, the hydrology of the region is such that the plains do not receive sufficient precipitation in a normal year to produce more than minor amounts of runoff, except in basins where the storage capacity is low. Runoff occurs during the snowmelt period and following intense or prolonged rains in spring. In the wetter years, runoff may be significant enough to produce floods in June or later in the summer.

DATA DEVELOPMENT

Precipitation

The distribution of rainfall depends on many local factors such as topography, elevation, proximity of forest covers, and lakes. Nevertheless, rainfall over an area can be considered to be normally distributed. Furthermore, as the number of measurement points increases, so does the reliability of the estimate of the precipitation amount. Thus, it is important to have as many measurement points as possible.

Various attempts have been made to develop techniques for estimating mean areal rainfall depth (Singh, 1989). Most of the recent techniques such as trend surface analysis, double Fourier series, analysis of variance etc. are complex and their application requires a certain amount of skill and judgement. In fact, regardless of the eloquence of the technique, the quality of the estimate is primarily a function of the quality of the data and the inherent diversity of the precipitation patterns. The simpler methods such as the isohyetal method or the Thiessen polygon method, are practically as good as the more complex ones (Singh, 1989).

In this study, the Thiessen polygon method was used for the computation of the average basin rainfall depth. The relevant precipitation stations were determined according to their proximity to the particular basin and whether or not they exhibited more or less similar rainfall patterns for any particular storm. It should be noted that due to the sparseness of the precipitation stations it was not possible to draw Thiessen polygons for all the basins. In some cases mean areal rainfall depths were calculated according to the distance weighted means of several gauges located close to the basin.

Runoff and Runoff Coefficients

The main difficulty in determining the runoff resulting from a particular rain event is the fact that the runoff lags the rain by at least several days, even for the smallest basins. A flood resulting from a particular rainstorm remains undistorted by runoff from previous (and subsequent) rainfall only if the interval between the rainfall events is greater than the flood time base. These cases are rare. In most instances it was necessary to use flood recession curves to analyse rainfall floods caused by a series of rainfall events.

A correlation diagram of discharges for adjacent time periods was used to express the recession curves. The recession curve representing the rainless period was determined from the hydrograph, it was transformed into a discharge correlation between adjoining time periods, and the envelop for a number of curves was adopted as the unified recession curve for the watershed. Then, the recession curve was transformed into a relationship between discharge and recession runoff volume by either a graphical or mathematical procedure.

From the watershed recession curve, the runoff volume for various floods was computed by plotting the average daily discharge and average daily rainfall as shown in Figure 2. The time of rise for each individual flood was determined by comparing the hydrograph with hyetograph and finally the runoff volumes of the various floods were computed. This complex procedure was not required for all the basins examined in the present study. For instance, the basins located in the southeastern region have very distinct and simple hydrographs.

Thus, runoff volumes from 450 events, covering the study region and over the period of record, were calculated. The runoff coefficients were obtained for each of these events by simply calculated the ratio of the volume (depth) of direct runoff to the volume (depth) of precipitation over the basin from the storm associated with the runoff event. These data are shown graphically in a variety of figures in the next section.

ANALYSIS

The runoff depths were plotted against the corresponding rainfall depths to provide an overview of the rainfall-runoff relationships. This is shown in Figure 3. The runoff from any particular rain event (5 to 220 mm) varies between zero and 64 mm. The upper bound of the data represents runoff occurring in basins with a significant antecedent moisture content. The slope of this line defines the maximum runoff coefficient under the prevailing hydrological, meteorological, and geophysical conditions. This suggests that the maximum percentage of precipitation that can run off is about 53%, regardless of the rain duration, intensity, or volume.

Considering rainfall and soil moisture as the prime factors governing runoff, two concepts for runoff have been developed in the literature. These are (1) runoff formation in excess of infiltration capacity (Horton, 1933) and (2) runoff generation related to the natural storage in the catchment, termed the variable contributing area concept (Hewlett and Hibbert, 1967). Infiltration excess runoff, also known as Hortonian runoff, is the runoff that occurs when rainfall exceeds the infiltration capacity. The soil at the lower part of the profile is always dry and the infiltration capacity is relatively low. It is almost impossible for most of the rainfall events to fill the profile to its field capacity. For small amounts of rainfall or in the time period shortly after the beginning of rainfall, the runoff coefficient, R/P, is close to zero. Both the runoff and the runoff coefficient increase with the amount of rainfall for situations with similar initial and final infiltration rates. For very large amounts of rainfall, the runoff coefficient tends to a constant value.

On the other hand in wet regions, the groundwater level is usually high because of an abundance of moisture. Moisture content in the lower part of the soil profile stays near the field capacity all year around, and only the upper part of the soil profile dries due to evaporation. Runoff begins immediately after the upper soil layer

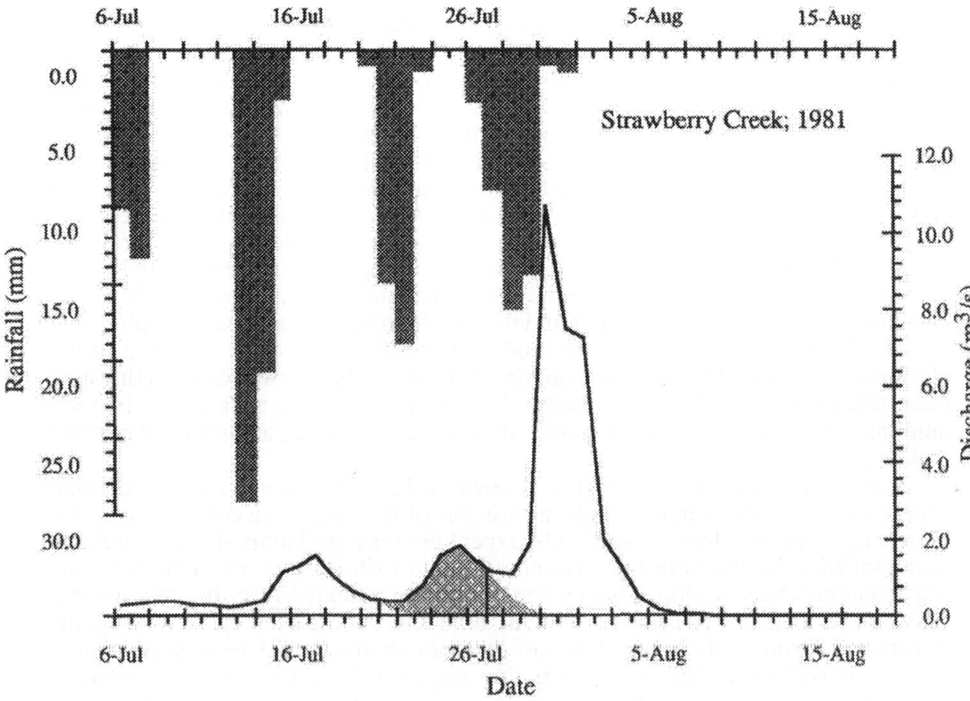

Figure 2. Representation of runoff volume corresponding to a single rainfall event.

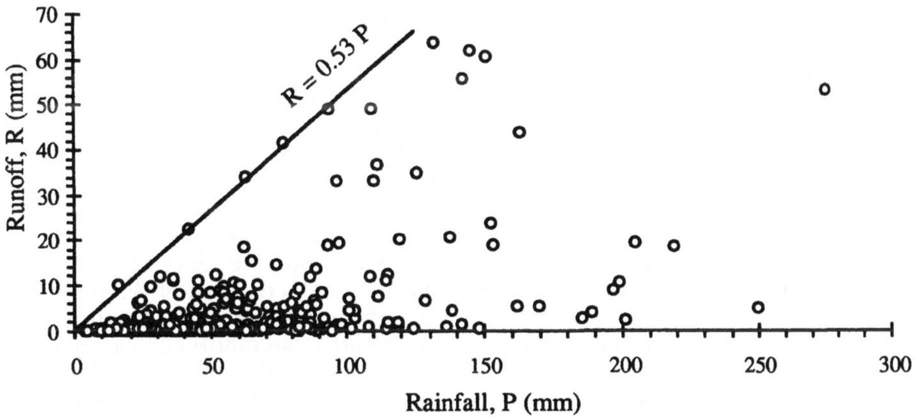

Figure 3. Relationship between the rainfall and runoff for all the basins.

becomes almost completely water saturated. During periods of high saturation, surface runoff may occur from rainfall of very low intensity. Due to the wide variation in groundwater levels and the hydrophysical properties and degree of saturation of the soil, the runoff process also develops nonuniformly. Runoff will begin at different times in different parts of the basin. When the amount of rainfall is small, runoff may be derived from only a very small part of the watershed. As rainfall amounts increase, more and more area contributes to runoff. For this condition the rate of change of runoff with precipitation is equal to the change in the fraction of the area from which runoff takes place. As the amount of rainfall increases, the contributing area increases, and a subsequent increase in rainfall gives rise to more and more intensive runoff. This phenomenon continues until the entire catchment area contributes to runoff and the runoff coefficient approaches unity.

The difference between the two modes of runoff formation is such that in the Hortonian case, rain intensity is the dominant factor. In the variable contributing area case, rain intensity plays a secondary role. The most important factor is the total amount of rainfall, to which is connected the process of complete saturation of the soil layer.

Sharp hydrograph peaks and the absence of long recessions suggest that the runoff formation mode in the south eastern part of the study area complies with the Hortonian concept. Hence, one should expect poorer correlation of runoff volume with rainfall volume for this region as compared to a similar plot for the more humid north and northeast region. This tendency can be evaluated from the relationship between runoff and rainfall volume in both the southeast (Figure 4) and the northwest (Figure 5). In Figure 4, runoff does not seem to correlate with the rainfall volume as well as it does in Figure 5. This suggests that the variable contributing area concept is probably more appropriate for the northwest as opposed to the drier southeast.

Figure 6 illustrates a graph of the runoff coefficients plotted against the rainfall depths. It is possible to draw an envelope curve for the entire study region. Furthermore, one can stratify the data so that a separate envelope curve can be drawn for a set of basins which are located in distinct climatic zones. These zones are differentiated by a type of climatic index defined as the ratio of the annual normal precipitation to the annual normal potential evapotranspiration. This is termed the wetness index, W, and is the inverse to the aridity index proposed by Bothe (1987).

The data suggest that there are at least three distinct zones within the prairie region. Each of these zones can be characterized by a distinct envelope curve which relates the maximum runoff coefficients to the precipitation amount. Curve 1 represents the envelope curve of the maximum runoff coefficients for watersheds with a wetness incex between 0.30-0.40. The runoff coefficients varies from almost zero to a maximum of 0.01. This subregion is located in the extreme southeast part of Alberta. Curve 2 represents the envelope curve of maximum runoff coefficients that could be expected due to rainfall in an area with a wetness index of 0.41-0.54. This subregion is located in the central southeast of Alberta, east of the foothills. For this subregion the runoff coefficient varies from near zero to a maximum of 0.25. Curve 3 represents the envelope curve of maximum runoff coefficients that could be expected for the watersheds with a wetness index of 0.55-0.75. This envelope curve represents the highest runoff coefficients (with the maximum value approaching 53%) for the entire plains area. The subregion extends along the

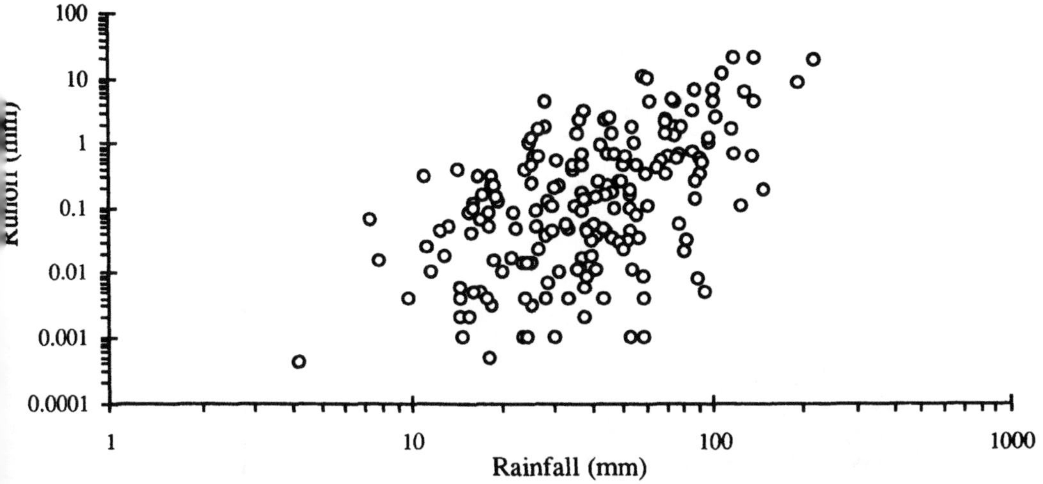

Figure 4. Rainfall-runoff relationship for basin in the southeastern part of Alberta.

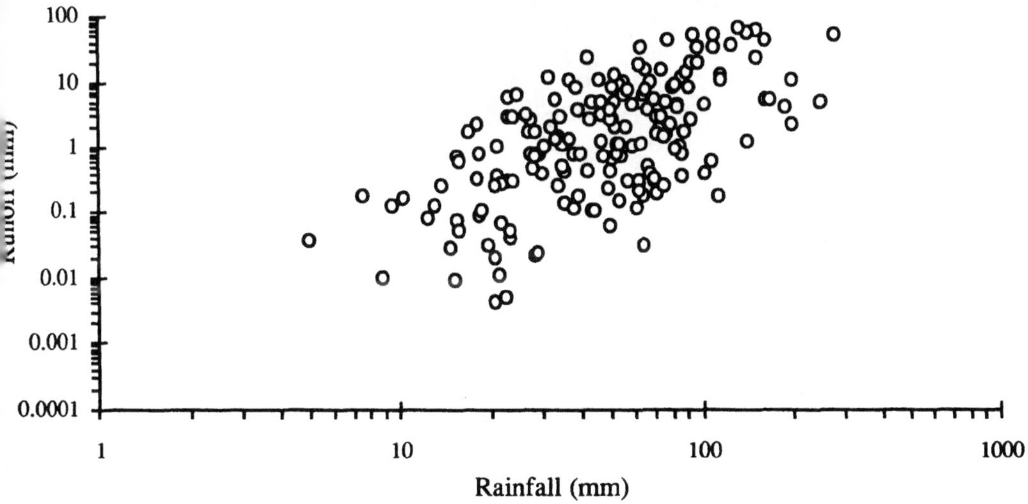

Figure 5. Rainfall-runoff relationship for basin in the western part of Alberta.

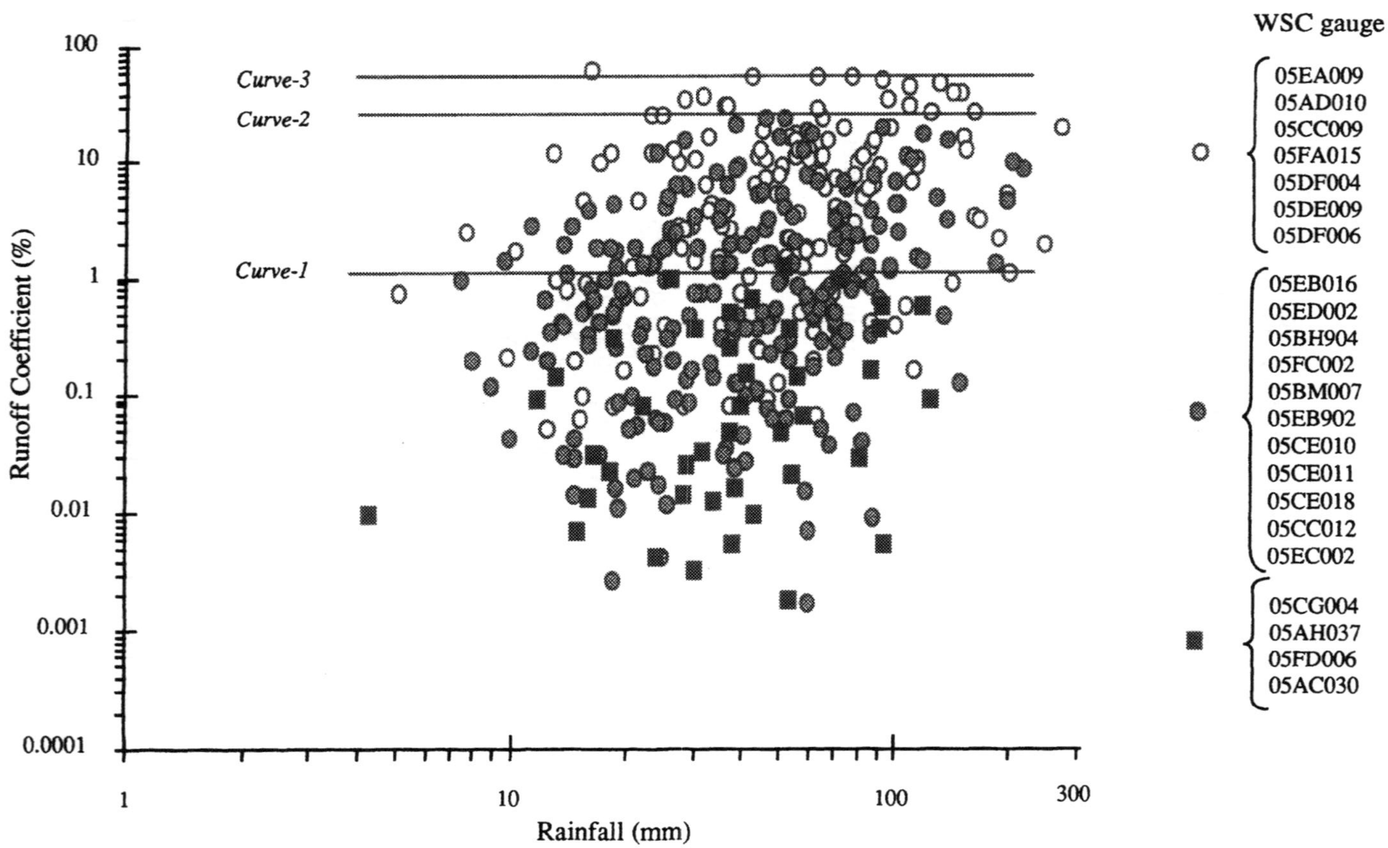

Figure 6. Envelop curves for maximum runoff coefficients.

eastern margin of the foothills and wastward across central Alberta. The three homogeneous subregions as defined by the envelope curves are shown in Figure 7.

CONCLUSIONS

This paper has provided a brief description of a hybrid approach that uses the basin runoff coefficient as the fundamental variable in determining runoff in ungauged, small prairie basins. A summary of the significant characteristics of 450 runoff events from 22 small prairie basins and a relationship between the upper limit of the runoff coefficients and the basin dryness index was also developed.

The analysis showed that the runoff coefficients varied between zero and 53% for precipitation events which ranged from 4 mm to 270 mm. There was a distinct upper limit to the runoff coefficients which varied from 1% in the southeast to 53% in the north and northwest. These upper limits were consistent with the relative aridity of the representative areas as defined by the ratio between the normal annual precipitation and potential evapotranspiration. This indicates that a design value of the runoff coefficient can be determined from regional analysis.

The data also suggest that it is inappropriate to characterize the runoff from the prairie basins as Hortonian. However, the scatter in the data also makes it difficult to explicitly define the runoff within the framework of the variable contributing area concept.

Efforts are currently being directed towards qualifying the response characteristics of these small prairie basins on the basis of some of their fundamental geomorphic characteristics. In addition, work is underway to evaluate the joint probability of the simultaneous occurrence of severe precipitation and high antecedent moisture conditions. This will lead to a determination of the frequency of the design flows determined by this hybrid approach.

ACKNOWLEDGEMENT

The report presented herein is part of a continuing research program to study hydrologic processes in Alberta. This program is carried out by the Environmental Research and Engineering Department of the Alberta Research Council under the auspices of the Alberta Cooperative Research Program in Transportation and Surface Water Engineering. Participants in the program are the Alberta Research Council and Alberta Transportation and Utilities.

The authors acknowledge the assistance of P. Mostert, J. Thompson, and B. Trevor for collecting and tabulating some of the data. Mr. Gerry Coles of the Survey Branch of Alberta Environment also provided some hydrologic data used in the analysis.

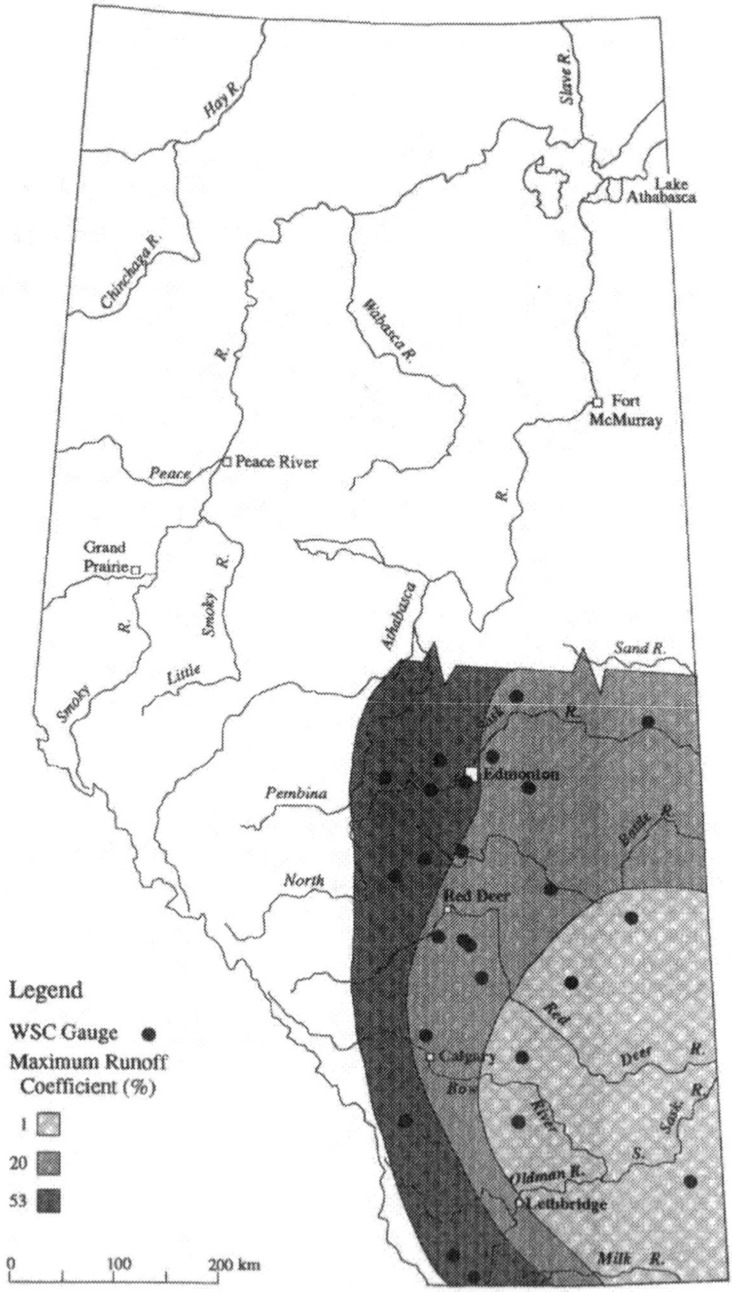

Figure 7. Regionalization of maximum runoff coefficients.

REFERENCES

Atmospheric Environment Service, (1990) "Rainfall Intensity, Duration, Frequency Values", Meteorology Division, Environment Canada.

Bothe, (1987) R.A., and Abraham, C., *"Evaporation and Evapotranspiration in Alberta"*, Hydrology Branch, Alberta Environment.

Hewlett, J.D., and Hibbert, A.R.(1967) "Factors Affecting the Response of Small Watersheds to Precipitation in Humid Areas", In: *International Symposium on Forest Hydrology* (Eds: W.E. Sopper and H.W. Lull), Pergamon Press, pp.275-290.

Horton, R.E. (1933) "The Role of Infiltration in the Hydrologic Cycle", *Transactions of the American Geophysical Union*, Vol.14, pp.446-460.

Hudson, H.R.(1989) *"Flood Regionalization: North Saskatchewan and Battle River Basins, Alberta"*, Alberta Research Council, Report SWE 89-07.

Laycock, A.H. (1957) "Precipitation and Streamflow in the Mountain and Foothill Region of the Saskatchewan River Basin", Prairie Provinces Water Board, Report #6.

Sabur, M.A. (1990) *"A Distributed Numerical Model for Watershed Hydrology"*, Ph.D. dissertation, Department of Civil Engineering, University of Alberta, Edmonton, Alberta.

Singh, V.P. (1989) *"Hydrologic Systems: Watershed Modeling"*, Vol.2, Prentice Hall, NJ.

Water Survey of Canada, (1990) *Historical Streamflow Summary of Alberta*, Environment Canada.

BASIN SCALE FOR DISTRIBUTED RUNOFF MODEL BASED ON CHANNEL NETWORK GEOMETRY

H. HAYAKAWA[1] and **M. FUJITA**[2]
[1]Department of Civil Engineering, Kitami Institute of Technology
Kitami, Hokkaido, 090, Japan
[2]Department of Civil Engineering, Hokkaido University
Sapporo, Hokkaido, 060, Japan

For prediction of the hydrologic response of ungauged basins and the changing response resulting from changes in land characteristics, physically based distributed runoff models are more advantageous than concentrated conceptual models. However, one of the still-unsolved problems in distributed models is the estimation of the appropriate subbasin scale. This paper focuses on how a single basin of moderate size is divided into sub-basins. This paper consists of two parts. One is to find an appropriate subbasin scale for distributed runoff models based on the hydrologic response of a basin by using quantitative analysis of the subbasin topography. The other is the application in runoff analysis of the theoretical generated channel network geometry. By this process, it is possible to generate a more complicated channel network from a skeleton one. The results of the study suggest that this proposed method is useful to estimate more accurately the runoff prediction of ungauged basins, which may be provided with a rough topographic map only.

INTRODUCTION

Recently, weather radars and geographic information systems (GIS) have been applied to runoff analysis. In Japan, many weather radars installed by the Ministry of Construction can yield rainfall informations with 3 (km)$\times$ 3 (km) mesh size. For making better use of these informations from weather radars, it may be more effective to divide the basin into reasonable small ones corresponding to a characteristic scale for rainfall-runoff phenomena. Accurate physically based distributed models will make an effective use of spatial variations in rainfall or other basin properties, such as topography, geology, vegetation and changing in land use.

The hydrologic response for a basin is broken down into a subbasin response and a channel network one. The scale of a subbasin or a subdivision of a basin must be considered as it affects both responses. Although the effects of a subbasin scale on the overall basin response have been discussed by Wood *et al.*(1988), few papers have discussed the appropriate subbasin scale considering the channel network geometry. Therefore, this paper discusses the appropriate subbasin scale of distributed runoff models based on the channel network geometry.

There are many models that have been proposed for relatively small catchments (SHE: Abbott *et al.*,1986; IHDM4: Beven *et al.*,1987; THALES: Grayson *et al.*,1992).

K. W. Hipel (ed.),
Stochastic and Statistical Methods in Hydrology and Environmental Engineering, Vol. 1, 355–368.

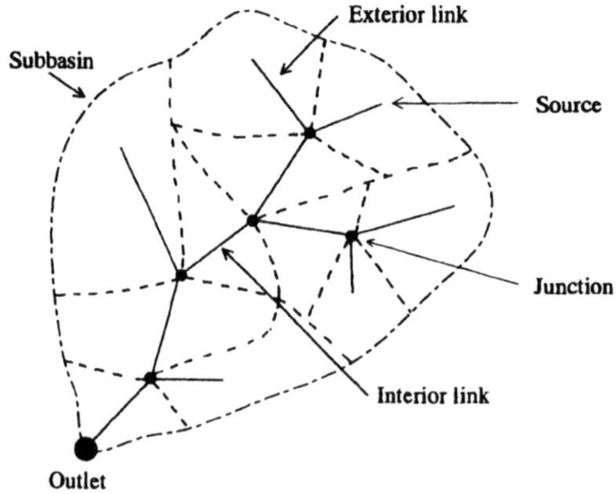

Figure 1. Schematic basin.

However, the calibration of physically based models is not a straightforward task due to the large number of model parameters involved and computational requirements of making multiple runs of such models. Taking into account these difficulties leads us to the establishment of a simplified distributed model that includes the merits of both the physically based distributed model and the concentrated conceptual model.

On the other hand, it is not so difficult to discuss the appropriate subbasin scale if detailed topographical maps or Digital Terrain Maps (DTM) of topography are on hand, although we are sometimes confronted with the difficulty of predicting runoff with no better than using a rough topographical map. This paper proposes the stochastic approach of estimating runoff by using poor topographical maps.

SIMPLIFIED DISTRIBUTED RUNOFF MODEL

A drainage basin is composed of a channel network and subbasins as shown in Figure 1, where a subbasin consists of one link and its associated hillslopes. Hillslopes are the main regions at which rainfall is converted into runoff, and a channel is the regions through which the runoff from hillslopes is transported to the outlet of subbasin. Channel network plays an important role in synthesizing and transporting the discharge from subbasins to the outlet. Therefore, the hydrologic response of a drainage basin is influenced not only by the subbasin response but also by the channel network response.

We consider that the following two requirements should be satisfied to develop the distributed models.

1. Entire catchment area is divided to many subbasins. Runoff models of single subbasin must be simple to save computation time. Runoff model which satisfies this requirement is a concentrated parameter model.

2. Each subbasin has different topography. Runoff model for each subbasin should be able to reflect the effects of different topography. It is considered that a runoff model which satisfies this requirement is a distributed parameter model.

The above two requirements are contradictory with each other. We employ the method which converts a distributed parameter runoff model to a concentrated one to avoid this contradiction. Consider the kinematic wave method as a distributed runoff model.

$$\frac{\partial h}{\partial t} + \frac{\partial q}{\partial x} = r \qquad 0 \le x \le \ell \tag{1}$$

$$q = \alpha h^m \tag{2}$$

where x is the flow distance, t time, h the flow depth, q the discharge per unit width, r the rainfall intensity, and ℓ hillslope length. After combining the Manning resistance equation with kinematic wave approximation, the following expressions for α and m are obtained:

$$\alpha = \sqrt{\sin\theta}/n \tag{3}$$

$$m = 5/3 \tag{4}$$

where θ is the hillslope angle and n is the Manning's roughness coefficient. Fujita(1980) derived the storage function method for rectangular hillslope plane from (1) and (2) as follows,

$$\frac{\partial S}{\partial t} + Q = r \tag{5}$$

$$S = KQ^P \tag{6}$$

$$K = \frac{m}{m+1}\left(\frac{1000^{m-2}\ell}{3.6\alpha}\right)^P, \qquad P = 1/m \tag{7}$$

where S(mm) is storage, Q(mm/hr) discharge, r(mm/hr) rainfall intensity, and K and P parameters of storage function method. Thus, parameter K is determined by the hillslope length ℓ(m), hillslope angle and Manning's roughness coefficient.

On the other hand, the channel flow routine for the channel network is composed under certain assumptions.

1. Channels play only a role in delaying the time that hydrograph from subbasins is propagated at end of channel; that is, lag time is $t_\ell = L_i/U_i$, where U_i is the propagation velocity of flood wave and L_i is the channel length.

2. Deformation of the hydrograph along the channel can be neglected.

3. The hydrograph is synthesized at each junction by the sum of the converging subbasins' hydrographs.

In this model, it is important to estimate the propagation velocity of a flood wave U_i(m/s). The empirical relationship between U_i and inflow peak discharge Q_{p_i}(m^3/s) is proposed by Yamabe (1971) as follow (8);

$$U_i = \frac{5}{3}\left(0.62 Q_{p_i}^{0.11}\right) \tag{8}$$

Therefore, it is possible to discuss the effects of changing subbasin scale because this model is the simplified distributed model based on subbasin topography and channel network geometry.

CHANNEL NETWORK GEOMORPHOLOGY

Since above runoff model can be used to estimate the effects of subbasin scale for hydrologic response, we focus on the relationship between subbasin scale and channel network geometry. Figures 2 and 3 show a channel network of the Shokotsu River at Takinoue and the Tatsuushi River, which is a right tributary of the Shokotsu River in Hokkaido, Japan, and each catchment area is about 237(km^2) and 160(km^2), respectively. These channel networks are extracted from 250(m) grid-based Digital Elevations Models(DEMs) of the Digital National Land Information, which is provided by the Ministry of Construction, using Michiguchi method(Michiguchi *et al.*,1990). This method creates a drainage network by setting some threshold (TH), which means a critical support area that defines the minimum drainage area required to a source of channel. The value of the threshold defines the resolution of obtained channel network. The larger the threshold is, the rougher the obtained channel network is; that is, the threshold corresponds to the topographical map scale. Although these

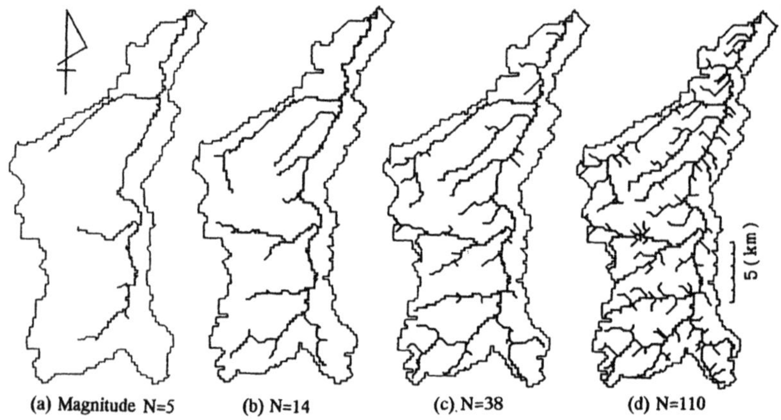

Figure 2. Channel network from DEM at the Shokotsu River
(a) Threshold(TH)=300, (b) TH=70, (c) TH=20 and (d) TH=8.

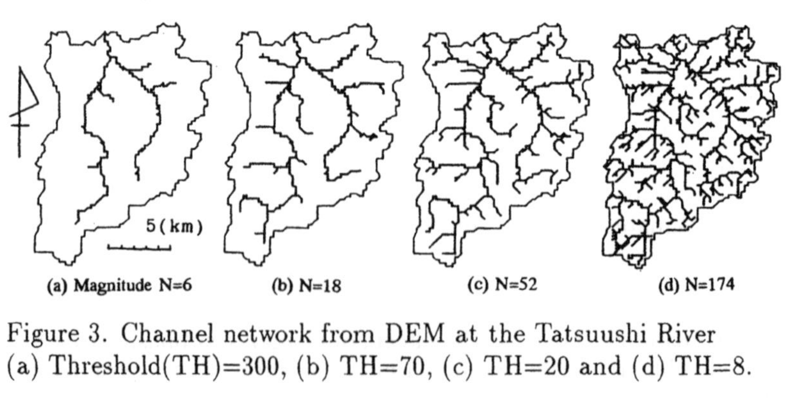

Figure 3. Channel network from DEM at the Tatsuushi River
(a) Threshold(TH)=300, (b) TH=70, (c) TH=20 and (d) TH=8.

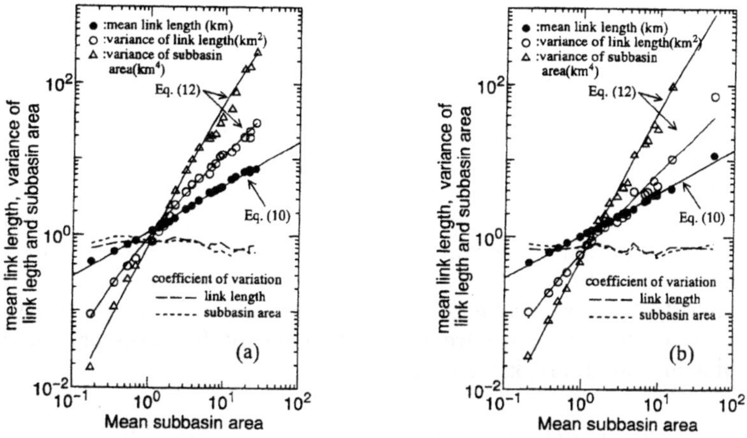

Figure 4. Relationship among mean link length, variance of link length and subbasin area, and mean subbasin area; (a) the Shokotsu River and (b) the Tatsuushi River.

channel networks depend on the threshold, the main structure of channel network remains same.

Next, Figures 4 (a) and (b) indicate properties of subbasin topography against mean subbasin area above both river basins, respectively. Mean link length ($\bar{L}$) is expressed in (9) by combining *Hack's equation* (Hack, 1956) and the expected number of tributaries flowing into mainstream, $\bar{M}_N$, by using topologically distinct channel network(TDCN) (Fujita, 1976).

$$\bar{L} = \frac{1.273 A_l^{0.6}}{\bar{M}_N + 1} = \frac{1.273 \bar{A}^{0.6} (2N-1)^{0.6}}{\bar{M}_N + 1} \tag{9}$$

$$= 1.004 \bar{A}^{0.6}, \quad N < 80, \quad \bar{L} : km, \bar{A} : km^2 \tag{10}$$

where $\bar{A}$ denotes mean subbasin area and N is the magnitude or the number of exterior links. Equation (10) is derived from the relationship $(\bar{M}_N + 1) \sim (2N - 1)$(Hayakawa *et al.*, 1990). By using the least squares regression for the Shokotsu River and the Tatsuushi River, the intercept and exponent of (10) are 1.055 and 0.560, and 1.114 and 0.599, respectively. Equation (9) or (10) indicates good agreement with simulated channel network from DEM.

Variance of link length and subbasin area against magnitude or mean subbasin area are derived by Hayakawa *et al.* (1991) as follows.

$$\sigma_{N_1}^2 = \sigma_N^2 \left(\frac{2N-1}{2N_1-1}\right) \left\{1 + \left(\frac{2N-1}{2N_1-1} - 1\right)\gamma\right\} \tag{11}$$

$$= \sigma_N^2 \left(\frac{\bar{A}_{N_1}}{\bar{A}_N}\right) \left\{1 + \left(\frac{\bar{A}_{N_1}}{\bar{A}_N} - 1\right)\gamma\right\}, \qquad N > N_1 \tag{12}$$

where σ_N^2 and $\sigma_{N_1}^2$ are the variances of subbasin properties such as link length and subbasin area, $\bar{A}_N$ and $\bar{A}_{N_1}$ are mean subbasin areas at the magnitude N and N_1, respectively. γ is the coefficient of the correlation between subbasins and reflects the characteristics of topographic feature. Figure 4 shows that Equation (12) agrees well with the results from the simulated channel network from DEM. By analyzing another basins, average value of γ is 0.7 for subbasin area and is 0.0 for link length, respectively. It has been proposed that the frequency distributions of link lengths and subbasin areas can be closely approximated by either a log normal or a gamma probability density function. We examined link length and subbasin area belong to a gamma distribution. As the shape and scale parameters of the gamma density, κ and η are related to the mean value($\bar{X}$) and the variance (σ_X^2) as it is written in (13) respectively, we can determine the frequency distribution for arbitrary magnitude or arbitrary subbasin scale (Hayakawa *et al.*, 1992).

$$\begin{aligned} g(X;\kappa,\eta) &= \eta^{\kappa}X^{\kappa-1}e^{-\eta X}/\Gamma(\kappa) \qquad \kappa,\eta>0 \\ \kappa &= \bar{X}^2/\sigma_X^2, \quad \eta=\bar{X}/\sigma_X^2 \end{aligned} \tag{13}$$

where variable X is subbasin topographic property, such as link length or subbasin area.

EFFECTS OF SUBBASIN SCALE FOR HYDROLOGIC RESPONSE

First of all, we applied this proposed model to some real basins by using simulated channel network from DEM at arbitrary threshold, and examined the sensitivity test of this model parameters. As shown in (7), storage coefficient K is a function of the hillslope length, the hillslope angle and Manning's roughness coefficient. Although the hillslope length and angle can be measured from the topographical maps, it is not easy to estimate Manning's roughness coefficient because surface-subsurface conditions of hillslopes affect it. The value of K is closely related to hydrologic response, the smaller K is, the more rapid runoff response is. Thus, it is possible to evaluate the effects of subbasin scale through K.

Now, assume that Manning's roughness coefficient is constant under the homogeneous surface condition in the whole basin. Figure 5 shows the observed and calculated direct runoff based on the simulated channel network from DEM at an arbitrary threshold. Observed direct runoff was separated by the variable slope method, which

TABLE 1. Outline of floods for runoff analysis

Basin (km^2)	Flood	Date	Total rainfall (mm)	Retention (mm)
Shokotsu R. ($236.7km^2$)	S1	Aug. 17, 1973	131.5	88.7
	S2	Aug. 22, 1975	135.3	90.7
Tatsuushi R. ($160.7km^2$)	T1	July 21, 1981	84.0	46.0
	T2	Aug. 5, 1981	114.0	38.0

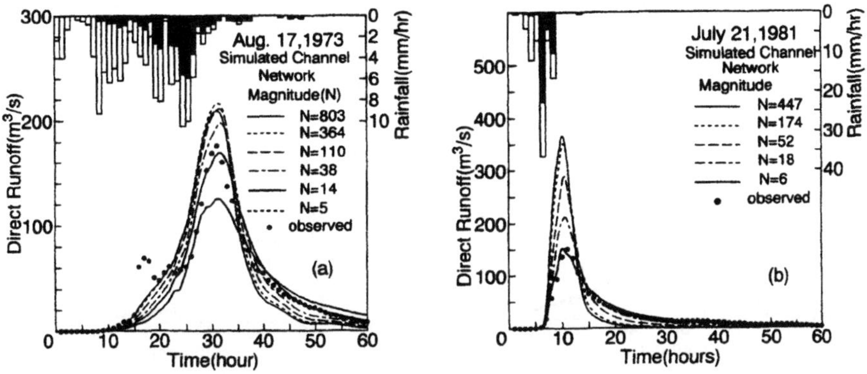

Figure 5. Calculated and observed hydrographs; (a) the Shokotsu River and (b) the Tatsuushi River, where n is constant $n = 1.0$.

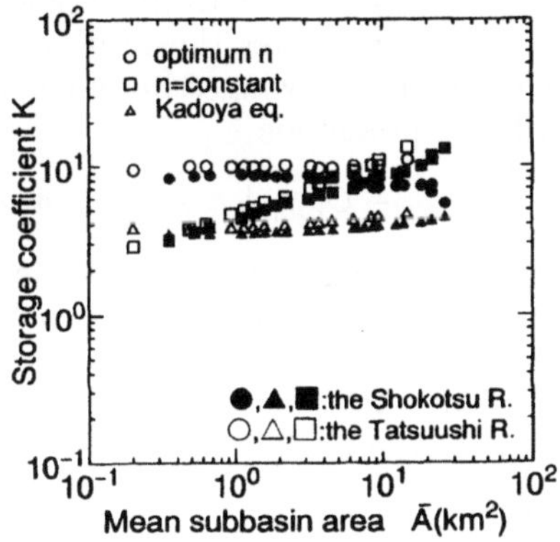

Figure 6. Relationship between Storage coefficient K and mean subbasin area $\bar{A}$ against Manning's roughness coefficient n; (a) n is constant, (b) optimum value of n and (c) empirical value of n (Kadoya *et al.*, 1978).

is one of baseflow separation techniques (V. T. Chow *et al.*, 1988), and effective rainfall was computed by the relation between retention and cumulated rainfall. Details of flood events used here are listed at Table 1. As the magnitude of channel network was increasing, the subbasin scale became smaller and the peak discharge of the hydrograph increased gradually. Finally the calculated hydrograph gradually converged critical one. The relationship between K and mean subbasin area is shown in Figure 6 for both basins. For smaller mean subbasin area, the value of K decreased gradually and approached a constant value. Generally speaking, the hydrologic response was depended on the variation of the subbasin scale under a constant Manning's roughness coefficient.

Next, there are some approaches of calibrating parameters no matter whether we can obtain abundant hydrologic data or not. As an ideal case, prepare detailed topographical maps or DTM and abundant rainfall-runoff data, the calibration of

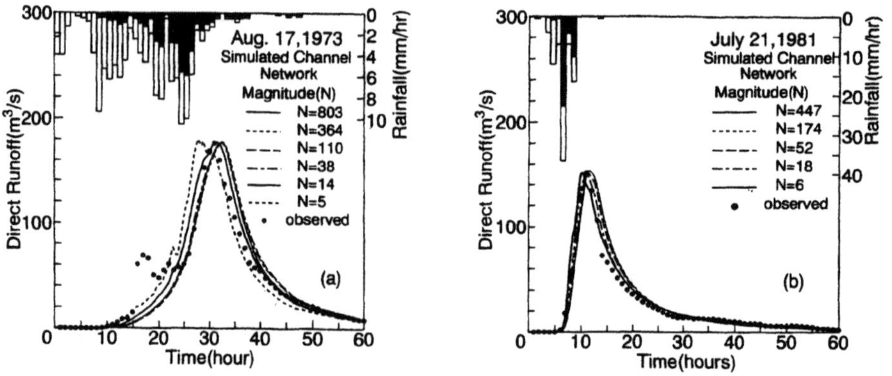

Figure 7. Calibrated and observed hydrographs; (a) the Shokotsu River and (b) the Tatsuushi River.

Manning's roughness coefficient becomes relatively simple and accurate runoff prediction can be obtained. Figure 7 indicates the results of calibrated direct runoff. This calibration was carried out as following criterion because of practically great demand for exact peak discharge prediction.

$$J = | Q_{ob,p} - Q_{cal,p} | \quad \longrightarrow \quad min \tag{14}$$

where $Q_{ob,p}$ and $Q_{cal,p}$ are the observed and calculated peak discharge, respectively. As shown in Figure 7, peak time of hydrographs is inclined to become faster by division of subbasins. The value of K showed just little fluctuation against subbasin scale as shown in Figure 6. Calibrated values of Manning's roughness coefficient for flood S1 and T1 were expressed as the following relation to hillslope length.

$$n = 27950\ \bar{\ell}^{-1.482} \quad \text{, for S1,} \quad \bar{\ell} : \text{mean link length(m)} \tag{15}$$

$$n = 7544\ \bar{\ell}^{-1.196} \quad \text{, for T1} \tag{16}$$

There is a well-known empirical equation in Japan proposed by Kadoya *et al.* (1978), and adopted into relatively small catchments relating n and ℓ_c:

$$n = 530\ \ell_c^{-1.08}, \quad \ell_c : \text{ concentration hillslope length(m)} \tag{17}$$

Figure 8 illustrates the calculated direct runoff from (17). The calculated hydrographs are greater than the observed in this flood event. From the relationship between K and $\bar{A}$ in Figure 6, it could be noticed that storage coefficient K was either a constant or values with slight fluctuating. Therefore, it is expected that the calculated hydrographs from both empirical equations are little dependent on subbasin scale.

For runoff prediction of the neighboring ungauged basins, this calibrated parameters are effective and can be used to estimate the storage coefficient for mean subbasin area only because the hillslope length and the angle are pretty well related with the subbasin area.

$$K = 8.64A^{-0.065} \quad \text{, for flood S1} \tag{18}$$

Figure 9 (a) is an example of the calculated hydrographs by applying (15) into flood S2 and (b) by applying (15) into flood T1. The difference between the calculated and observed results can be reduced by dividing the subbasin into more smaller subbasins even if the optimum value of model parameters could not be prepared.

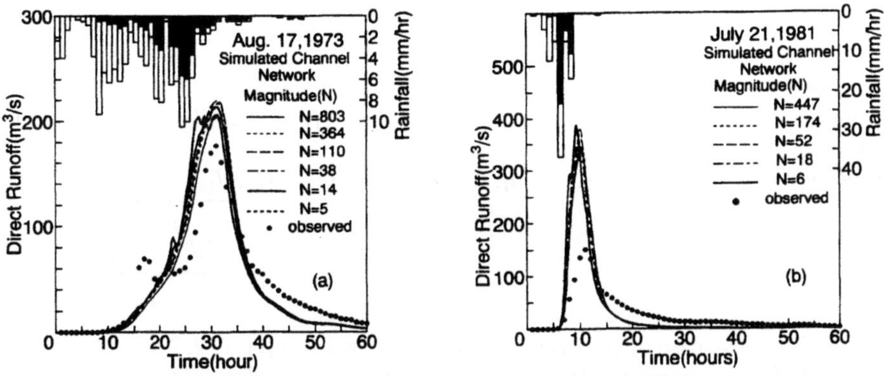

Figure 8. Calculated and observed hydrographs; (a) the Shokotsu River and (b) the Tatsuushi River, where Manning's roughness coefficient is given by empirical equation (Kadoya *et al.*, 1978).

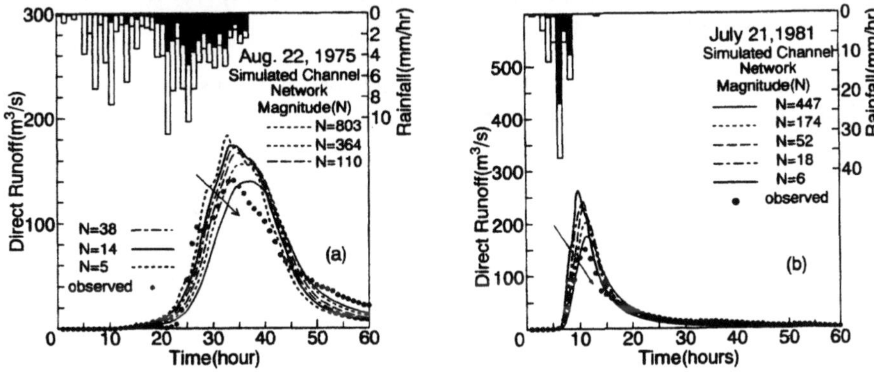

Figure 9. Examples of calculated hydrographs by (15); (a) the Shokotsu River in flood S2 and (b) the Tatsuushi River in flood T1.

SIMPLE GENERATION PROCESS OF CHANNEL NETWORK

It is not so difficult to discuss the appropriate subbasin scale if detailed topographical maps or DTM are on hand, and we are sometimes confronted with the difficulties to obtain such data. As shown in Figures 2 and 3, it is clear that the channel network geometry becomes more complicated as the threshold is decreasing. Let's consider a channel network with magnitude N. It is convenient to express channel network pattern by a sequence of N numbers, $C_N(J_1, J_2, \ldots, J_i, \ldots)$, where J_i denotes the number of links at equidistant positions and i-th links from outlet, as shown in Figure 10 (Werner *et al.*, 1973). Now, assume that every links generate j links due to scale-up of map under the condition of homogeneous characteristics of basin, the new channel network pattern can be shown as follows;

	Original Channel Network	New Channel Network
Magnitude	N	$\Rightarrow N' = N + j(2N-1)$
Total number of links	$2N-1$	$\Rightarrow 2N'-1 = (2j+1)(2N-1)$
Channel network pattern	$C_N(J_1, J_2, \ldots, J_i, \ldots)$	$\Rightarrow C_{N'}(J_1, \underbrace{2J_1, \ldots, 2J_1}_{j}, \ldots \ldots\ldots, J_i, \underbrace{2J_i, \ldots, 2J_i}_{j}, \ldots)$ (19)

For example, in the case of the original channel network $C_4(1,2,2,2)$ and $C_4(1,2,4,0)$, the generated new channel network pattern with $j = 1$ is denoted as follows.

$$C_4(1,2,2,2) \Rightarrow C_{11}(1,2,2,4,2,4,2,4) \Rightarrow C_{32}(1,2,2,4,2,4,4,8,2,4,4,8,2,4,4,8) \Rightarrow C_{95}(1,2,2,4, \ldots\ldots\cdot) \Rightarrow, \ldots\ldots \quad (20)$$

$$C_4(1,2,4,0) \Rightarrow C_{11}(1,2,2,4,4,8,0,0) \Rightarrow C_{32}(1,2,2,4,2,4,4,8,4,8,8,16,0,0,0,0) \Rightarrow C_{95}(1,2,2,4, \ldots\ldots\cdot) \Rightarrow, \ldots\ldots \quad (21)$$

Figure 11 shows an example of new channel network patterns with $j = 1$ in which stem from the original channel network $C_4(1,2,2,2)$ and $C_4(1,2,4,0)$.

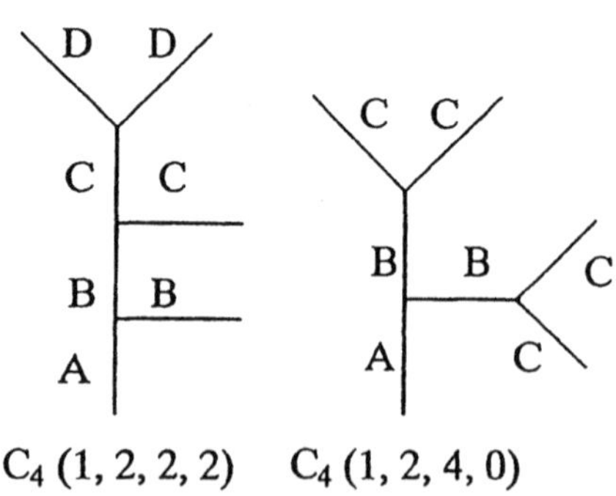

Figure 10. Schematic network in the case of magnitude 4.

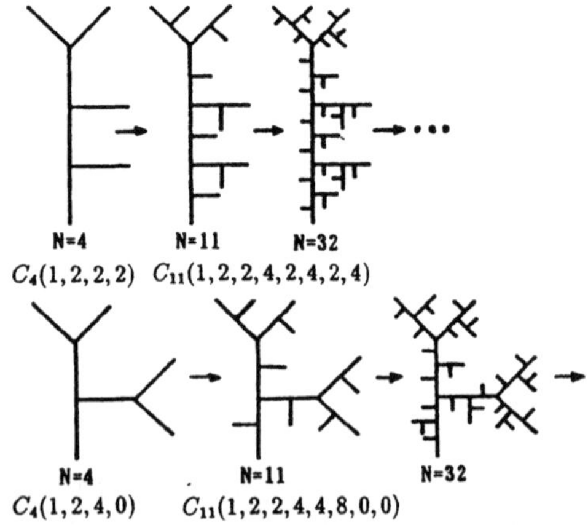

Figure 11. Process of generated channel network geometry.

APPLICATION OF GENERATED CHANNEL NETWORK GEOMETRY

The above process is applied to the runoff prediction of real basins with poor topographic data. At first, let consider the original channel network in the Shokotsu River, which has a magnitude of 5 as shown in Figure 2. Assume that a basin consists of a channel network with equivalent link length and equivalent subbasin area and new one channel ($j = 1$) stems from the center of the original link and the original associated area is divided into three equal parts. Here, this generating process repeats for 4 times. Manning's roughness coefficient n is calculated through flood S1 by using (15), and the result is applied to runoff analysis for flood S1 and S2. Figure 12 indicates the obtained result, which is compared with observed data for flood S1 and S2. There is a little difference between the calculated and the observed data even if the basin is divided into the smaller subbasins. The effect of subbasins scale for hydrologic response is not clear. The difference of the observed and calculated results is due the variations in the subbasins topography being not considered in this case.

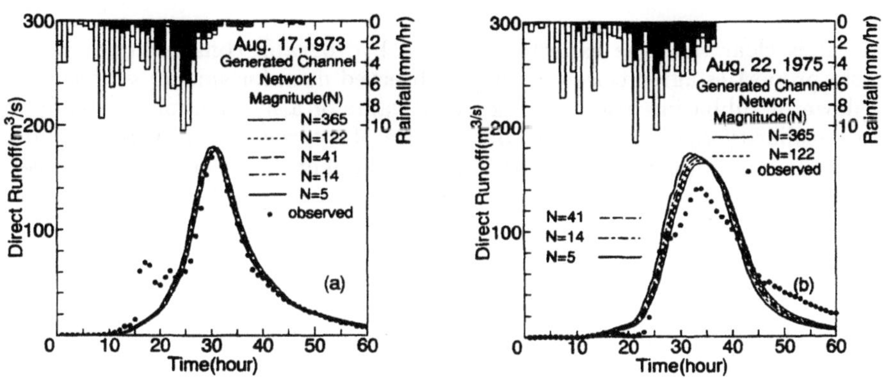

Figure 12. Calculated and observed hydrographs based on the generated channel network with equivalent link length and equivalent subbasin area at the Shokotsu River; (a) flood S1 and (b) flood S2.

Next, consider the variations in the subbasin topography for this runoff analysis. As discussed above, the variations of the link length and the subbasin area obey a gamma distribution, and the relationship between subbasins scale and the properties of subbasin topography are prepared in (9) $\sim$ (13). The coefficient of variation is indicated in Figure 4. The relationships are difficult to study by theoretical approaches, so that a simulation method was adopted here. In practice, one simulated channel network is combined with generated channel network, and simulated link length and subbasin area which are produced by using random numbers of a gamma distribution. Since the link length and the subbasin area are correlated as (9), it can be adopted depending on the link length and the subbasin area. The results which are obtained by having the variations of subbasin properties be considered are illustrated in Figure 13. This figure illustrates the expected value and the variance of calculated hydro-

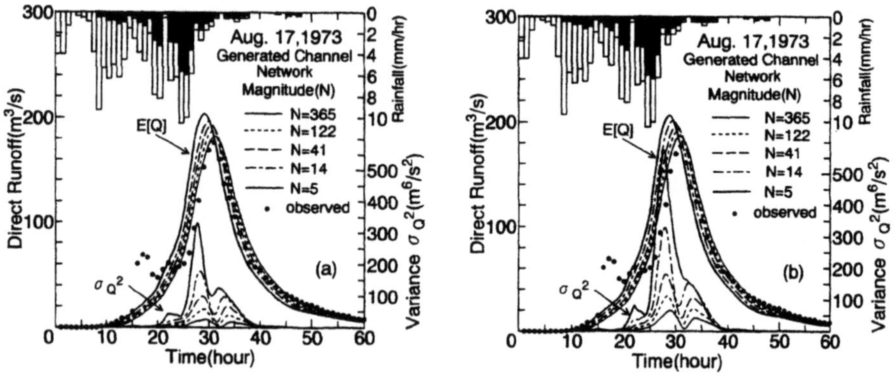

Figure 13. The expected value and variance of calculated hydrograph based on the generated channel network with considering of variation of subbasin topographic property at the Shokotsu River; (a) coefficient of variation of subbasin area $(\rho) = 0.8$ and (b) $\rho = 1.2$.

graph. It is clear that the expected value of hydrograph considering variations of subbasin properties agree closely with the observed data for smaller subbasin scales. Scale-down of subbasins brings the expected value of a hydrograph to steady state. It is a peculiarity that peak values of variance of hydrograph occurred in with the rising and falling limbs of the hydrograph. Consequently, the difference between the observed and the calculated hydrograph using equivalent link length and subbasin area, can be minimized by considering variations in subbasin topography.

CONCLUSIONS

This paper has discussed the effects of subbasin scale for distributed runoff models. A distributed runoff model was established based on the channel network geometry. By using the simulated channel network of a river valley from DEM, the relationship between subbasin scale and hydrologic response with changing Manning's roughness coefficient can be made clear.

In addition, this paper has proposed a new method to estimate runoff accurately with poor topographic data. The process of generated channel network geometry is derived and applied into runoff analysis for real basins. The results of this approach is summarized as the following;

1. It is possible to predict runoff with mean values of subbasin topographic properties but with a slight overestimation.

2. The differences between the observed and calculated data can be compensated by considering variations in the subbasin topographic properties.

3. Effects based on subbasin scale on hydrologic response provide results similar to those of calculations based on channel networks from DEM.

Consequently, this method appears to be valid for runoff analysis of basins where topographic and hydrologic data are insufficient. For more accurate runoff prediction, it should be better to divide the subbasin into more smaller subbasins based on the process of generated channel network geometry.

REFERENCES

Abbott, M. B., Bathurst, J. C., Cunge, J. A., O'Connell, P. E. and Rasmussen, J. (1986) " An introduction to the European Hydrological System—Systéme Hydrologique Européen,"SHE",1: History and philosophy of a physically-based, distributed modelling system ",J. of Hydrology, 87: 45-59.

Abbott, M. B., Bathurst, J. C., Cunge, J. A., O'Connell, P. E. and Rasmussen, J. (1986) " An introduction to the European Hydrological System—Systéme Hydrologique Européen,"SHE",2: Structure of a physically-based, distributed modelling system ",J. of Hydrology, 87: 61-77.

Beven, K. J., Calver, A. and Morris, E. M. (1987) " The Institute of Hydrology distributed model ", Rep. 81, Inst. of Hydrology, Wallingford, Oxon, England.

Chow, V. T., Maidment, D. R. and Mays, L. W. (1988) Applied Hydrology, McGraw-Hill, New York.

Fujita, M. (1976) " Study on characteristics of tributary distribution in channel networks ", Proceedings. of JSCE, 246:35-45 (in Japanese).

Fujita, M. (1980) " The impact of variations of slope length on storage-discharge real relationships ",Proceedings. of JSCE, 314: 75-86 (in Japanese).

Grayson, R. B., Moore, I. D. and McMahon, T. A. (1992) " Physically based hydrologic modeling 1. A terrain-based model for investigative purposes ", Water Resour.,Res.,28: 2639-2658.

Hack, J. T. (1956) " Studies of longitudinal stream profiles in Virginia and Maryland ", U. S. Geol. Surv. Prof. Pap., 294-B.

Hayakawa, H. and Fujita, M. (1990) " Study on hydrologic response via channel network geomorphology ", Hydraulic Engrg. Proceedings of the 1990 National Conference, HY Div/ASCE, 239-244.

Hayakawa, H. and Fujita, M. (1991) " Runoff characteristics based on channel network structure ", Proceedings of Hydraulic Engrg., JSCE, 35: 155-160 (in Japanese).

Hayakawa, H. and Fujita, M. (1992) " Development of distributed runoff modeling based on channel network property ", Proceedings of the Sixth IAHR Inter. Sympo. on Stochastic Hydraulics, Taipei, 671-678.

Kadoya, M., Fukushima, A. and Sagoo, J. (1981) " Modeling of natural hilly watershed and optimum parameters of kinematic flood runoff model ", Annuals of Disaster Prevention Res. Inst. of Kyoto Univ., 21(B-2):1-15 (in Japanese).

Michiguchi, T. and Fujita, M. (1990) " Simulation technique of river networks based on National Land Digital Information ", the Proceeding of JSCE Hokkaido Branch, 46: 275-280 (in Japanese).

Tateya, K. and Fujita, M. (1989) "Study on the relation of storage-discharge model to channel network characteristics ", Journal of Japan Society of Hydrology and Water Resources, 2(2): 41-48 (in Japanese).

Yamabe, K. (1971) " Study on the hydraulic geometry of stream channel ", Water Temperature Research, Res. Inst. for Water Temperature, 15(2): 31-37 (in

Japanese).
Werner, C. and Smart, J. S. (1973) " Some new methods of topologic classification of channel networks ", Geographic Analysis, 5:271-295.
Wood, E. F., Sivapalan, M., Beven, K. and Band, L. (1988) " Effects of spatial variability and scale with implications to hydrologic modeling ", J. of Hydrology, 102: 29-47.

PART VII

ENTROPY AND FRACTAL ANALYSIS

AGGREGATING HYDROLOGICAL DATA FOR RISK ANALYSIS

N.C. LIND and V. SOLANA
Institute for Mathematics and Fundamental Physics.
National Research Council of Spain, CSIC. Serrano 123, Madrid, E-28006 Spain

A method to estimate hydrological extremes is proposed, relaxing the usual assumption that the distribution type is knowable. The weaker assumption is made that physically related distributions are affine, different only in scale and location. This permits aggregation of data from different series of observations and yields a more accurate definition of the distribution tail. Entropy-based estimation, used with quantile constraints, facilitates the use of historical information and has many other advantages, including invariance under strictly monotonic transformations, invariance under conditioning of the variables and invariance under equivalent ways of updating. The method of Relative Entropy with Fractile constraints (REF) encodes the data as quantile constraints (rather than as moments) and minimizes the relative entropy between a reference distribution and the posterior distribution. For illustration the annual maximum discharge in a river is analyzed, using short series of observations of the flow in four unregulated rivers in the Bay of Biscay watershed of the Iberian Peninsula. Robustness against assumptions of reference distribution type is shown. It is concluded that aggregation and REF modelling can offer important advantages for practical engineering hydrology.

INTRODUCTION

Civil engineers must design specifically with a view to rare natural events. Dams, dikes and bridges over rivers, in particular, must function under highly uncertain conditions of hydraulic demand. Analysis procedures borrowed from the empirical sciences may be expedient but have no empirical justification. In particular, events of high return periods do not admit of any frequentist interpretation; probabilities are quantities that support decisions that aim to maximize expected utility. "Estimation" in this context means modelling.

In conventional parametric statistics the first step is to assume a distribution type. If a particular distribution type has been found generally to fit well to available data, this type is accepted to model other streams, and for return periods well beyond observation. For rare natural phenomena this has no scientific basis. Data are scarce and distribution shapes are unknown. Conventional statistics methods are inappropriate for the analysis of engineered system that must be highly reliable. This "tail sensivity problem" has commanded attention in civil engineering reliability research for decades. Many methods are now available to help cope with the uncertainties of rare combinations of high demand and low capacity of systems that are free of gross error (Madsen *et al.* 1986). However, these methods are either approximations or they assume that the distributions are known. Much work is still required to develop adequate ways to deal with the uncertainty arising from scarcity of data and the fact that the probability distributions are always unknown, indeed unknowable.

K. W. Hipel (ed.),
Stochastic and Statistical Methods in Hydrology and Environmental Engineering, Vol. 1, 371–380.

The distribution function of random variable in the real world is unknown (indeed, it doesn't even exist). This fact is important when studying extremes in nature. Then it may be natural to fall back to the position that the distribution function model is of the same type for sufficiently similar variables. Families of variables having distributions of unknown shape but could not reasonably be presumed not be similar are common in engineering practice. Examples are the strength of concrete made with one kind of cement but differing in the water/cement ratio, wind speeds at neighboring locations in a geographically homogeneous region, or wind set-up and wave heights on a coastline. Such *serial data* are considered in this paper.

In studies of serial data is usual to assume common distribution type as known, but to analyze each data set in insolation. This paper considers instead a collective analysis of physically related families of data, using the weaker assumptions that all distributions within a family are *affine*, i.e. similar except for location and scale. The distribution type is assumed unknown; it is assumed only to be the same for all series in the family. This allows aggregation of data from different test series, and leads to a more accurate definition of the distribution tail. In turn, this allows a surer determination of extreme quantiles.

If the distribution in a family are affine, they are all related to one distribution prototype by a linear mapping. Many different statistics could be related between family members and the prototype. Many statistical procedures are available to do the required mapping (eg. NERC 1975, USWCR 1982, Hosking and Wallis 1986, 1988). This task requires robust methods. Mean and bias-corrected standard deviation are usually good statistics for the purpose.

When there is a reason to presume that the distributions are mixtures, as is true of the precipitation in an area subject to temperate rains and occasional tropical storms, then it is better only to assume affinity in the high tail. To ensure an accurate mapping into the prototype the first moment of this region of interest could be used instead of the standard deviation.

The purpose of this paper is to illustrate aggregation and entropy-based estimation (i.e., probabilistic modelling) of related hydrological data. The method of relative entropy with fractile constraints (REF) is adapted to improve the estimation of probabilities of rare events. A method is proposed to estimate hydrological extremes in a region from observations and historical information.

Regional hydrological estimation procedures that employ systematic records have been studied by many (NERC 1975, Potter 1987). Incorporation of historical data was considered by Hosking and Wallis (1986) who found it unlikely to be useful in practice, and by Jin and Stedinger (1989) who reached the opposite conclusion. The latter used a mixed moment-based and likelihood-based procedure that employs only threshold exceedance information from the historical data, discarding some information. These methods assume that the distributions from the more or less homogeneous sites are independent and of the same, known type.

ENTROPY-BASED ESTIMATION

Entropy-based estimation was pioneered by Jaynes (1957) and has found numerous applications in science and engineering. See the Maximum Entropy and Bayesian Methods (MAXENT) Series in References. Almost all entropy-based estimation assumes the data are in the form of moment constraints. Used instead with quantile constraints in the REF method, entropy-based estimation facilitates the use of historical information, which is encoded in the same way as other data.

The REF method has other advantages that have been documented elsewhere (Lind and Solana 1990, Solana 1990, Solana and Lind 1993, Lind 1994, Solana 1994). The REF method yields a unique solution when other methods fail. The posterior distribution function satisfies a combinatorial condition of the problem (the sample rule) unlike common parametric solutions. The objective function, the relative entropy, has a transparent *rationale*. REF solutions are invariant under monotonic transformations of the random variable domain, invariant under conditioning of the random variable domain and invariant under equivalent probabilistic updating. These desiderata are fundamental for consistency, and they are satisfied only by the REF inference procedure which provides exactly the same results as the Logical inference procedure derived using logical probabilities in Solana (1993).

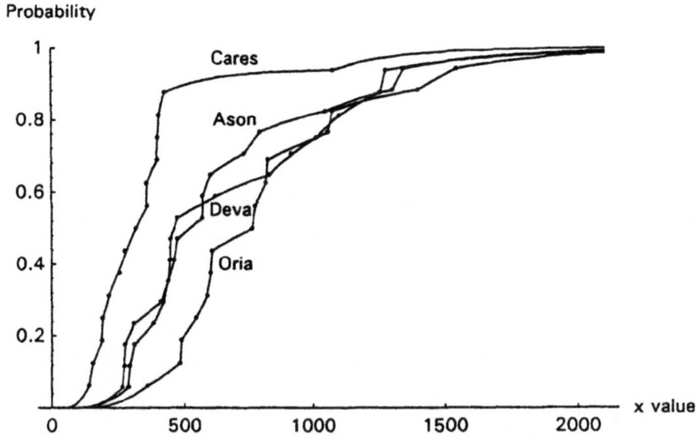

Figure 1. Example 1: Data plotted as $[x_{(i)}, i/(n+1)]$ and the Posterior Distribution Functions having Minimum Entropy Lognormal Reference Distributions.

The REF algorithm is the same for all sets of candidate distributions. The estimation process introduces a minimum of information and thus approaches objectivity in a quantified sense. The method identifies a least information distribution that is identical to the Maximum product of spacings (MPS) estimate.

The REF method (Relative Entropy with Fractile constraints) encodes the data as prescribed quantiles that subdivide the domains of the random variables into intervals (Lind and Solana 1990). The REF method minimizes the relative entropy, or cross entropy (Shore and Johnson 1980) between two distributions, the reference distribution and the posterior distribution. The reference distribution is selected form a sufficiently wide set of candidate types. The REF method selects the element of this set that best fits the data in the sense of minimizing the information about the intervals. The posterior is the distribution closest to the reference (in the sense of minimizing the relative entropy) that satisfies the quantile constraints.

The REF method has been described in the literature only for a single real-valued random variable. Extension to a finite set of independent variables is described in the sequel and follows because of the additive property of the relative entropy. Historical data may be encoded simply as prescribed quantiles, in the same fashions as systematic records. All data can be weighted according to judged quality. Robustness against assumptions of reference distribution type is high, as the example below illustrates.

AGGREGATION ANALYSIS

The approach to aggregation is easily explained when the unknown X is real-valued; extension to more dimensions is obvious. If a probability distribution $F = F(x)$ is affine with another probability distribution $F_0 = F_0(x)$, which may be a prototype, then for some a and b and for all x

$$F(a + bx) = F_0(x). \tag{1}$$

F is thus defined by F_0 and the parameters of the mapping (1), which may be taken as the mean m and standard deviation s of $F(.)$. The proposed approach is, first,to estimate m and s for each series in the family, second, to map all data into the space of F_0, third, to aggregate and plot the data to fit a model for F_0 and four, to invert the mapping to get the model for each series.

TABLE 1. Observed Annual Maxima Discharge of four Unregulated Iberian Peninsula Rivers of the Bay of Biscay Watershed, m^3/s.
Note: (*) Missing observations are indicated by an asterisk.

Observations, 1971-1988						Natural Logarithm of the Observations (Sorted)			
Year	Ason	Cares	Deva	Oria		Ason	Cares	Deva	Oria
1971	*	273	440	1012					
1972	*	187	264	825					
1973b	422	397	*	775		5.662	*	5.575	*
1974	572	357	448	764		5.673	4.927	5.598	5.886
1975	574	253	832	1273		5.730	5.023	5.609	6.182
1976b	308	315	474	*		5.945	5.231	5.726	6.190
1977b	794	426	1072	*		6.045	5.247	6.016	6.304
1978b	1046	403	410	*		6.084	5.365	6.086	6.383
1979b	734	*	624	1099		6.135	5.533	6.098	6.403
1980b	1298	*	1057	819		6.163	5.609	6.104	6.411
1981b	439	*	917	609		6.349	5.752	6.161	6.638
1982b	382	190	*	484		6.352	5.877	6.436	6.652
1983	1340	138	1397	1253		6.401	5.877	6.723	6.708
1984	603	357	273	604		6.598	5.983	6.821	6.715
1985	288	1073	307	488		6.677	5.988	6.963	6.919
1986	475	214	445	592		6.952	5.998	6.977	7.002
1987	291	152	270	547		7.168	6.054	7.242	7.133
1988	462	399	1542	360		7.200	6.978	7.340	7.149
Sample Mean:						6.321	5.696	6.342	6.578
Sample Standard Deviation:						0.476	0.497	0.578	0.358
Unbiased Estimator of StdD.:						0.492	0.514	0.597	0.371

In some contexts it may be desirable to restrict coefficient a in (1) to be zero. On physical grounds this may be the case in hydrology. Then all distributions in the family are similar except for a change of scale. The effect of this restriction on the resulting models is illustrated in the example.

When mapping to and from the prototype distribution it is appropriate to use the conventional correction of bias in the sample variance. Thus, the sample standard deviation for each series is multiplied by the factor $[n/(n-1)]^{1/2}$, where n is the series sample size, as if to estimate the population parameter.

A plot of the aggregated data will suggest the best way to model F_0 and determinate the desired information. Quantiles may be found by interpolation or extrapolation, or an appropriate distribution model may be derived. In any case the model for each series is determinated by the inverse mapping. The REF method is illustrated in the example below.

TABLE 2. Synopsis of Methods of Analysis

1. Isolated Modelling-Ason Data:	
(a) MOM	Method of Moments;
(b) MPS	Maximum Product of Spacings Method ≡ Least Information Reference Distribution;
(c) TEA	Tail Entropy Approximation;
(d) REF+MOM	REF Method, MOM reference distribution (from case a);
(e) REF+MPS	REF Method, MPS reference distribution (from case b).
2. Aggregated Modelling-All Data:	
(f) TEA*AGG*MOM	Tail Entropy Approximation applied to mapped data;
(g) AGG*REF	REF Method, similar least-information reference distributions (MPS) selected from a candidate set of all normal, gamma, Gumbel, lognormal and Cauchy distributions;
(h) MPS*AGG	Maximum Product of Spacings Method applied to the aggregated data;
(i) REF*AGG	Minimum relative entropy estimation of aggregated data.
3. Regional-Local Decomposition-All Data:	
(j) ALG*RLD*MOM	Algebraic convolution of local and regional components;
(k) SYN*RLD*MOM	Method of moments applied to synthesized data combining local and regional components.

REGIONAL-LOCAL DECOMPOSITION

A recent publication (Klemes et al 1992) suggests a novel method of decomposition of a hydrological time series that leads to estimation of extreme quantiles. The idea is to construct a large sample of synthetic values of seasonal maxima of daily precipitation totals in a location, X, by assuming that it is the sum of two mutually independent components, the orographic precipitation, Y, and the synoptic precipitation, Z. Z is approximated by the observations Z_i, in a neighboring location, while Y is defined by the observed differences $(X-Z)_i$. The approach reflects the thought, widely held among hydrologists and meteorologists that extreme precipitation events mostly occur because of rare *combinations* of (not so rare) individual components of precipitation. Thus, from n simultaneous observations X_i and Z_i Klemes et al. synthesize n^2 observations $X_{ij} = Y_i + Z_j,\ i,j \in \{1,2,\ldots,n\}$. They then use the empirical distribution to estimate extreme quantiles.

The idea of Klemes et al. (1992) can be applied to regional aggregation of data to estimate extremes of a hydrological variable, X. Let Z denote the regional average of the same variable, but excluding X. Let Y denote the difference $X - Z$, and assume that Z and Y are independent. Suppose that there are m observations of Z and n observations of Y, and suppose that these data do not justify rejection of the hypothesis of independence between Y and Z. Then X can be estimated in classical ways from the mn observations $X_{ij} = Y_i + Z_j,\ i,j \in \{1,2,\ldots,n\}$. Alternatively, Y and Z may be estimated individually by any method and convoluted to estimate X. These approaches are also explored in the following example .

EXAMPLE

For illustration of the various methods, consider the problem to estimate high extremes of the maximum discharge of the Ason, an unregulated river in Northern Spain. The data are 18 years of observations of the annual maximum discharge of the rivers Ason, Cares, Deva and Oria in northern

Spain shown in Table 1. The observations in Columns 2-5 are plotted in Figure 1. The data are transformed logarithmically and sorted in columns 6-9 in Table 1. For comparison, estimates of the 100-year and 1000-year discharges of the river Ason are produced by several different approaches. The operations are indicated by three-letter abbreviations listed and briefly explained in Table 2.

TABLE 3. Selection of Reference Distribution for the Aggregated Data

Type	River	Parameters		Minimum Relative Entropy
Normal	A	628.185	380.672	3.35548
	C	372.576	225.776	
	D	713.568	432.413	
	O	689.685	417.941	
Gamma	A	3.21123	201.358	2.75754
	C		110.401	
	D		216.962	
	O		244.231	
Cauchy	A	487.182	160.263	3.01272
	C	301.872	99.32	
	D	451.692	148.613	
	O	681.695	224.287	
Lognormal	A	6.32501	0.563399	2.59336
	C	5.69968		
	D	6.34690		
	O	6.57683		
Gumbel	A	477.734	265.191	2.77186
	C	259.679	144.148	
	D	515.639	286.232	
	O	591.069	328.103	

The REF method allows a comparison between any finite number of types of reference distributions that could be candidates. Here, all normal, gamma, Gumbel, lognormal and Cauchy distribution functions were compared. The analysis was done first for each of the four rivers in isolation. The lognormal distribution provides the best fit in all cases except one, for which a Gumbel distribution fits marginally better. Next, the aggregated data was analyzed for similar reference distributions. The lognormal distribution again provides the best fit.

Table 3 shows the resulting parameters and the relative entropy for the aggregated case. The lognormal gives the best fit in the sense that the least information lognormal reference distribution yields the minimum relative entropy. This justifies the choice of lognormals over a broad set of alternatives to model the distributions. In the analyses following, therefore, only lognormal distributions are examined. Results for river Ason are shown in Table 4, lines (e) and (i).

The least information reference distribution is identical to the maximum product of spacings (MPS) distribution. The resulting estimates of the 100-year and 1000-year discharges of the river Ason are listed in Table 4, line (b).

Conventionally fitted distributions may also be used as reference distributions to derive a posterior distribution for each river (Lind and Solana 1990). In Figure 2 the reference distribution (e.g. the MPS distribution or a MOM-fitted longnormal) is denoted by $P(x)$. The greatest of n observations, $x_{(n)}$ is labelled (n) and plotted at $(x_{(n)}, n/(n+1))$. The density function of the posterior that minimizes the cross-entropy relative to $P(x)$ is proportional to the density $p(x)$ of $P(x)$; the factor of proportionality α_n equals $(n+1)p_n$, where $p_n = 1 - P(x_{(n)})$, such that $Q(X)$ passes through point (n). Figure 3 shows the posterior distributions for the Ason data and illustrates the great discrepances between at the levels of the return periods important in hydrological engineering.

Table 4 lists the factors α_n for the various lognormal models for Ason. It is observed, for example, that the largest extreme for river Ason is much greater than predicted by the reference distribution $P(x)$ used in case (d), and that the factor α_n is correspondingly large. Conversely, in case (e) $P(x)$ passes very close to point (n), and α_n is near unity. Figure 2 shows how the factor α_n is used to calculate all quantiles of $Q(x)$ greater than $x_{(n)}$: the $1-f$ quantile falls at $x = P^{-1}((n+1)p_n f)$.

A normal distribution was also fitted to the transformed data by the conventional method of moments (MOM). This distribution function corresponds to a lognormal distribution correctly fitted to the original data (Table 4, line (a)).

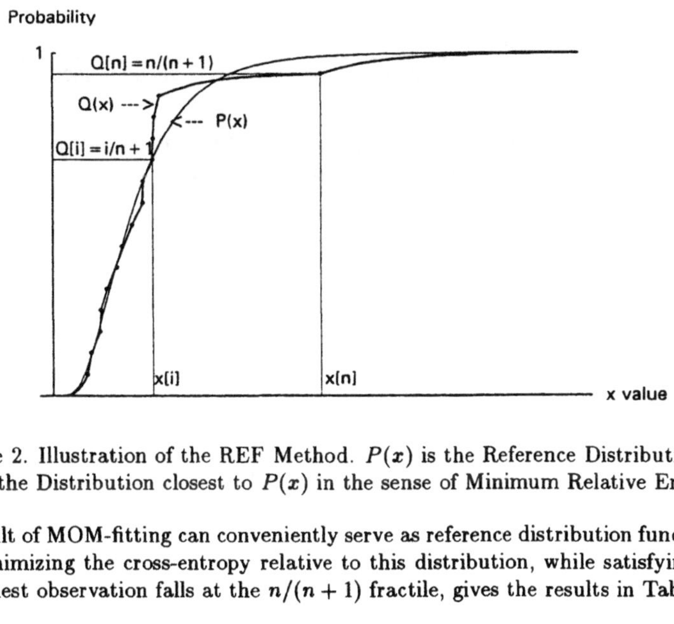

Figure 2. Illustration of the REF Method. $P(x)$ is the Reference Distribution, $Q(x)$ is the Distribution closest to $P(x)$ in the sense of Minimum Relative Entropy.

The result of MOM-fitting can conveniently serve as reference distribution function for the REF method. Minimizing the cross-entropy relative to this distribution, while satisfying the constraint that the highest observation falls at the $n/(n+1)$ fractile, gives the results in Table 4, line (d).

Perhaps the simplest of all methods of analysis is Tail Entropy Approximation (TEA) (Lind and Hong 1991). For a two parameter distribution, such as the lognormal distributions used here, the TEA method consists in selecting two quantiles and adjusting the distribution function parameters to satisfy these constraints. One good choice is the sample mean and $x_{(n)}$. Results are shown in line (c) in Table 4 for the Ason data and line (f) for the aggregated data.

Aggregated Modelling was done by the method of moments, which assigns a pair of coefficients (a, b) to the data for each river in isolation (eg. Table 4, line (a)). After standardizing the data, it was aggregated for all rivers and mapped back to the mean and variance of each river. The result is then treated as a new data set for any of the rivers. Analysis by the TEA, MPS and REF methods give the results in lines (f) to (i) in Table 4.

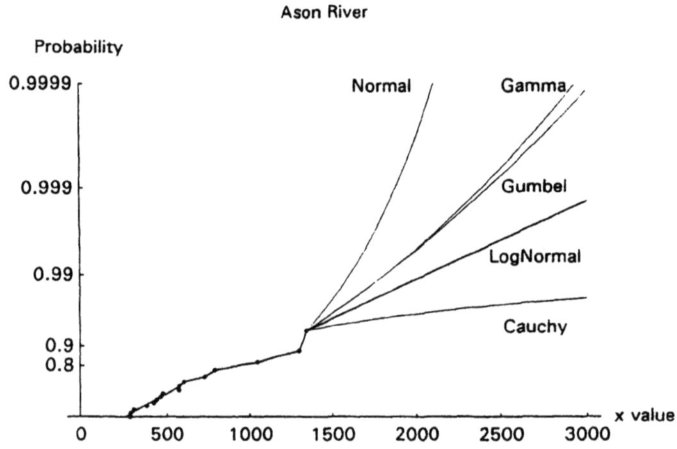

Figure 3. Ason Observations and Posterior Distribution Functions for the References listed in Table 2.

Regional-Local Decomposition (RLD). Table 1 shows ten years in which maxima are available for Cares, Deva and Oria. The means, Z_i, of the logarithms for these years and rivers used as observations to estimate Z by the MOM, giving $(m, s)_Z = (6.212, 0.485)$. There are eight years of simultaneous observations of Z and X, the maximum for Ason, yielding eight observations $Y_j = X_j - Z_j$.

One way to continue is to estimate Y by MOM from these eight observed values, giving $(m, s)_Y = (0.071, 0.430)$. Algebraic convolution to sum Y and Z then gives $(m, s)_X = (6.212, 0.513)$ and thus the estimated quantiles in line (j) of the Table 4.

Combining instead each Z_i with each Y_j yields 80 synthetic data points that can be treated by any of the methods described above. For example, Table 4, line (k) shows the results of REF analysis.

DISCUSSION

It should remembered that in professional engineering practice it is not appropriate to use statistical methods blindy, particularly for series of data as short as in the present example. Statistics at best can be only a support for careful judgment and practical experience. Classical statistics, in particular, tends to pursue regular features of phenomena and ignore "outliers". When, as in the present case, the extreme is the essence of the problem, it should not be surprising if conventional statistics fares poorly. Historical data are often among the best indicators of rare natural phenomena. The REF method offers the advantage that historical data may be encoded just as simply as systematic records, i.e., as prescribed quantiles.

TABLE 4. Models and Estimated Annual Maxima of Total Daily Discharge of River Ason, m^3/s. (Data: Table 1).

Method (see Table 2)	Tail Parameters			Return Period, years	
	m	s	α_n	100	1000
(a) MOM	6.321	0.492	1	1749	2548
(b) MPS	6.325	0.555	1	2031	3103
(c) TEA	6.321	0.561	1	2055	3155
(d) REF*MOM	6.321	0.492	1.585	1900	2722
(e) REF*MPS	6.325	0.555	1.025	2042	3146
(f) TEA*AGG*MOM	6.321	0.571	1	2102	3251
(g) REF*AGG*MOM	6.321	0.492	2.492	2053	2899
(h) MPS*AGG	6.325	0.563	1	2069	3181
(i) REF*MPS*AGG	6.325	0.563	1.019	2061	3173
(j) ALG*RLD*MOM	6.212	0.513	1	1647	2437
(k) SYN*RLD*MOM	6.212	0.485	1	1543	2235

A single example can merely illustrate and suggest, not prove. With this *caveat* it is observed in Table 4 that the popular method of moments gives conspicuously low quantiles in the unconservative direction (cases (a) and (k)). Case (j) also stands out by low quantiles; this shows a weakness of the decomposition approach that may be a general property. Case (d) can be considered as a correction in the high tail of the method of moments; the result is much closer, but still evidently low in relation to all other results, which are effectively identical. Tail entropy approximation (cases (c) and (f)), in particular, seems to perform like the other methods, although much simpler than even the method of moments.

CONCLUSIONS

Conventional statistical estimation can perform poorly in fields such as hydrology, hydraulic meteorology, civil engineering hydraulics and in other fields where the data are scarce and the distributions not known. Quantile-based entropy methods are available and may be preferable. Historical data, which are often important in engineering hydrology, may be taken into account in a simple fashion; they are encoded naturally as prescribed quantiles in the same way as systematic records. The REF method involves roughly the same amount of computation as maximum likelihood estimation, i.e., more computation than the method of moments. This is not so important, since it is easily programmed and does not require mainframe capacity. Tail entropy approximation can perform as well as other entropy methods, although even simpler than the method of moments.

If the modeller believers that the data for a set of related hydrological variables in a region are all of the same distribution type, differing at most in mean and standard deviation, then regional aggregation is indicated. The modelling operation can be conceived of as the estimation of the mapping parameters for each variable. Regional aggregation is advantageous because the aggregation provides a better perspective on the distribution tails, of vital importance for design. In particular, regional aggregation casts a light on "outliers" which, in this context, may be of greatest importance.

Decomposition of a local variable into the sum of a regional base variable and an independent local perturbation variable, here called "local-regional decomposition", can permit the generation of large sets of synthetic data. However, such decomposition is not likely to reveal the possibility of unusual extremes of low probability if they are not present in the local observations.

ACKNOWLEDGMENTS

Work reported in this paper was carried out in part with the financial support of the Dirección General de Investigación Científica y Técnica of Spain as research project PB-89077 and the Natural Sciences and Engineering Research Council of Canada. Special thanks are due to Confederación Hidrográfica del Norte de España who provided the data for the example.

REFERENCES

Hosking, J.R.M. and Wallis, J.R. (1986) "The value of historical data in flood frequency analysis," Water Resources Research, Vol. 22, No. 11, pp. 1606-1612.

Hosking, J.R.M. and Wallis, J.R. (1988) "The effect of intersite dependence on regional flood frequency analysis," Water Resources Research, Vol. 24, No. 4, pp. 588-600.

Jaynes, E.T. (1957) in E.T.Jaynes: Papers on Probability, Statistics and Statistical Physics, Rosenkrantz R.D. (ed.) (1983) D. Reidel, Dordrecht, The Netherlands.

Jin, M. and Stedinger, J.R. (1989) "Flood frequency analysis with regional and historical information," Water Resources Research, Vol 25, No. 5, pp. 925-936.

Klemes, V., Nikleva, S. and Chin, W.Q. (1992) "Probability of a PMP - A feasibility study," CDSA-CANCOLD Joint Conference, Safety of Dams and the Environment, Quebec City, Quebec.

Lind, N.C. (1994) "Information Theory and maximum product of spacings estimation," Journal of the Royal Statistical Society, Series B, Vol. 56, No. 1.

Lind, N.C. and Hong, H.P. (1991) "Tail Entropy Approximations", Structural Safety.

Lind, N.C., Hong, H.P. and Solana, V. (1989) "A Cross-Entropy Method for Flood Frequency Analysis," Stochastic Hydrology and Hydraulics, Vol. 3, No. 3, pp. 191-202.

Lind, N.C. and Solana, V. (1990) "Fractile constrained entropy estimation of distributions based on scarce data," Civil Engineering Systems, Vol. 7, No. 2, pp. 87-93.

Madsen, H.Q., Krenk, S. and Lind, N.C. (1986) Methods of Structural Reliability, Prentice-Hall.

Natural Environmental Research Council (NERC), (1975) Flood studies report, Vol. 1-5, London, UK.

Potter, K.W. (1987) "Research on flood frequency analysis: 1983-1986," Rev. Geophys., Vol. 25, No. 2, pp. 119-124.

Shore, J. and Johnson, R.W. (1980) "Axiomatic Derivation of the Principle of Maximum Entropy and the Principle of Minimum Cross-Entropy", IEEE Transactions on Information Theory, Vol. IT-26, No. 1, pp. 26-37.

Solana, V. (1991) "Consistency Principle for Data-based Inference," in Maximum Entropy and Bayesian Methods in Science and Engineering, Fougère, P.F. (ed.), Fundamental Theories of Physics, Vol. 39, Kluwer, Dordrecht. pp. 41-52.

Solana, V. (1993) "An Invariance Principle for Updating Probabilities," in Maximum Entropy and Bayesian Methods, Mohammad-Djafari, A. (ed.), Fundamental Theories of Physics, Vol. 53, pp. 99-112, Kluwer, Dordrecht.

Solana, V. (1994) "Entropy versus Bayesian Inference. Why Bayesian Inference for an Uncertainty Quantity Fails", in Maximum Entropy and Bayesian Methods, G. Headbreder (ed.), Fundamental Theories of Physics, Kluwer, Dordrecht.

Solana, V. and Lind, N.C. (1989) "A Monotonic property of distributions based on entropy with fractile constraints" in Maximum Entropy and Bayesian Methods, J. Skilling (ed.), Fundamental Theories of Physics, pp. 481-490, Kluwer, Dordrecht.

Solana, V. and Lind, N.C. (1993) "Relative entropy method using tail fractiles," Reliability Engineering and System Safety, Vol. 40, pp. 61-67.

U.S. Water Resources Council (USWRC), (1982) "Guidelines for determining flood frequency," Bulletin No. 17B, Hydrol. Comm., Washington, DC,

RETURN PERIODS FOR RAINFALL BY FRACTAL ANALYSIS

Susumu OGAWA
Water Quality Management Center
Waterworks Bureau, Tokyo Metropolitan Government
2-7-1 Hongo, Bunkyo-ku, Tokyo 113
Japan

The current design storm profiles are derived from a rainfall depth-duration-frequency relationship, and differ from the original profiles. Fractal models can be used to describe irregular time series and the fractal dimension can be an element to characterize storm profiles. Rainfall profiles are determined by four factors, fractal dimensions, duration time, peak position and peak intensity. These factors except peak intensity distributes normally with the annual maximum rainfall rate data. Fractal dimensions range between 1.5 and 2.0, 1.85 on average. However, these three factors for the current design storm profiles are far from the mean values. Therefore, the return periods of the design storm profiles may not be accurately evaluated.

INTRODUCTION

The design storm profiles for flood control are determined by synthetic profiles corresponding to different return periods derived from a rainfall depth-duration-frequency relationship. In Tokyo we adopted the single lumped profiles with a peak at half of the duration, 24 hours. However, doubts concerning the representativity of such profiles led several authors to suggest an arbitrary rearrangement in the rainfall intensity for successive time increments [Hawken, 1921; Judson, 1933]. Thus, the return periods for the storm profiles were analyzed by fractal.

FRACTAL AND SHAPE FACTORS

A rainfall depth-duration-frequency relationship can be given by calculating a Gumbel distribution for the observed depth within duration time. The profiles have other parameters such as peak positions, the maximum duration time and irregular temporal variation, which are

K. W. Hipel (ed.),
Stochastic and Statistical Methods in Hydrology and Environmental Engineering, Vol. 1, 381–384.

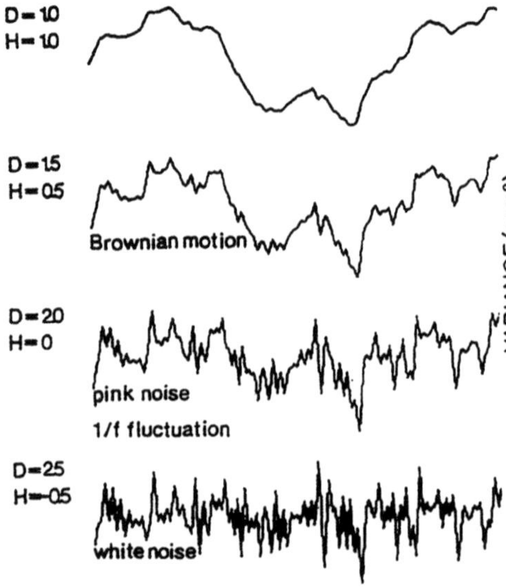

Figure 1. Fractional Brownian functions.

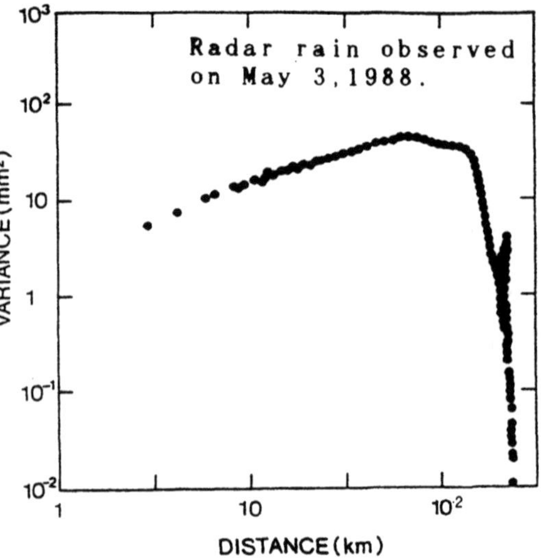

Figure 2. Semivariogram for rainfall.

not considered in this frequency analysis. Among these shape factors of the profiles, irregular temporal variation can be modeled by fractional Brownian functions characterized by fractal dimensions as shown in Fig.1. The fractal dimensions for storm profiles are calculated by the slope semivariograms as shown in Fig.2. the relation between the dimensions and the slope is expressed as,

$$D=2-H \tag{1}$$

where D is fractal dimension and H is Hurst exponent, half of the slope of the semivariogram.

Since fractals have no characteristic length, two different shape factors are needed in time series. They are duration time and peak rainfall rate. Moreover, peak position is very important for flood control. Therefore, four shape factors should be selected. However, the current design storm profiles are constructed from only the return periods of peak rainfall rate. This might be right if the rest of shape factors should distribute uniformly.

The rainfall data have been observed in Tokyo from 1926 to 1988. The population of the rainfall data was divided into 1 hour annual maximum storm profiles and 1 day annual maximum ones, which have quite different profiles.

RESULTS

The fractal dimensions calculated from the annual maximum rainfall intensity distribute normally ranging between 1.5 and 2.0, 1.85 on average. The distributions of the 1 hour annual maximum storm profiles and the 1 day

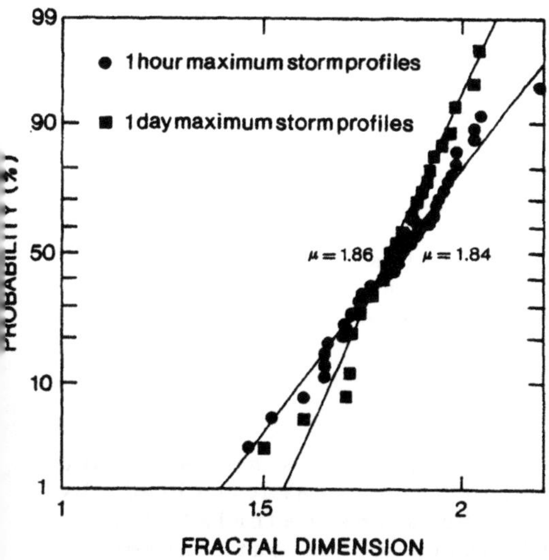

Figure 3. Distribution of fractal dimensions.

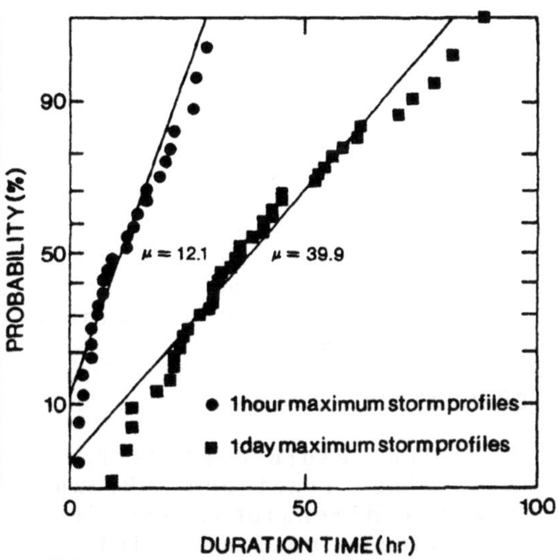

Figure 4. Distribution of duration time.

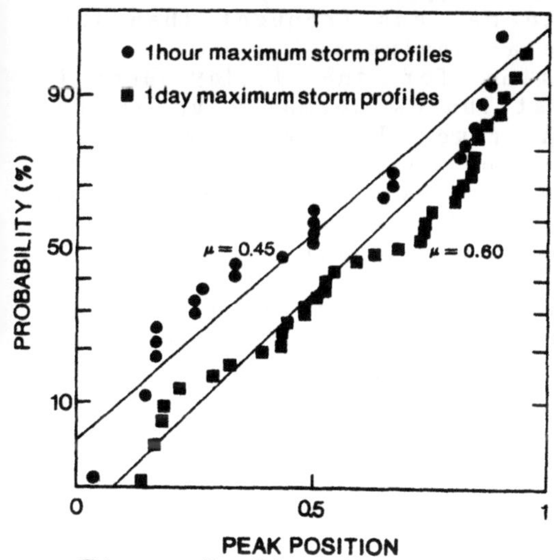

Figure 5. Distribution of peak position.

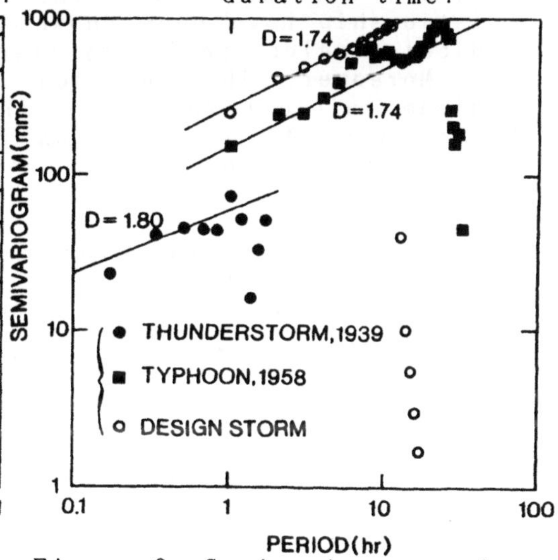

Figure 6. Semivariograms of thunderstorm, typhoon and design storm.

annual maximum ones are almost the same. However, the fractal dimensions of the design profiles converge on 1.75 despite different return periods. As the fractal dimensions increase, the component of the high frequency waves increases as shown in Fig.1. It means that the mean observed profiles fluctuate more violently than the design storm profiles and have more high frequency waves.

Moreover, the duration time also distributes normally up to 89 hours, with 40 hours on an average for the 1 day annual maximum rainfall rate and 12 hours on an average for the 1 hour annual maximum. The peak position that is the ratio of peak time to the duration distributes normally between 0 and 1, the mean 0.45 for the 1 hour annual maximum storm profiles and 0.60 for the 1 day annual maximum storm profiles.

Therefore, the design storm profiles in Tokyo might be less frequent than the indicated return period and should be reconstructed concerning profiles shape characterized by fractal dimensions.

CONCLUSION

The return periods for the storm profiles are analysed with the fractional Brownian function characterized by fractal dimensions. The fractal dimensions calculated from the annual maximum rainfall intensity distribute normally ranging between 1.5 and 2.0, 1.85 on average. The design profiles derived from a rainfall depth-duration-frequency relationship have been shown to be less frequent than the indicated return period by fractal analysis.

Moreover, the duration time for the 1 day annual maximum rainfall rate also distributes normally up to 89 hours, with 40 hours on an average. This mean value is about twice as long as the one for design storm profiles, 24 hours. The peak position distributes normally, with 0.45 on an average for the 1 hour annual maximum rainfall rate and 0.60 on an average for the 1 day annual maximum. Both of the means shift from the center. The latter is more dangerous than the former because the saturated area of soil moisture in a basin increases with duration time till the peak, which often causes flood disasters.

The return periods for storm profiles can be calculated with four factors; fractal dimensions, duration time, peak position and peak rainfall rate. The current design storm profiles should be reconstructed with these shape factors.

REFERENCES

Hawken,W.H.(1921) "An analysis of maximum runoff and rainfall intensity", Trans.Instn.Engrs.Austr.,2,193-215.

Judson,C.C.(1933) "Runoff calculations, a new method", Proc.Instn.Munic.Co.Engrs.,59,861-7.

Lovejoy,S.(1982) "Area-perimeter relation for rain and cloud area", Science,216,9,185-6.

Ogawa,S.,Hirano,M.,Moriyama,T.,and Ando,T.(1993) "Fractal analysis of temporal and spatial distributions for radar rain", Proc.Hydrau.Engr.,JSCE,37,21-26.

AUTHOR INDEX

SUBJECT INDEX

Water Science and Technology Library

1. A.S. Eikum and R.W. Seabloom (eds.): *Alternative Wastewater Treatment.* Low-Cost Small Systems, Research and Development. Proceedings of the Conference held in Oslo, Norway (7–10 September 1981). 1982 ISBN 90-277-1430-4
2. W. Brutsaert and G.H. Jirka (eds.): *Gas Transfer at Water Surfaces.* 1984 ISBN 90-277-1697-8
3. D.A. Kraijenhoff and J.R. Moll (eds.): *River Flow Modelling and Forecasting.* 1986 ISBN 90-277-2082-7
4. World Meteorological Organization (ed.): *Microprocessors in Operational Hydrology.* Proceedings of a Conference held in Geneva (4–5 September 1984). 1986 ISBN 90-277-2156-4
5. J. Němec: *Hydrological Forecasting.* Design and Operation of Hydrological Forecasting Systems. 1986 ISBN 90-277-2259-5
6. V.K. Gupta, I. Rodríguez-Iturbe and E.F. Wood (eds.): *Scale Problems in Hydrology.* Runoff Generation and Basin Response. 1986 ISBN 90-277-2258-7
7. D.C. Major and H.E. Schwarz: *Large-Scale Regional Water Resources Planning.* The North Atlantic Regional Study. 1990 ISBN 0-7923-0711-9
8. W.H. Hager: *Energy Dissipators and Hydraulic Jump.* 1992 ISBN 0-7923-1508-1
9. V.P. Singh and M. Fiorentino (eds.): *Entropy and Energy Dissipation in Water Resources.* 1992 ISBN 0-7923-1696-7
10. K.W. Hipel (ed.): *Stochastic and Statistical Methods in Hydrology and Environmental Engineering.* A Four Volume Work Resulting from the International Conference in Honour of Professor T. E. Unny (21–23 June 1993). 1994
    10/1: Extreme values: floods and droughts ISBN 0-7923-2756-X
    10/2: Stochastic and statistical modelling with groundwater and surface water applications ISBN 0-7923-2757-8
    10/3: Time series analysis in hydrology and environmental engineering ISBN 0-7923-2758-6
    10/4: Effective environmental management for sustainable development ISBN 0-7923-2759-4
    Set 10/1–10/4: ISBN 0-7923-2760-8
11. S. N. Rodionov: *Global and Regional Climate Interaction: The Caspian Sea Experience.* 1994 ISBN 0-7923-2784-5
12. A. Peters, G. Wittum, B. Herrling, U. Meissner, C.A. Brebbia, W.G. Gray and G.F. Pinder (eds.): *Computational Methods in Water Resources X.* 1994 Set 12/1–12/2: ISBN 0-7923-2937-6

Kluwer Academic Publishers – Dordrecht / Boston / London

MIX
Papier aus verantwortungsvollen Quellen
Paper from responsible sources
FSC® C105338

If you have any concerns about our products,
you can contact us on
ProductSafety@springernature.com

In case Publisher is established outside the EU,
the EU authorized representative is:
Springer Nature Customer Service Center GmbH
Europaplatz 3, 69115 Heidelberg, Germany

Printed by Libri Plureos GmbH
in Hamburg, Germany